High Temperature Coatings

High Temperature Coatings

Sudhangshu Bose

ELSEVIER

AMSTERDAM • BOSTON • HEIDELBERG • LONDON
NEW YORK • OXFORD • PARIS • SAN DIEGO
SAN FRANCISCO • SINGAPORE • SYDNEY • TOKYO

Butterworth-Heinemann is an imprint of Elsevier

Butterworth–Heinemann is an imprint of Elsevier
30 Corporate Drive, Suite 400, Burlington, MA 01803, USA
Linacre House, Jordan Hill, Oxford OX2 8DP, UK

⊗ Recognizing the importance of preserving what has been written, Elsevier prints its
books on acid-free paper whenever possible.

Library of Congress Cataloging-in-Publication Data
Bose, Sudhangshu.
  High temperature coatings/Sudhangshu Bose.
    p. cm.

  1. Refractory coating.
  2. Heat resistant materials.   I. Title.
  TS695.9.B67 2007
  671.7′3—dc22

                                                              2006035922

British Library Cataloguing-in-Publication Data
A catalogue record for this book is available from the British Library.

ISBN 13: 978-0-7506-8252-7

Transferred to Digital Printing, 2012

This book is lovingly dedicated to my wife Juthika and our two sons, Krishnangshu and Jayanta. They allowed me the luxury of not attending to many chores for the excuse of becoming a book author.

Table of Contents

About the Author

Dr. Sudhangshu Bose is a Fellow and Manager at Pratt & Whitney, the manufacturer of Gas Turbine and Rocket Engines. He is also an Adjunct Professor at the Hartford, Connecticut campus of Rensselaer Polytechnic Institute, Troy, New York. He holds a Ph.D in Materials Science and Engineering from University of California, Berkeley, having previously obtained B.Sc (Honors) and M.Sc in Physics from Ranchi University, Ranchi, India. Dr. Bose has taught undergraduate and graduate level courses in Physics and conducted research in Materials Characterization by x-ray diffraction prior to completing the doctoral degree. For the last twenty-nine years at Pratt & Whitney and its sister divisions, Dr. Bose has conducted and managed research and development in advanced materials and processes including oxidation and corrosion in fuel cells and gas turbine engine, catalysis, high temperature coatings, super-alloys, intermetallics, and ceramic matrix composites. He holds 14 patents and continues to teach high temperature alloys and coatings courses within Pratt & Whitney and at Rensselaer.

Preface

The idea of a book on high temperature coatings arose out of a need for teaching material for a semester long course for engineering students. Additionally, it would provide a resource of reference material for practicing professionals. It took the author several years during his forays into teaching to assemble the material in a balanced and logical format.

High temperature coatings belong to a technologically important, economically lucrative, intellectually fertile, and fast evolving field. It encompasses basic understanding of materials, principles of physical metallurgy and ceramics, environmental interactions, processing sciences and manufacturing techniques. There is a wealth of information available on various aspects of these coatings. However, the information is dispersed in myriads of technical journals, conference proceedings, handbooks, book chapters, published reports from the manufacturers in the industry, and users of coatings, as well as from National Laboratories and other organizations in a host of countries. The author had the fortune to have access to many of these resources, in addition to his own continuing experience in the aerospace industry spanning over a quarter of a century.

For the users of this book, a rudimentary knowledge of physics and chemistry is assumed. The fundamental concepts, which cover thermodynamic and kinetic principles as well as phases and phase diagrams, are described in the first two chapters. Basic understanding of structural alloys used in high temperature processes and equipment, notably gas turbine engines, which require high temperature coatings, is introduced in a separate chapter. Environmental interactions including oxidation and high temperature corrosion are discussed in Chapters 4 and 5. Metallic coatings for protection against oxidation and corrosion are covered in Chapter 6 in reasonable detail including processing, characterization, and properties. A number of processes traditionally used in the electronic industry to deposit thin films have been included because of their unique characteristics and slow but progressive adaptation to deposit thicker coatings in special cases. Thermal Barrier Coatings (TBC) for thermal protection are addressed in Chapter 7. For better flow and ease of understanding, this chapter is physically separated into three sections starting with an introduction in the beginning, plasma sprayed TBC in the middle, and Electron Beam TBC at the end, together with the literature references for the whole chapter. Non destructive inspection, coatings repair, and coating experience in the field are treated in the last three chapters. Extensive references are given for the readers to consult the original sources.

Obviously, the author benefited immensely from the work of many of the pioneers from the past and the present. One such individual is Dr. Bill Goward. The author is deeply grateful to Dr. Goward for his thoughtful review and constant guidance during the writing of the book. Professors Fred Pettit of University of Pittsburgh, Professor David Clarke of University of California, Santa Barbara, Professor Maury Gell of University of Connecticut, and Dr. Bob Miller of NASA Glenn Research Center, all of whom are prolific contributors to the knowledge base in the field of high temperature coatings, were very kind to review parts of the manuscript and provide suggestions for improvement. My special appreciation goes to Professor Ernesto Gutierrez-Miravete of Rensselaer, Hartford for his encouragement.

Elsevier's Joel Stein has been of immense help from the very beginning of this project, providing advice, encouragement, and the words of wisdom of a seasoned editor. I owe him my sincere thanks. Special thanks are also due to Shelly Palen of Elsevier for her editorial help and to Julie Ochs for her superb production of the manuscript.

The author has made every effort to correct inaccuracies and acknowledge the original sources including authors and publishers. He regrets any omissions.

My appreciation goes to the following sources and copyright owners for their permission to use excerpts, data, figures, tables, and art works:

Elsevier, John Wiley & Sons, Inc., Springer, Taylor and Francis, Maney Publishing, Blackwell Publishers, The Minerals, Metals, and Materials Society (TMS), American Society of Materials (ASM) International, The American Ceramic Society, American Society of Mechanical Engineers (ASME) International, American Institute of Physics, Science and Technology Letters, Annual Reviews, National Association of Corrosion Engineers (NAC), The Electrochemical Society, Praxair Technologies, Sulzer-Metco, Professor Jogender Singh (Pennsylvania State University), Drs. Jim Smialek and Bob Miller, both of NASA Glenn Research Center, Dr. Thomas Strangman of Honeywell, Department of Energy, United State Airforce, and the National Aeronautic and Space Administration.

Finally, I am thankful to Pratt & Whitney for permission to use examples from its collection of coatings, coated engine hardware, publications, and the photo of a turbine blade which appears on the cover of this book.

SUDHANGSHU BOSE

Manchester, Connecticut, USA
Sudhabose@gmail.com
2007

Chapter 1

INTRODUCTION

1.1 HIGH-TEMPERATURE ENVIRONMENT

A large number of industrial processes operate in very aggressive environments characterized by high temperature, increased temperature gradients, high pressure, large stresses on individual components, and the presence of oxidizing and corroding atmosphere, as well as internally created or externally ingested particulate material, which induces erosion and impact damage. A few representative examples of such processes are shown in Fig. 1.1 (Stroosnijder et al., 1994). Machines include aircraft gas turbine engines, steam turbines, industrial gas turbines, coal conversion, petroleum refining, and nuclear power generation. The generation of large amounts of heat and associated high component temperature lie at the heart of all of these processes. For example, in jet engines, fuel is mixed with highly compressed air and the mixture ignited. As a result of the heat generated, the air expands and works on the turbine to rotate it. The turbine in turn forces the compressor to rotate, which compresses the incoming air. The exiting exhaust gas creates thrust for propulsion. The gas temperature in modern gas turbine engines could well exceed 1650°C (3000°F) in the turbine section, with cooled parts reaching temperatures as high as 1200°C (2200°F). In coal gasifiers, coal reacts with steam at high temperatures to convert it into usable gas, which can be transported by pipelines and fed directly into processing plants. Process temperatures could be as high as 1650°C (3000°F) with component temperatures reaching 1090°C (2000°F). Process temperature is also high in petroleum refining as well as nuclear power generation. In petroleum refining, crude petroleum is catalytically cracked at high temperatures and fractionated into usable petroleum products such as liquid petroleum gas, gasoline, kerosene, diesel, heavy oils, plastics, asphalt, and coke. In nuclear power plants the heat from controlled fission of fuel elements, such as uranium and thorium, is used to produce steam, which is in turn fed to turbines to generate electrical power.

All these processes require materials of construction with high-temperature capability under load to meet performance and durability requirements. During operation, the structural materials of individual components degrade. In addition to fatigue and creep damage of structurally loaded components, the materials undergo oxidation, corrosion, and erosive wear. Typical temperatures seen in some of the industrial processes are compared in Fig. 1.2, benchmarked against melting points of some basic structural materials.

The strength ranges of a few potential materials of construction for high-temperature processes as a function of temperature are shown in Fig. 1.3 (Meetham, 1988). Properties of many of these engineering materials such as tensile, creep, and fatigue strength are generally optimized for maximum load-carrying capability, with less emphasis on the environmental resistance. As an example, turbine blades for jet engines are made of precipitation-strengthened nickel base superalloys, the precipitates being the gamma prime phase in a gamma phase matrix. One of the constituents of the alloys is aluminum, which participates in the formation

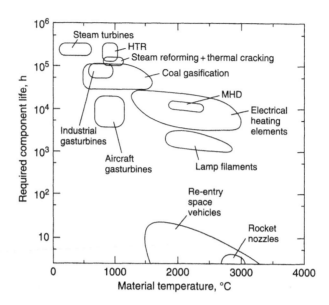

Figure 1.1 High-temperature processes with component temperatures and required lives (M. F. Stroosnijder, R. Mevrel, and M. J. Bennett, The interaction of surface engineering and high temperature corrosion protection, *Materials at High Temperatures*, 1994, 12(1), 53–66). Reprinted with permission from Science and Technology Letters.

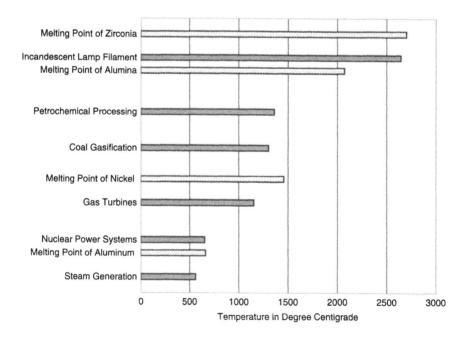

Figure 1.2 Typical temperatures for some industrial processes.

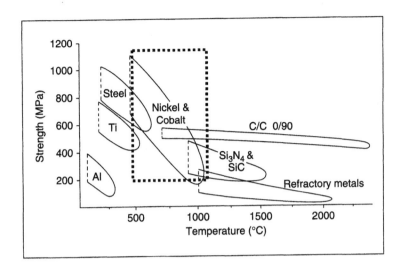

Figure 1.3 Temperature capabilities of classes of materials (G. W. Meetham, Requirements for and factors affecting high temperature capability, *Materials and Design*, 1988, 9(5), 247). Reprinted with permission from Elsevier.

of strengthening precipitates as well as in providing for oxidation resistance. Although higher aluminum content of the alloys increases high-temperature oxidation resistance, it is kept at a level below 6% to maximize creep strength. If such bare alloys are exposed to the environment of high-pressure turbine of modern gas turbine engines, they will degrade fast by several of the processes discussed below.

- *Oxidation*: The alloys do not have adequate levels of such critical elements as Al and Cr to impart oxidation resistance for the life of the parts. Increasing Al, therefore, would apparently be a logical solution to increase oxidation capability. However, a high Al level will lower the melting point of the alloys and their creep resistance, reducing, in turn, the load-bearing capability, a "must have" property of turbine blades and other turbine hardware.

- *High-Temperature Corrosion*: The Cr content of the majority of the alloys used in the turbine section of the gas turbine engine is below the requirement for corrosion resistance. Again, the reason for lower Cr content is to have other critical elements necessary to meet the structural capability requirement to carry load.

- *Heat Damage*: Two of the consequences of high-temperature exposure are oxidation and corrosion, as explained earlier. Additional damages due to cyclic exposure to high temperature in the presence of fluctuating stresses come about by the process of thermal fatigue. The strengthening of the alloys helps to some degree in developing resistance to fatigue cracking. The overall protection against heat damage requires surface treatment of the alloys. For oxidation and corrosion, surface treatment by coatings is the only choice. The coatings provide barriers between the alloys and the outer environment. These coatings can be tailored to meet requirements for environmental resistance, for example, by increased Al and Cr content because they do not need to have load-bearing capability. Chapter 6 addresses oxidation- and corrosion-resistant coatings. An additional avenue to reduce heat damage to the alloys is to lower their temperature without sacrificing the performance of the turbine.

This is addressed for hot components by thermally insulating them. A class of coatings called thermal barrier coatings (TBCs) in combination with active cooling accomplishes this function.

Nondestructive testing of the coating is an essential tool to monitor quality of new coatings and progressive degradation during their use. Understanding of this technologically important area is important.

Repair of coatings is another technologically important area, which also has financial components due to the high cost of coatings and coated hardware.

This book addresses most of the aspects of coatings discussed earlier as a single resource. For better grasp of the subject matter, introductory materials essential to the understanding of the science and technology of coatings are covered in Chapters 2 and 3. Knowledge of how the coatings have fared in the real environment is described in Chapter 10 to tie into the earlier chapters on their processing, structure, and properties.

REFERENCES

Meetham, G. W., Requirements for and factors affecting high temperature capability, *Materials and Design*, 1988, 9(5), pp. 244–252.

Stroosnijder, M. F., R. Mevrel, and M. J. Bennett, The interaction of surface engineering and high temperature corrosion protection, *Materials at High Temperatures*, 1994, 12(1), pp. 53–66.

Chapter 2

Background in the fundamental concepts of thermodynamics, kinetics of reactions, diffusion, crystal structure, phase equilibrium, and phase diagrams of alloys is essential in order to understand the selection, processing, and behavior of high-temperature coatings. These concepts are discussed briefly in this chapter.

2.1 THERMODYNAMIC CONCEPTS

The principles of thermodynamics are used to determine the spontaneity of many chemical and metallurgical phenomena and the direction in which they proceed. A few examples of practical applications of thermodynamics include processes for extraction of metals from ores, formation of alloys, precipitation and grain growth, and degradation of materials in service. In order to understand these processes, a number of thermodynamic parameters need to be defined. The concepts, discussed next, will be specifically helpful in understanding the processes of oxidation and corrosion, phase and microstructure changes of high-temperature coatings, and their interactions with substrate alloys.

Enthalpy

Enthalpy H is the measure of the combination of internal energy E and the product of pressure P and volume V of a system. Thus $H = E + PV$. For example, if the system consists of an inflated balloon made of a thermally insulating rubber, the internal energy is the energy of the gas molecules inside the balloon, and the pressure is that of the ambient atmosphere. If a small amount of heat dQ is added to this system, it will increase internal energy by dE and induce a change in volume dV while the pressure remains constant at P. Thus, $dQ = dE + PdV$. For constant pressure, $dQ = d(E + PV)$, which leads to $dQ = dH$. This relationship indicates that enthalpy is a measure of heat in the system. Change in enthalpy is related to the heat capacity at constant pressure, $(dH/dT)_p = (dQ/dT)_p = C_p$, where dT is the change in temperature. Integration gives

$$H = H_0 + \int C_p \, dT,$$

in which H_0 is a constant and T is temperature in absolute (K) units.

Entropy

Entropy is a measure of the order or randomness of a system. For example, if red and white billiard balls are arranged alternately in an array on a tray, the arrangement is highly ordered with low entropy. If the tray is shaken sufficiently, the arrangement will become randomized, with a red ball as likely to have a red neighbor as a white neighbor. In all processes, entropy maximizes. The quantitative aspect of entropy can be more easily understood with the balloon example. If we add heat dQ to the balloon kept at temperature T, the entropy change is given by $dS = dQ/T = C_p \, dT/T$. Integration provides $S = S_0 + \int C_p \, dT/T$, in which S_0 is a constant.

Free Energy

Free energy or, more appropriately, Gibbs free energy (after the inventor of the concept, J. Willard Gibbs) is a composite thermodynamic concept involving both enthalpy H and entropy S. It is given by $G = H - TS$. If a small change occurs in a system kept at constant temperature, the resulting free energy change is given by $\Delta G = \Delta H - T \Delta S$.

Free energy change determines the direction in which a process such as a chemical reaction proceeds. For example, at constant temperature free energy decrease (ΔG negative) for the reaction $4/3Al + O_2 = 2/3Al_2O_3$ indicates that the reaction will go from left to right; that is, aluminum will spontaneously oxidize when in contact with oxygen. On the other hand, for the reaction $2FeO = 2Fe + O_2$, the free energy increases (ΔG is positive), which means that the reaction will *not* go from left to right. In other words, iron oxide will not decompose spontaneously into elemental iron and oxygen. However, for the reaction $2Fe + O_2 = 2FeO$, at some temperatures, the free energy decreases (ΔG is negative), which indicates that iron will spontaneously oxidize to iron oxide. Thermodynamics does not predict the rate at which the oxidation or, for that matter, any reaction, proceeds. The rates are controlled by kinetics of the process. Kinetics will be discussed in a later section. Equilibrium is defined as a state of constancy in which changes do not occur over time. For example, in the hypothetical reaction $mA + nB = pC + qD$, at equilibrium, as many atoms of A and B react to form C and D as the latter two react to reform A and B. In thermodynamic calculations, the absolute values of enthalpy, entropy, and free energy are never known and are seldom required. The calculations are based on changes in these parameters. In order to assess changes, a "standard state" is used as a basis relative to which all changes are measured. The standard state is defined as the state in which the "pure" substance (solid, liquid, or gas) exists at a pressure of 1 atmosphere and temperature of 298°K (25°C). The absolute values of the parameters at standard state are generally indicated with a superscript such as $G°$.

Equilibrium Constant

The concept of equilibrium constant is adopted from chemistry and is used in calculations relating to free energy changes. For the reaction $4/3Al + O_2 = 2/3Al_2O_3$, the equilibrium constant is given by $K = [Al_2O_3]^{2/3}/[Al]^{4/3}[O_2]$, where [ ] denotes concentration. Note that the concentration is raised to a power equal to the number of molecules (or atoms) participating in the reaction. The free energy change of the reaction at standard state is related to the equilibrium constant through the equation $\Delta G = \Delta G° + RT \ln K$ where R is the gas constant, "ln" is the traditional symbol of natural logarithm, ΔG is the free energy change in an arbitrary state, and $\Delta G°$ is the free energy change involving reactants and products in their standard state. When the reaction is in equilibrium, $\Delta G = 0$. This gives $\Delta G° = -RT \ln K$. The convention used in assessing concentration in reactions defines *pure* materials in either reactants or products as having a concentration of unity. Thus $[Al_2O_3]$ and $[Al]$ are each unity, resulting in $\Delta G° = -RT \ln(1/[O_2]) = RT \ln[O_2]$.

Activity Coefficient

In the foregoing discussion, the concentrations $[Al_2O_3]$ and $[Al]$ each have been taken as unity because they are assumed to be in their standard state as pure solids. However, this assumption is not valid in many instances where Al exists in an alloy or a solution. In such cases the concept called *activity* is used instead of concentration. Thus, for a Ni base alloy containing 6% aluminum, knowing that the initial oxidation product is Al_2O_3, the oxidation reaction needs to be written as $4/3Al$ (alloyed in Ni) $+ O_2 = 2/3Al_2O_3$ with the equilibrium constant $K = [Al_2O_3]^{2/3}/[a_{Al}]^{4/3}[O_2]$, where a_{Al} is the activity of aluminum in the alloy. Because the product oxide is still pure Al_2O_3, its concentration (or activity) is unity. The free energy change is now given by $\Delta G° = -RT \ln K = RT \ln[O_2] + 4/3RT \ln a_{Al}$. This relationship defines activity, which is the "effective concentration" of the species in the alloy. A more complete definition of activity is given by $a_{Al} = P_{Al \text{ in alloy}}/P_{Al}$, the ratio of vapor pressure of Al over the alloy divided by the vapor pressure of Al over pure Al (which is its standard state) at the same temperature. The activity of gaseous species is represented by the partial pressure. We therefore replace $[O_2]$ by P_{O2}. The difference between the oxidation of a pure substance and an alloy can be elucidated by the following example of oxidation of Si:

$$Si \text{ (pure)} + O_2 = SiO_2, \Delta G° = -RT \ln\{[SiO_2]/[Si] \cdot [O_2]\} = RT \ln P_{O2},$$

because $[SiO_2]$ and $[Si]$ are each unity because both the metal and the oxide are pure solids in their standard state;

$$Si \text{ (in alloy)} + O_2 = SiO_2, \Delta G° = -RT \ln\{[SiO_2]/[a_{Si}] \cdot [O_2]\} = RT \ln a_{Si} + RT \ln P_{O2},$$

since $[SiO_2]$ is unity because the oxide is pure solid and in its standard state. However, Si is not in its standard state as pure solid but is in the alloy. The concentration is therefore replaced by its activity in the alloy.

2.2 CONCEPT OF KINETICS

Thermodynamic analysis through estimation of free energy changes predicts whether processes such as oxidation, dissolution, precipitation, or grain growth are possible. However, prediction of the rates at which these processes occur is outside the purview of thermodynamics. The principles of "kinetics" determine whether the processes take place at reasonable rates. A simple example of the limitation of thermodynamics is demonstrated by the relationship between diamond and graphite. The reaction C (as diamond) = C (as graphite) at ambient temperature and pressure has associated with it a negative free energy change. Therefore, diamond should spontaneously convert to graphite. However, we know that this does not occur. The underlying reason involves kinetics or rate of the reaction. Each atom in diamond is attached to four atoms at the corners of tetrahedra, whereas in graphite each atom in a plane is attached to three atoms with a weak bond between parallel planes. Thus, conversion from diamond to graphite would need extensive breaking and making of bonds with atomic rearrangement in between, all of which require expense of energy. In absence of this energy, the thermodynamic prediction is therefore not realized.

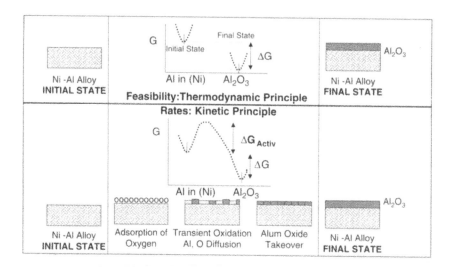

Figure 2.1 Illustration of the concept of activation energy.

Activation Energy

The example of diamond to graphite conversion, or lack of it, emphasizes the role of energy in the processes. This energy is called activation energy. The concept of activation energy is illustrated for oxidation of Ni–Al alloy in Fig. 2.1 by comparing the thermodynamic principle on the top half, which predicts feasibility, and kinetic principle at the bottom half, which predicts rates of the process.

As a constituent of the alloy, Al has a free energy (plotted on the vertical axis, the horizontal axis being known as the "reaction" coordinate axis) corresponding to the minimum of the parabola identified as "Initial State". The parabola reflects the dependence of free energy on the composition of the alloy. The final product of oxidation is Al_2O_3, which has associated free energy represented by another parabola identified as "Final State". The change in free energy in the process of oxidation is given by ΔG, the difference between the "Final State" and the "Initial State". The intermediate steps in the process play no role in the thermodynamic concept. Thermodynamics tell us that the oxidation reaction will occur because ΔG is negative.

Now let us look at the process in more detail as depicted in the bottom portion of the figure. Between the initial and the final sates there are several intermediate states, which include such steps as oxygen molecules dissociating into atoms, adsorption (and occasional desorption) of the dissociated oxygen atoms on the surface of the alloy, motion of the adsorbed atoms to seek low-energy sites, formation of transient oxides, diffusion of Al to the surface, and conversion of the transient oxides to the final oxide of Al. The slowest of the steps controls the rate of the overall process. The intermediate steps require external energy input. The magnitude of this energy required is represented by the height of the "hill," ΔG_{Activ}, between the two parabolas discussed earlier. In order to be successful in the oxidation process, the system needs to go from the "Initial State" to the "Final State" by climbing over the energy hill. This hypothetical hill is called the activation barrier. Thermal energy ($\sim k_B T$, where k_B is the Boltzmann constant) in the system provides energy to surmount the activation barrier. As the temperature of the system is increased, more thermal energy is available to overcome the barrier, and therefore the oxidation rate increases.

The rate of oxidation is the rate at which the barrier is climbed, which in turn is given by $k = k_1 \exp(-\Delta G_{Activ}/RT)$, in which k is called the rate constant, ΔG_{Activ} is the free energy of activation, and k_1 is a constant. In line with thermodynamic relationships, the free energy of activation is expressed in terms of enthalpy and entropy of activation, $\Delta G_{Activ} = \Delta H_{Activ} - T\Delta S_{Activ}$. Thus,

$$k = k_1 \exp(-\Delta G_{Activ}/RT) = k_1 \exp(\Delta S_{Activ}/R) \exp(-\Delta H_{Activ}/RT).$$

Combining the temperature independent and the temperature dependent terms separately, the rate constant is given by

$$k = k_0 \exp(-\Delta H_{Activ}/RT),$$

in which ΔH_{Activ} is called the activation energy and $k_0 = k_1 \exp(\Delta S_{Activ}/R)$ is independent of temperature. The exponential dependence on temperature is known as the *Arrhenius equation*. Taking the natural logarithm of both sides, we get

$$\ln k = \ln k_0 - (\Delta H_{Activ}/R)(1/T).$$

A simple method to determine the activation energy is to plot $\ln k$ at several temperatures against $1/T$, where T is temperature in Kelvin. The data should follow a straight line of negative slope $(\Delta H_{Activ}/R)$, where R is the gas constant. An example of rate constant plot as a function of temperature is given by Fig. 2.2 for oxidation of several metallic materials.

The concepts of rate constants and activation energy are used in Chapter 4 to study oxidation.

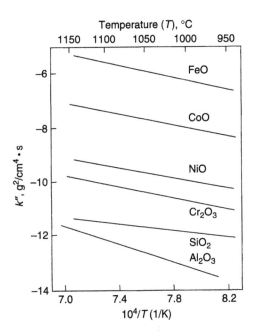

Figure 2.2 Rate constants of parabolic oxidation as function of temperature, $k'' = k (8/V)^3$, where V is the equivalent volume of oxide (James L. Smialek and Gerald H. Meier, High-temperature oxidation, in *Superalloys II*, Eds. C. T. Sims, N. S. Stoloff, and W. C. Hagel, Wiley, 1987). Reprinted with permission from John Wiley & Sons, Inc.

Diffusion

Diffusion is the process of mass transport through matter. Our interest is in mass transport through solids. Generally, there are three major mechanisms for diffusion in solids. In the first, known as volume diffusion, atoms diffuse by migrating from atomic sites through vacancies. The second mechanism is a modification of the first in which atoms migrate through defect sites such as dislocations, surfaces, and grain boundaries. The third mechanism involves movement through interstitial atomic sites. The driving force for diffusion is the reduction in free energy. In most practical cases this translates into the existence of a concentration gradient and diffusion occurring from higher concentration C_i to lower concentration C_f (Fig. 2.3).

The rate of diffusion is governed by Fick's first and second laws (Askill, 1970). These laws are represented by two equations.

The first law of diffusion: Under steady-state conditions, that is, when the concentration at any point does not change with time, the flux of diffusing species is given by Fick's first law,

$$J = -D\nabla c$$

with J being the flux (atoms/m^{2}s), D the constant of proportionality called the diffusivity or diffusion coefficient (m^2/s), and ∇ the differential operator (called "del") given by $i\partial/\partial x + j\partial/\partial x + k\partial/\partial x$. Here i, j, and k are unit vectors along the orthogonal x, y, and z directions, respectively. Thus, ∇c is the concentration gradient, c being the concentration (atoms/m^3). In one dimension, with the concentration gradient dc/dx, the first law reduces to

$$J = -D\ dc/dx.$$

The second law of diffusion: Under non-steady-state conditions, where the concentration at any point changes with time, the rate of change of concentration is given in three dimensions by

$$dc/dt = D\left(d^2c/dx^2 + d^2c/dy^2 + d^2c/dz^2\right).$$

In one dimension this equation reduces to

$$dc/dt = D\ d^2c/dx^2.$$

The diffusion coefficient D is strongly temperature dependent and can be expressed in the form of the Arrhenius equation,

$$D = D_0\exp(-\Delta H/RT)$$

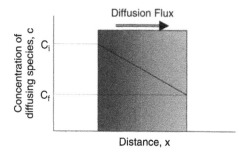

Figure 2.3 Diffusion along the concentration gradient.

where D_0 is a constant, which involves, among other factors, the jump frequency of atoms, and ΔH is the activation energy of the diffusion process. Typically, the relative values of diffusion coefficients rank in the following order:

$$D_{\text{surface}} > D_{\text{grain boundary}} > D_{\text{volume}}$$

or in terms of activation energy,

$$\Delta H_{\text{surface}} < \Delta H_{\text{grain boundary}} < \Delta H_{\text{volume}}.$$

Here the subscripts "surface," "grain boundary," and "volume" indicate that the diffusion paths are predominantly over surfaces, along grain boundaries, or within the grain, respectively. The comparative order of diffusion coefficients is very important in understanding the behavior of coatings. Coatings are typically fine grained with more grain boundaries compared with substrate materials. Diffusion rates in coatings are therefore expected to be faster than in substrates of similar composition.

2.3 CRYSTAL STRUCTURE

The concepts of crystal structure and phases are equally important in understanding coating behavior. Figure 2.4 illustrates some of these concepts. Of the three states of matter, our interest is predominantly limited to solids, which come in two forms, amorphous such as glass and crystalline such as diamond.

Crystalline materials exhibit regular periodic arrangements of "motifs" of atoms, ions, or molecules in a three-dimensional array known as a *lattice*. The smallest unit of this arrangement in a lattice is called a unit cell. Amorphous materials, on the other hand, are characterized

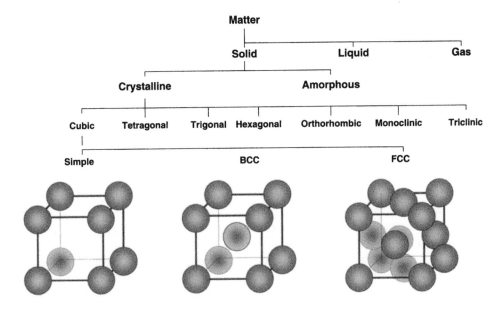

Figure 2.4 Crystal structure and atomic arrangement.

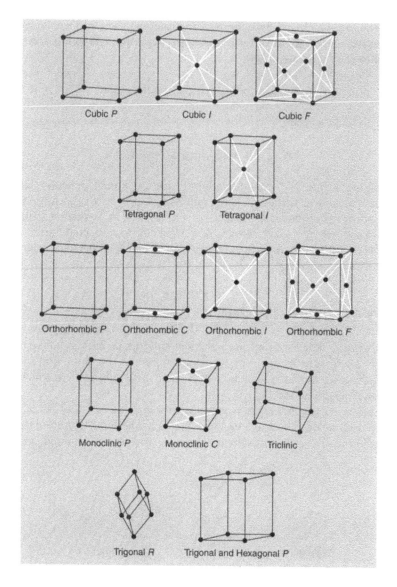

Figure 2.5 The seven crystal systems and 14 Bravais lattices (Charles Kittel, *Introduction to Solid State Physics*, 5th ed., Wiley, 1976). Reprinted with permission from John Wiley & Sons, Inc.

by atomic arrangements with only short-range order. Based on the size and shape of the unit cells, crystalline materials are classified into one of seven crystal systems, some of which may have lattice points at the center of the unit cell or on the face centers, forming the 14 "Bravais lattices" shown in Fig. 2.5.

Most metallic systems belong to the cubic crystal system in which the unit cells are cubes. In a few metals and many ceramics, the atoms are arranged in noncubic unit cells such as hexagonal or orthorhombic systems. In the cubic system of atomic arrangement, there are three possibilities shown at the bottom of Fig. 2.4. The arrangements are known as the simple cubic (sc), body-centered cubic (bcc), and face-centered cubic (fcc). In the simple cube the atoms are arranged at the corners of the cubic lattice. In the bcc structure, in addition to the corners there are atoms at the center of the cubic lattices. The fcc structure contains atoms at the center of each face in addition to those belonging to the corners. The hexagonal structure of some metals contains an atom at each corner of the hexagon, an atom at the center of each of the end faces, and three atoms located in specific positions within the hexagon.

In the simple cube each unit cell has eight atoms at the corners. Each corner is shared among eight neighboring cells. Thus, each unit cell effectively "owns" only one eighth of each atom, totaling to one. In the case of a bcc unit cell, the corner atoms are each shared among eight unit cells, while the atom in the body center belongs to only one cell. Adding these gives two atoms per unit cell. Similar analysis shows that an fcc unit cell has four atoms.

Defects in Crystals

Although ideal crystals of metal would consist of a three-dimensional periodic arrangement of atoms, real crystals always contain many types of defects, some of which are in thermodynamic equilibrium. Occasionally atoms would be missing from their ideal position, creating "vacancies" that are called point defects. The vacancies are an essential part of the crystal structure helping in the process of diffusion. Dislocations are another group of imperfections, called line defects, in which the crystalline lattice is distorted around a line, as shown in Fig. 2.6 (Callister, 1995). Dislocations originate during solidification of crystals. They are also created by condensation of vacancies as well as by the process of plastic deformation. As Fig. 2.6 shows, there are two types of dislocations (Hull, 1969), the edge and the screw. The edge dislocation can be visualized as formed by the insertion of an extra half plane of atoms. The screw dislocation, on the other hand, may be imagined as the line created by shear stress (Fig. 2.6), above which the upper part of the crystal has been moved one lattice unit to the right relative to the bottom part of the crystal. Relative ease of movement of dislocations provides metals

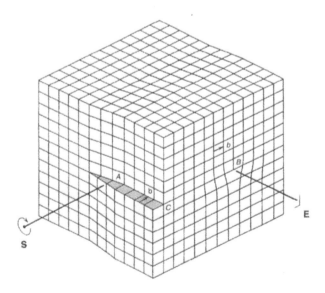

Figure 2.6 Schematic of edge (E) and screw dislocations (S). Burgers vector is a measure of displacement associated with dislocations (William D. Callister, Jr., *Materials Science and Engineering, An Introduction*, 3rd ed., Wiley, New York, 1995). Reprinted with permission from John Wiley & Sons, Inc.

with ductility and capacity to plastically deform under stress. In addition to point and line defects, crystalline materials also contain surface defects such as grain boundaries, which are the interfaces between adjacent grains.

2.4   EQUILIBRIUM PHASES

Pure metals are crystalline and conform to specific crystal structures, which may change with temperature. For example pure iron has a bcc structure which changes to fcc at temperatures between 912 and 1394°C (1674 and 2541°F). It changes back to bcc above 1394°C (2541°F). However, at a constant temperature, iron, like many other pure metals, exhibits the same crystal structure. When metals are alloyed with other metals or non-metals, the crystal structure may change. Let us consider Ni as an example. It crystallizes in an fcc structure. If we alloy Ni by adding Al, several significant changes occur. For up to about 4 wt % Al, there is no change in the atomic arrangement of Ni except an occasional Al atom replacing a Ni atom randomly, as in Fig. 2.7(a). The structure remains essentially like fcc Ni, called the γ phase. As the Al content is increased, it starts selectively replacing the corner atoms (Fig. 2.7(b)), while the atoms on the cube faces remain Ni. The share of atoms for each cell is 1 Al and 3 Ni, giving the composition Ni_3Al, which is known as the γ' phase. If the alloying addition is continued, eventually at about 25 wt % Al, the crystal structure changes so that the corner atoms remain Ni, while Al enters the center of the cube. The central Al atom belongs to this cell, while the corner atoms are each shared with eight cells. The overall share is therefore 1 Al and 1 Ni, providing the composition NiAl, known as the β phase, Fig. 2.7(c).

Binary Phase Diagram

In the Ni–Al alloy system, as the Al content is varied, other phases in addition to the γ, γ', and β appear. A map of all of these phases formed as a function of composition and temperature is called the (equilibrium) phase diagram, shown in Fig. 2.8 (based on *Metals Handbook,* 1973). Equilibrium here refers to the fact that the phases in the map remain stable over time when held at appropriate temperatures. Because the diagram shown here describes an alloy of two components only, Ni and Al, it is called a binary phase diagram. The two phases bounding the diagram (in this case fcc Al and Ni) are called the terminal phases.

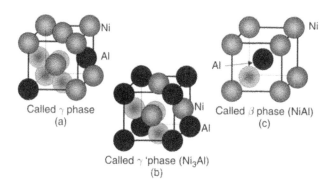

Figure 2.7   Effect of alloying on crystal structure.

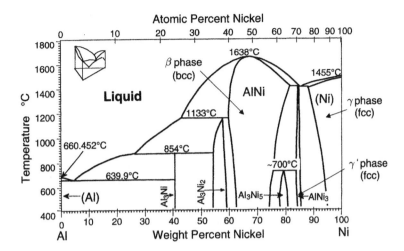

Figure 2.8 The Ni–Al phase diagram (*Metals Handbook*, 8th ed., Vol. 8, ASM, Metals Park, OH, 1973). Reprinted with permission from ASM International.

Additional information contained in the diagram includes the melting temperatures as a function of composition and the range of composition within which multiple phases coexist in equilibrium. Quantitative estimation of phases can also be made. The composition is represented in weight percent or atomic percent. Some features of the Ni–Al phase diagram are obvious from Fig. 2.8. The γ phase dissolves up to 4 wt % Al below 400°C (752°F). The solubility increases as the temperature is raised. The melting point of the resulting alloy decreases as Al content is increased in this phase. The γ' phase, on the other hand, has a narrower phase field, which means that the composition does not depart significantly from Ni$_3$Al. The β field is very broad, indicating that in this phase the concentration of Al can depart widely from the stoichiometric composition NiAl. Also, the melting point of the β phase is much higher than that of pure Ni. These characteristics of the β phase are very favorable to have in high temperature coatings. The high Al concentration helps in providing a large reservoir of Al for oxidation protection through the formation and replenishment of Al$_2$O$_3$ scale. The change in microstructure of a β-phase-containing coating on thermal exposure can also be analyzed by the use of the phase diagram.

Ternary Phase Diagram

Practical alloys and coatings are seldom limited to two components. Alloys with three components require a ternary phase diagram to map the phases. The ternary diagrams are actually planar slices of three-dimensional plots, shown as inset in Fig. 2.8, with the sides of the triangle representing concentrations of the three components and the vertical axis denoting temperature. Each planar slice corresponds to a particular temperature. Figure 2.9 shows an Ni–Cr–Al ternary diagram at 850°C (1562°F). The individual phase fields, as well as fields, in which multiple phases coexist, are generally clearly indicated.

The elemental composition corresponding to a particular point such as at A in the ternary can be clculated in the following way:

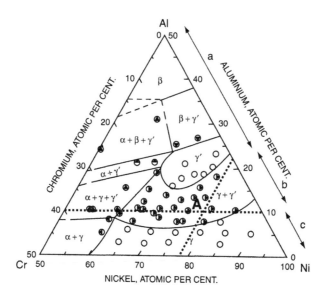

Figure 2.9 Isothermal Ni–Cr–Al ternary phase diagram showing the various phases as a function of composition at 850°C (A. Taylor and R. W. Floyd, The constitution of nickel rich alloys of the nickel–chromium–aluminum system, *J. Inst. Metals*, 1952–53, 81, 451–464). Reprinted with permission from Maney Publishing.

- Two lines are drawn through the point A, each line being parallel with one of the two sides of the triangle.

- The three intercepts *a*, *b*, and *c* are measured. Their ratios follow the ratios of elemental contents. One has to remember that *a*, *b*, and *c* are respectively on opposite sides of (or across from in the case of Cr) Ni, Cr, and Al:

$$\mathrm{Ni} : \mathrm{Cr} : \mathrm{Al} = a : b : c$$

As more alloying elements are added, the phase diagram becomes more complex. Such complex diagrams can be calculated for specific alloy systems using thermodynamic data and computer programs like Computer Coupling of Phase Diagram, acronym CALPHAD (www.calphad.com) and Thermo-Calc (www. thermo-calc.com).

REFERENCES

Askill, J., *Tracer Diffusion Data for Metals, Alloys, and Simple Oxides*, IFI/Plenum, New York, 1970.

Callister, W. D., Jr., *Materials Science and Engineering: An Introduction*, 2nd ed., Wiley, New York, 1995.

Hull, D., *Introduction to Dislocations*, 1st ed., p. 64, Pergamon Press, New York, reprinted 1969.

Kittel, C., *Introduction to Solid State Physics*, John Wiley & Sons, Inc., 5th Ed, 1976.

Metals Handbook, 8th ed., Vol. 8, pp. 262, Metals Park, Ohio, ASM, 1973.

Smialele, J. L., and G. H. Meier, in Superalloys II, Eds, C. T. Sims, N. S. Stoloff, and W. C. Hagel, John Wiley & Sons, New York, 1987.

Taylor, A., and R. W. Floyd, The constitution of nickel rich alloys of the nickel-chromium-aluminum system, *J. Inst. Metals*, 1952–53, 81, pp. 456.

Chapter 3

SUBSTRATE ALLOYS

The role of materials with high-temperature capability in many load-carrying components in industrial applications has already been introduced in Chapter 1. Although nonmetallic materials such as monolithic ceramics and ceramic composites are making inroads into some of these applications, the demand for materials is met predominantly by metallic alloys. These alloys are selected because of high melting point, strength (tensile strength, creep strength, fatigue strengths), ductility and toughness, and low density wherever possible. The environmental resistance of the alloys is seldom adequate. High-temperature coatings, the primary focus of this book, are therefore used extensively to protect the alloys from oxidation, corrosion, heat, and other processes of degradation.

3.1 TEMPERATURE CAPABILITY OF METALS AND ALLOYS

Although alloys of many metals find specialized applications in several fields, such as the electrical industry, electronics, medical equipment, and prosthetic devices, four groups of alloys based on the metals aluminum, titanium, iron, and nickel and cobalt support the bulk of the demand in industrial processes. The range of density-corrected strengths of these classes of alloys as a function of temperature is shown schematically in Fig. 3.1. It is clear from the figure that aluminum, with its low melting point, 660°C (1220°F), and its commercially available alloys with temperature capability limited to about 150°C (300°F), does not satisfy requirements for high-temperature application. The only alloys relevant to our interest include the nickel and cobalt base alloys and to some extent the titanium alloys and steels. The mechanisms by which these alloys are strengthened are briefly discussed next.

3.2 STRENGTHENING MECHANISMS

Pure metals are weak and do not have adequate environmental resistance. Therefore, they are seldom used without strengthening by various mechanisms and protected from the environment by surface treatments. Table 3.1 summarizes a number of processes by which metals are strengthened.

Solid-solution strengthening and precipitation hardening are effective at high temperatures and, therefore, are used in strengthening high-temperature alloys. Of the other mechanisms listed in Table 3.1, grain size control is used in conjunction with solid-solution and precipitation hardening. The oxide dispersion strengthening (ODS) process is sparsely used in selected cases for moderate-temperature applications. Martensitic transformation and work hardening are not effective at temperatures of interest here.

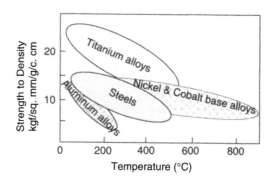

Figure 3.1 Temperature capabilities of several classes of alloys.

Table 3.1 Strengthening Mechanisms for Metals

Mechanisms	Effective Temperature
Solid-solution strengthening	High temperature
Precipitation (or age) hardening	High temperature
Grain size control	Moderate temperature
ODS	Moderate temperature
Martensitic transformation strengthening	For specific metals and at low temperature
Work hardening	Low temperature

Strengthening by solid solution and precipitation involves impeding motions of dislocations in the metallic structure. In a solid solution, appropriate alloying elements (called the solute) are dissolved in a metal (called the solvent). The solute atoms randomly substitute for the solvent atoms (Fig. 3.2a) without altering the phase; that is, the crystal structure. In the vicinity of the solute atoms, compressive or tensile stresses are generated depending respectively on whether the substituting atom is larger or smaller than the solvent atom. Motions of dislocations are impeded by the presence of the stress fields. The size differences between the solvent and the solute atoms as well as the volume fraction of the solute determine how effective the strengthening would be. For this mechanism to be useful, the solute atoms should have reasonable solubility in the solvent metal, and the atomic size difference should be significant.

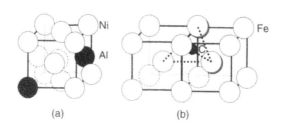

Figure 3.2 Schematic of solid solution (a) substitutional (Al dissolved in Ni), (b) interstitial (C dissolved in Fe).

In addition to substitutional solid solution in which solute atoms randomly replace solvent atoms, the solute atoms may also enter the solvent in the interstitial sites (Fig. 3.2b). For example, carbon, being a small atom, forms an interstitial solid solution with iron. The stress fields associated with the interstitial atoms also interfere with dislocation motion, thereby influencing strength.

In precipitation hardening, the solute atoms participate in creating a fine and uniformly dispersed second phase. To achieve this condition, the solvent metal needs to form a supersaturated solid solution at high temperature. On cooling, the solubility decreases and the solute atoms are rejected from solution in the form of fine precipitates. These precipitates impede dislocation motion. The size, strength, "coherency," and volume fraction of the precipitates control the strength achieved by this process. These parameters are controlled through appropriate heat treatment of the alloys.

3.3   TITANIUM ALLOYS

Titanium and its alloys (Donachie, 2000) have very attractive structural properties. The metal has a low density, $4.54\,g/cm^3$, about 60% of that of steel and nickel and cobalt base superalloys, a high melting point, 1668°C (3035°F), and excellent corrosion resistance. Titanium exhibits a hexagonal close packed (hcp) structure called α which transforms into a body-centered cubic (bcc) structure β on heating to above 883°C (1621°F). Titanium alloys are very strong and, on a density-corrected basis, form some of the strongest alloys available. Because of their high strength and low density, titanium alloys are extensively used in gas turbine engine as fan blades, compressor blades and vanes, disks, and cases. One of the limitations of titanium alloys is their susceptibility to interstitial formation during processing or application above 600°C (1100°F) stabilizing the α phase. This occurs by the spontaneous capturing of oxygen (as well as carbon and nitrogen). The α phase forms a brittle skin known as the "α case". The formation of the α case debits structural properties of the alloys. Prior to any processing, the α case, therefore, needs to be chemically or mechanically removed.

Titanium alloys fall in one of three families depending on the predominant phases: the alpha alloys, the alpha–beta alloys, and the beta alloys. In the alpha alloys by the α phase is stabilized by the addition of Al, which raises the β transus, the transus being the temperature boundary between the single-phase and two-phase regions (Fig. 3.3a). The beta alloy, on the other hand, contains elements such as V, Nb, Mo, Cr, W, Fe, Co, and Si, all of which stabilize the β phase by lowering the β transus (Fig. 3.3b). In the alpha–beta alloy both the phases exist together. A few elements such as Zr, Sn, and Hf are neutral in that they do not stabilize either of the phases but contribute to property improvement. The alpha alloy derives its strength from solid solution strengthening due to Al and Sn. Every 1% Al addition increases strength by about

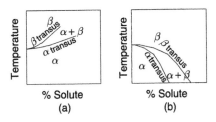

Figure 3.3   Schematic of (a) α and (b) β alloy stabilization.

Table 3.2 Some commercially available titanium alloys.

Alloy	Year of Introduction	Phase Composition	Maximum Use Temperature
Ti–6Al–4V	1954	$\alpha\beta$	315°C (600°F)
Ti–6Al–2Sn–4Zr–2Mo	1967	$\alpha\beta$	510°C (950°F)
Ti–6Al–2Sn–4Zr–2Mo–Si	1974	$\alpha\beta$	565°C (1050°F)
Ti–5.5Al–4Sn–4Zr–0.3Mo–1Nb–0.5Si	1984	Near α	593°C (1100°F)

8 ksi (55 MPa), while a 4 ksi (28 MPa) increase comes from every 1% addition of Sn. The alpha alloys are insensitive to heat treatment due primarily to the lack of second phases. The beta alloys always contain small amounts of the α phase. The size and distribution of the α phase are controlled by a combination of thermal and mechanical treatments. The improved strength is derived from this second phase precipitate. The most useful titanium alloys, however, consist of both α and β phases. These alloys are amenable to heat treatment, which is beneficially used to control both solid solution strengthening and precipitate hardening. A number of commercially available titanium alloys are shown in Table 3.2. Figure 3.4 (Donachie, 2000) shows temperature dependence of creep strength of several Ti alloys. The data in Table 3.2 as well as in Fig. 3.4 clearly show that the maximum use temperature of these titanium alloys is limited to 593°C (1100°F).

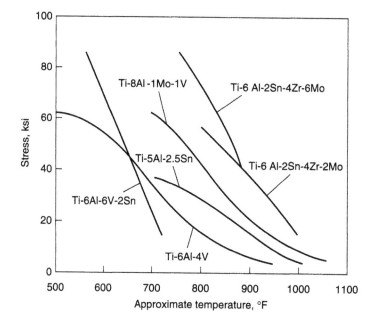

Figure 3.4 Creep strength (0.1% creep in 150 hrs) of selected titanium alloys as a function of temperature (Mathew J. Donachie, Jr., *Titanium: A Technical Guide*, 2nd. ed., ASM International, 2000). Reproduced with permission from ASM International.

When exposed to an oxygen-containing atmosphere at elevated temperature, titanium alloys not only dissolve oxygen and form α case but also oxidize. The product of oxidation is a TiO_2 (titania) scale. Oxygen easily diffuses through titania and, therefore, unlike alumina on many aluminum-containing alloys, titania is nonprotective against continued oxidation.

To protect titanium alloys against oxidation at elevated temperatures, several coating systems such as platinum aluminides (Gurrappa and Gogia, 2001), magnetron sputtered Ti–Al (Leyens et al., 1996), and Ti–Al–Cr (Leyens et al., 1997) have been investigated. Some of these coatings processes are covered in chapter 6.

3.4 STEELS

Steels (Honeycombe, 1982) are among the most versatile, widely used, and complex structural materials. Density-corrected strengths of steels are competitive with nickel base superalloys at lower temperatures as shown in Fig. 3.1, and at much lower cost. However, at higher temperatures their strengths fall off below those of the superalloys. The structure of steels is based on the iron–carbon phase diagram (Fig. 3.5). Although this is not a true equilibrium phase diagram because of the nonequilibrium cementite (Fe_3C) as a terminal phase, for all practical thermal exposures, the diagram faithfully explains the microstructure and phase behavior. To achieve a wide range of structural and physical properties, other alloying constituents

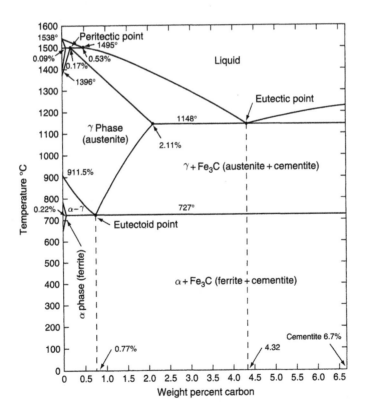

Figure 3.5 The iron–iron carbide metastable phase diagram (J. Chipman, *Metall. Trans.*, 1972, 3). Reprinted with permission from ASM International.

besides carbon are also added to steels. Depending on the carbon content, steels are classified into low-carbon (<0.25 wt % C), medium carbon (0.25 to 0.60 wt % C), and high carbon (>0.60 wt % C) steels. Each of the classes may contain additional alloying elements such as Cr, Mo, Ni, V, Nb, Ti, and Al to form what are known as alloy steels. As shown in Fig. 3.5, pure iron at room temperature has a bcc structure called α iron or "ferrite." Above 911.5°C (1673°F) it converts to a fcc γ phase called "austenite." On further heating it changes back to a bcc phase called "δ ferrite" at 1396°C (2546°F), finally melting at 1538°C (2800°F). As Fig. 3.5 shows, the sizes of the various phase fields vary with the carbon content.

Steels are strengthened by a combination of solid solution strengthening by interstitial elements such as carbon and nitrogen, substitutional solid solution either in the α or the γ phase, precipitation hardening from carbides of Nb and V, grain size control, and processing-induced texture of grains such as by rolling. Different alloying elements stabilize different phases. The degree of solid-solution strengthening by substitution is controlled by expanding (with the addition of C, N, Cu, and Zn), contracting (with the addition of Ta, Nb, and Zr), opening (with the addition of Ni, Mn, Co, Ru, Rh, Pd, Pt, etc.), or closing (with the addition of Si, Al, Be, and P) the γ phase field of Fig. 3.5.

The austenite phase in Fig. 3.5 can dissolve as much as 2.11 wt % ($\sim$10 at %) carbon in a solid solution. If austenite is quenched very fast to a relatively low temperature in the vicinity of room temperature, instead of changing to ferrite and cementite as the phase diagram Fig. 3.5 shows, it transforms into a phase called "martensite" supersaturated with carbon. The high carbon content distorts the crystal structure to body-centered tetragonal (bct). The unique feature of this transformation is that it *does not involve diffusion of atoms*. Instead, the phase forms by shear or local displacive transformation at rates close to the velocity of sound in the alloy. The maximum temperature below which martensitic transformation occurs is known as the martensite start temperature M_s. The martensite grains, with plate or needle-like morphology, are hard and brittle. The presence of martensite in steels increases strength significantly.

To achieve the optimum combination of microstructure, strength, and ductility of steels, heat treatment regimens have been developed using TTT (time temperature transformation) diagrams (Honeycombe, 1982).

Some of the commercially available steels include the following:

High Strength Low Alloy Steel (HSLA): These low carbon (0.03 to 0.15 wt % C) steels derive their strengthening from a solid solution (C, Mn, Si), precipitation hardening (carbides of Nb, Ti, and V), grain size control stabilized by grain boundary carbides, and processing-induced texture. Typical tensile strength and failure strains are 700 MPa (100 ksi) and 12 to 18%, respectively.

Bainite Steel: Such steels consist of a fine dispersion of cementite in a strained ferrite matrix. Bainite steels are produced by the suitable isothermal heat treatment of low carbon (<0.05 wt % C) steels using a TTT diagram between 250°C (482°F) and 550°C (1022°F). Typical tensile strength and failure strains are 600–1200 MPa (86–171 ksi) and 15–20%, respectively.

Dual-Phase Steels: These low carbon (0.1 to 0.2 wt % C) steels contain two phases, a fine dispersion of 10 to 20 vol % martensite in a matrix of ferrite. The microstructure and phase contents are achieved by heating to the $\alpha + \gamma$ phase field (Fig. 3.5) to form 10 to 20 vol % austenite followed by rapid quenching to convert to martensite. These steels have moderate to high tensile strength with high ductility.

TRIP (Transformation Induced Plasticity) Steels: In these steels, austenite (γ phase) is retained at room temperature by fast quenching followed by mechanical deformation during processing. During use or testing under load, the austenite phase transforms to martensite, providing the strengthening mechanism. Tensile strengths in the range of 600 to 1300 MPa (86 to 171 ksi) and failure strains of 25–40% are not uncommon.

The temperature capability of steels in stressed applications is controlled by strength as well as environmental resistance and is generally limited to 650°C (1200°F).

3.5 NICKEL–IRON ALLOYS

This is a class of alloy (Brown and Muzyka, 1987) containing 15 to 60% iron and 25 to 60% nickel with the fcc structure of γ austenite as a matrix and strengthened by precipitates with additional benefits from solid solution and grain boundary strengthening. These alloys are less expensive than Ni and Co base superalloys. Ni–Fe alloys may be classified into five groups.

Group 1: This group is relatively Fe rich with 25 to 35% Ni and less than 2% Ti. The strengthening phase is the coherent fcc γ' (called gamma prime, to be discussed in Section 3.7) precipitates. The use temperature is limited to 650°C (1200°F). Commercial alloys in this group include Tinidur, V-57, and A-286.

Group 2: This group is rich in Ni (> 40%) with increased contribution from a solid solution. Strengths of this group of alloys exceed those of the iron rich group. Commercial members of this group of alloys are Inco X-750 and Inco 901.

Group 3: This group is rich in Ni and owes its strength to coherent bct (body-centered tetragonal) γ'' (called gamma double prime) precipitates. Inco 706 and 718 are examples of this group, with the latter being one of the most used alloys of this class. The temperature capability of these alloys extends from cryogenic to 650°C (1200°F).

Group 4: This group is an Fe-rich Fe–Ni–Co alloy with low thermal expansion achieved through removal of Cr and Mo, which stabilize ferrite in the structure. The strength of this group of alloys is derived primarily from the coherent fcc γ' precipitates. Commercial members of this class are Inco 903 and 909 with temperature capability limited to 650°C (1200°F). Elimination of Cr makes the alloys more susceptible to oxidation and corrosion.

Group 5: The unique feature of this Ni-rich group is the absence of coherent precipitates. Some of the members of this group derive strength from precipitates of carbides, nitrides, and carbonitrides. Others (Hastelloy X and N-155) are solid-solution strengthened. The application temperature of the latter for nonstressed application is limited to 1093°C (2000°F).

3.6 NICKEL AND COBALT BASE SUPERALLOYS

This is a unique class of complex alloys based on Ni and Co that not only exhibits extraordinarily high strength but also maintains strength across a wide elevated temperature range, hence the name "superalloys." Some of these alloys are used in load-bearing application at ≥80% of their melting point! The two metals are attractive because of their high melting point and crystal structures amenable to extensive alloying. Whereas Ni exhibits fcc crystal structure, Co is hexagonal close packed (hcp) at room temperature. Alloying element additions are generally

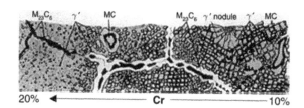

Figure 3.6 Microstructure evolution of Ni base alloy and Cr content change (adapted from C. T. Sims, *Superalloys: Genesis and character*, in *Superalloys II,* Eds. C. T. Sims, N. S. Stoloff, and W. C. Hagel, Wiley, 1987). Reprinted with permission from John Wiley & Sons, Inc.; C. R. Brooks, *Heat Treatment, Structure and Properties of Nonferrous Alloys,* ASM, 1982; Mathew J. Donachie, Jr., and Stephen J. Donachie, *Superalloys: A Technical Guide,* 2nd. ed., ASM, 2002). Reprinted with permission from ASM International.

used to stabilize Co in the fcc form. A number of strengthening mechanisms are utilized to produce practical alloys with Ni and Co. These are depicted in Fig. 3.6 (Brooks, 1982; Sims, 1984; Donachie and Donachie, 2002) and include:

1. *Strengthening of the austenitic γ phase by solid solution:* The γ phase constitutes the matrix of the alloys. Several metallic elements dissolve in the γ matrix (Cr, Co, Fe, Mo, Ta, W, Re in Ni base alloys and Ni, Cr, Mo, W, Nb, Ta in Co base alloys), randomly replacing the matrix atom (Fig. 3.2a). Because of their size difference relative to the matrix, the substituting solute atoms create stress fields that impede dislocation motion by interacting with the stress fields of dislocations. Slowing dislocation movement increases strength. This mechanism operates both in Ni and Co base superalloys.

2. *Precipitates of various carbides within the grains and at the grain boundaries:* Carbon, which is invariably present in the matrix, forms carbides with some of the alloying elements (MC type carbide with Ti, Ta, W, Mo, Hf, Nb, and $M_{23}C_6$ type carbide with Cr, W, and Mo in Ni base alloy; in Co base alloys MC type carbide with Ti and $M_{23}C_6$ type carbide with Cr) during processing and heat treatment. Carbides within grains impede dislocation motion, thereby increasing strength. Grain boundary carbides pin boundary movement, discouraging grain growth. They also impede grain boundary sliding. The carbides together with the solid solution provide the primary mechanism for strengthening of Co base superalloys. For Ni base superalloys these mechanisms are secondary to γ′ and γ″ precipitation strengthening.

3. *Coherent fcc γ′ and bct γ″ precipitates:* Precipitates of the ordered intermetallic compound of composition Ni_3Al, known as gamma prime, γ′, provide the primary mechanism for strengthening Ni base superalloys. Supplemental strengthening is achieved from solid solution and carbide precipitates. In the γ′ crystal structure, the Ni sites may also contain Co, Cr, and Mo, while the Al sites may contain Ti and Nb. The unique characteristic of the γ′ structure is that it is coherent with the γ matrix. Figure 3.7 compares schematically the matrix–precipitate interface for coherent and incoherent precipitates. Coherent interfaces have low energy. The precipitates have, therefore, little driving force for growth in size, and therein lies the clue to the strength retention of γ′-strengthened superalloys over a wide high-temperature range. These fine and hard precipitates impede dislocation motion, resulting in a spectacular increase in strength. Figure 3.8 (DeLuca and Annis, 1995) shows cuboidal γ′ precipitates in a γ matrix.

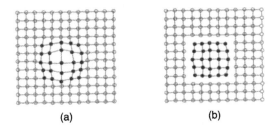

(a)   (b)

Figure 3.7 Schematic of (a) a coherent precipitate with large strain fields indicated by distortion of the lattice, (b) incoherent precipitate.

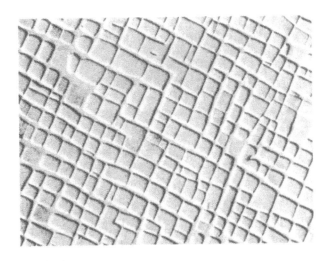

Figure 3.8 Electron micrograph showing gamma prime precipitates (size 0.35 to 0.6 μm) in a gamma matrix. Courtesy of Daniel P. DeLuca.

The size and the volume fraction of γ′ are critical parameters affecting creep and fatigue strength of the alloys. The former is controlled by the Al + Ti content and ranges from 0.2 for wrought Fe–Ni superalloys to greater than 0.6 in Ni base superalloys. The size ranges between 0.2 and 0.5 μm. Fine-size γ′ precipitates have spherical morphology, whereas larger sizes tend to be cuboidal.

In some superalloys containing Fe such as Inco 718, the predominant strengthening phase is γ″, which has a bct structure with composition Ni_3Nb. The γ″ phase is coherent with the γ matrix and provides high strengths in the low to moderate temperature regime. However, above 650°C (1200°F), γ″ decomposes to either γ′ or δ, resulting in a significant drop in strength.

The range of creep rupture strengths of superalloys strengthened by several of the mechanisms discussed earlier is shown in Fig. 3.9 (Donachie and Donachie, 2002).

For some creep-strength-critical applications, such as rotating turbine blades in modern gas turbine engines, Ni base superalloys in polycrystalline form are not acceptable because they tend to fail at grain boundaries oriented transversely to the centrifugal force of rotation, which is along the blade axis. These blades are produced by the process of casting, which involves melting of appropriate alloys in a furnace and pouring the melt into a ceramic shell.

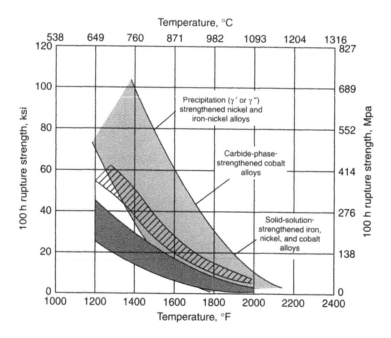

Figure 3.9 Superalloy creep rupture strengths as function of temperature (Mathew J. Donachie, Jr. and Stephen J. Donachie, *Superalloys: A Technical Guide,* 2nd. ed., ASM, 2002). Reprinted with permission from ASM International.

The creep properties of the alloys can be significantly improved by eliminating transverse grain boundaries and encouraging columnar grains oriented parallel to the blade axis. Such grain structures are achieved through "directional solidification" by modifying the casting process so that heat is removed from the melt by slowly pulling the ceramic shell from the furnace. Further increases in creep properties have been made possible by eliminating grain boundaries altogether in single-crystal superalloys. This is achieved by using a grain selector or appropriate seeds in the casting process. Figure 3.10 (Gell et al., 1987) shows the structure of polycrystalline, columnar, and single-crystal turbine blades developed by Pratt & Whitney.

3.7 NEED FOR COATINGS

The focus of the development of substrate alloys just discussed is generally to achieve high strength, ductility, and producibility. Environmental resistance toward oxidation and high-temperature corrosion does not go hand in hand with those goals. For example, increased Al and Cr provide improved oxidation and corrosion resistance. However, beyond a certain level, these elements reduce creep strength of the resulting alloys. To achieve both strength and resistance to the environmental degradation, the two functions are separated. The load capability is provided by the substrate alloy, while oxidation and corrosion resistance is achieved by the application of thin coatings with adequate Al and Cr. The thickness of the coating is so adjusted that it does not carry any significant load.

Depending on the temperature of use, each of the alloys discussed in this chapter may require coatings compatible with its composition and structural (modulus) and thermal

Figure 3.10 Turbine blades made of nickel base alloys with various grain structures (M. Gell, D. N. Duhl, D. K. Gupta, and K. D. Sheffler, Advanced superalloy airfoils, *JOM*, July 1987). Reprinted with permission from the Minerals, Metals & Materials Society.

(coefficient of thermal expansion) properties. The coating processes discussed in Chapters 6 and 7 are applicable to all of them.

REFERENCES

Brooks, C. R., *Heat Treatment, Structure and Properties of Nonferrous Alloys*, ASM, 1982.

Brown, E. E., and D. R. Muzyka, Nickel–iron alloys, in *Superalloys II,* Eds. C. T. Sims, N. S. Stoloff, and W. C. Hagel, John Wiley & Sons, New York, 1987, pp. 165–188.

Leyens, C., M. Schmidt, M. Peters, W. A. Kaysser, Sputtered intermetallic Ti–Al–X coatings: Phase formation and oxidation behavior, *Mater. Sci. Eng.*, 1997, A239–240, 680–687.

Chipman, J., Thermodynamics and phase diagrams of the Fe–C system, *Met. Trans.*, 1972, 3, 55.

DeLuca, D. P., and C. Annis, Fatigue in single crystal nickel superalloys, *Pratt & Whitney Report FR-23800*, 1995, p. 18.

Donachie, Matthew J., Jr., *Titanium: A Technical Guide*, 2nd ed., ASM International, 2000.

Donachie, Matthew J., Jr., and Stephen J. Donachie, *Superalloys: A Technical Guide*, 2nd ed., ASM, 2002.

Gell, M., D. N. Duhl, D. K. Gupta, and K. D. Sheffler, Advanced superalloy airfoils, *JOM*, July 1987, pp. 11–15.

Gurrappa, I., and A. K. Gogia, Development of oxidation resistant coatings for titanium alloys, *Mater. Sci. Technol.*, 2001, 17, 581–587.

Honeycombe, R. W. K., *Steels: Microstructure and Properties*, Edward Arnold and ASM, 1982.

Leyens, C., M. Peters, D. Weinem, and W. A. Kaysser, Influence of long-term annealing on tensile properties and fracture of near-alpha titanium alloy Ti–6Al–2.75Sn–4Zr–0.4Mo–0.45Si, *Metall. Trans.*, 1996, 27A, 1709–1717.

Sims, C. T., Superalloys: Genesis and character, in *Superalloys II,* Eds. C. T. Sims, N. S. Stoloff, and W. C. Hagel, John Wiley & Sons, New York, 1987, pp. 3–26.

Chapter 4

OXIDATION

Oxidation is an environmental phenomenon in which metals and alloys (and other materials) exposed to oxygen or oxygen-containing gases at elevated temperatures convert some or all of the metallic elements into their oxides. The oxide can form as a protective scale if it remains adherent, and reduces further oxidation, or may continually spall off, exposing fresh metal. The latter results in progressive metal loss. Additionally, internal oxidation may occur. The technological implications of oxidation lie in the loss of load-bearing capability of the original metal or alloy component, eventually resulting in component failure.

Oxidation is studied by controlled exposure of metals to oxidizing atmosphere such as O_2, CO_2, H_2O, NO_2, and combinations thereof, at high temperature, for various lengths of time. Continuous weight change may be monitored in microbalances, or metal thickness loss may be monitored microstructurally. Although all of the gases just listed induce oxidation in metals and alloys, the oxide composition, morphology, adherence, and spallation characteristics strongly depend on the gas composition.

4.1   OXIDATION PROCESS

Most metals oxidize readily because the free energy change ΔG associated with the reaction is negative. An example is aluminum oxidizing to form aluminum oxide:

$$4/3Al + O_2 = 2/3Al_2O_3;$$

$\Delta G^\circ = -251.8$ kcal/mole of oxygen at room temperature and 1 atmosphere (atm) pressure.

Temperature Effects

When a metal in standard state (pure solid at 298 K, 77°F) reacts with oxygen in standard state (as a gas at 1 atm pressure and 298 K) to form an oxide in standard state (pure solid at 298 K), the free energy release is given by ΔG°. The free energy change is related to the enthalpy and entropy change by $\Delta G^\circ = \Delta H^\circ - T\Delta S^\circ$. Variation of ΔG° with temperature is shown in the "Ellingham" diagram of Fig 4.1 (Richardson and Jeffes, 1948).

Several characteristics of oxidation are evident from the details of Fig 4.1:

- Noble metals are at the top of the diagram indicating that they are less reactive and their oxides are easily reducible.

- The slope of the lines is given by $d(\Delta G^\circ)/dT = d(\Delta H^\circ - T\Delta S^\circ)/dT \approx -\Delta S^\circ$ since ΔH° is not very temperature sensitive. Let us take the example of aluminum oxidation,

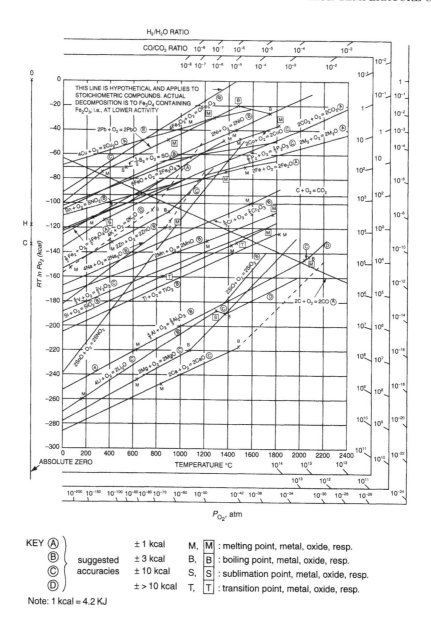

Figure 4.1 Ellingham diagram of free energy of formation of oxides as function of temperature (F. D. Richardson and J. H. E. Jeffes, *J. Iron Steel Inst.*, The Metals Society, 160, 1948; J. C. B. Alcock and E. Easterbrook, Thermodynamics and kinetics of gas–metal systems, in *Corrosion*, Vol. 1, Eds. L. L. Shreir, R. A. Jarman, and G. T. Burstein, Butterworth Heinemann, 3rd ed. 1994). Reprinted with permission from Elsevier.

$$\Delta S^\circ = S(2/3Al_2O_3) - S(4/3Al) - S(O_2),$$

where S is the absolute entropy.

We can make some approximations here. Because both aluminum and aluminum oxide are solids and, therefore, highly ordered, they may be assumed to have small (compared with gaseous oxygen) and roughly equal entropy values that cancel each other, leaving $\Delta S^\circ = -S(O_2)$. Thus, the slope of the lines in the Ellingham diagram is proportional to the entropy of 1 mole of oxygen. This argument is true as long as the metals are oxidizing to form solid oxides. Other characteristics of oxidation in Fig. 4.1 include the following:

- Melting has a very small effect on the slope since the associated change in entropy is small.

- Transition through vaporization or sublimation has a strong effect on the slope because of a significantly large change in S°.

- Exceptions to positive slope are exhibited in reactions in which both reactants and products involve gases as in the following reactions:

$$C + O_2 = CO_2,$$

$$2C + O_2 = 2CO.$$

For the first reaction, $\Delta S^\circ = S(CO_2) - S(C) - S(O_2) \approx S(CO_2) - S(O_2)$ because for carbon, being a solid, $S(C) \approx 0$. Assuming that both CO_2 and O_2, being gases, have roughly equal entropy which cancel out leaving $\Delta S^\circ \approx 0$. As the slopes of the lines are proportional to $-\Delta S^\circ$, the line representing this reaction will be close to the horizontal. For the second reaction, similar arguments give $\Delta S^\circ = 2S(CO) - 2S(C) - S(O_2) \approx 2S(CO) - S(O_2) \approx S(CO)$, which indicates a negative slope.

- A positive slope in the Ellingham diagram indicates that most oxides become less stable at elevated temperatures.

Partial Pressure Effects

Let us take the case of aluminum reacting with one mole of oxygen to form aluminum oxide. The reaction proceeds in standard state because the change in free energy is negative. The reaction "equilibrium" constant is given by:

$$K = [Al_2O_3]^{2/3}/[Al]^{4/3} P_{O2}.$$

[ ] represents concentration, and P is the partial pressure. For pure Al and Al_2O_3, $[Al_2O_3]$ and $[Al]$ are assumed to be unity. This gives $K = 1/P_{O2}$. Therefore, $\Delta G^\circ = -RT \ln K = RT \ln P_{O2}$ or $P_{O2} = \exp(\Delta G^\circ/RT)$.

At $1000\,K$ ($1340°F$), from Fig. 4.1, ΔG° for Al_2O_3 is about $-215,000\,cal/mole$ of oxygen. Inserting in the value of ΔG° gives $P_{O2} = 6 \times 10^{-48}$ atm. What is the significance of this number? It suggests that at $1000\,K$, pure solid aluminum will not oxidize if the partial pressure of oxygen is less than 6×10^{-48} atm. Because the vacuum environments in all practical equipment have partial pressure of oxygen greater than this value, it is not possible to protect aluminum from oxidizing. The reason that aluminum seems to exist in metallic form is due to the fact that the oxide formed is a thin dense scale that becomes practically impervious to further diffusion of oxygen at ambient temperature. This is a kinetic phenomenon, which will be considered in the latter part of this section.

Copper, a more noble metal than aluminum, however, can be protected against oxidation. Copper oxidizes by the following reaction:

$$4Cu + O_2 = 2Cu_2O,$$

$$K = [Cu_2O]^2/[Cu]^4 P_{O2}.$$

For pure Cu forming pure solid oxide, $[Cu_2O]$ and $[Cu]^4$ are unity, giving $K = 1/P_{O2}$ or

$$P_{O2} = \exp(\Delta G^\circ / RT).$$

At 1000 K (1340°F), from Fig. 4.1, ΔG° for Cu_2O is about $-46,600$ cal/mole of oxygen, giving $P_{O2} = 6 \times 10^{-11}$ atm. The significance of this number is that if oxygen partial pressure is kept greater than 6×10^{-11} atm, copper will oxidize. In contrast, copper oxide can be reduced if heated at 1000 K in an environment with oxygen partial pressure less than 6×10^{-11} atm. This can be achieved by heating in a mixture of H_2/H_2O:

$$2H_2O = 2H_2 + O_2,$$
$$K = P_{H2}{}^2 P_{O2}/P_{H2O}{}^2 = \exp(-\Delta G^\circ / RT).$$

Using ΔG° at 1000 K $-95,750$ cal/mole of oxygen, we get $P_{H2}{}^2 P_{O2}/P_{H2O}{}^2 = 10^{-20}$ atm. To reduce copper oxide, $P_{O2} < 6 \times 10^{-11}$ atm is needed. Inserting this value into the equation above gives $P_{H2}/P_{H2O} > 10^{-5}$ atm. Thus, steam containing 10 parts per million hydrogen at 1000 K will reduce copper oxide to metallic copper.

Nickel will oxidize on heating in oxygen according to the reaction

$$2Ni + O_2 = 2NiO.$$

The partial pressure of oxygen below which this oxidation to NiO can be prevented can be calculated:

$$K = [NiO]^2/[Ni]^2 P_{O2}.$$

For pure Ni forming pure solid oxide, $[NiO]$ and $[Ni]$ are unity, giving $K = 1/P_{O2}$ or

$$P_{O2} = \exp(\Delta G^\circ / RT).$$

Using the value of ΔG° at 1000 K, the equilibrium pressure comes to be 7×10^{-11} atm. Thus, if nickel is heated in an atmosphere with P_{O2} less than 7×10^{-11} atm, it will not oxidize.

Composition Effects

In the examples just discussed, we have assumed pure metal oxidizing; that is, the thermodynamic activity a_{Al}, and $a_{Cu} = 1$. However, for an alloy containing aluminum or copper, particularly as a minor constituent, these assumptions are not valid. For an alloy with aluminum activity a_{Al}, the oxidation reaction equilibrium constant will be:

$$K = [Al_2O_3]^{2/3}/[Al]^{4/3} P_{O2} = 1/a_{Al}{}^{4/3} P_{O2},$$
$$\Delta G^\circ = -RT \ln K = RT \ln[a_{Al}{}^{4/3} P_{O2}], \text{ or}$$
$$P_{O2} = [\exp(\Delta G^\circ / RT)]/a_{Al}{}^{4/3}.$$

The foregoing equation shows that the lower the metal activity (dilute), the higher the oxygen partial pressure required to oxidize the alloy. The details of alloy oxidation will be covered in a later section.

Kinetics of Oxidation

Although the thermodynamic consideration of free energy change predicts the various stable oxidation products, it does not provide the rates at which oxidation progresses. The individual steps leading to the formation of the oxide are also not addressed by thermodynamics. These steps can be qualitatively summarized as follows:

- Oxygen molecules are adsorbed on the metal surface.
- Molecular oxygen dissociates into atomic oxygen.
- Oxygen atoms migrate to low-energy sites on the metal surface.
- Atomic oxygen ionizes and forms bonds with the metal atoms of the surface.
- Multiple adsorbed layers build up.
- Oxide islands grow and overlap to form transient oxide film.
- Oxygen/metal ions diffuse through the film to enable the formation and continued growth of stable oxide.

Oxide Scale Protectiveness

The oxide scale could be either protective against further oxidation or porous, allowing oxygen permeation and continuation of the oxidation process. A general indicator of the protectiveness of the oxide scale is given by the Pilling-Bedworth Ratio (PBR). It is defined by PBR = Volume of oxide formed/Volume of metal consumed = $1/n[(m/r')/(M/R')]$, where m and M are the molecular and atomic weights of the oxide and metal respectively, r' and R' are the respective densities, and n is the number of metal atoms in the oxide molecule. PBR values of some metal oxide systems and the protectiveness of the oxide scale are listed in Table 4.1.

If PBR $<< 1$, the volume of oxide formed is less than the volume of the metal consumed. In order to conform to the underlying metal, the oxide needs to be porous and hence nonprotective. If PBR ~ 1, the scale is protective. If PBR $>> 1$, the oxide volume is greater than the volume of the metal replaced. The oxide scale is highly compressed, resulting in buckling and spall. Exposure of a fresh metal surface perpetuates continued oxidation. Thus, for an oxide scale to be protective, PBR should be close to unity. However, there are many exceptions to the PBR

Table 4.1 PBR of Some Metal Oxide Systems

System	PBR	Protectiveness
Al/Al_2O_3	1.28	P
Cr/Cr_2O_3	1.99	P
Si/SiO_2	2.15	P
Ca/CaO	0.64	NP
Ta/Ta_2O_5	2.47	NP

P, protective; NP, nonprotective

predictions due to geometry (edge) effects, volatility of oxides, and chemical effects of alloying elements.

4.2 OXIDATION TESTING AND EVALUATION

Oxidation Rates

The rate at which a metal or an alloy oxidizes depends on its composition, its real surface area as opposed to geometrical area, the temperature, the gas composition, and the nature of the oxidation cycle. The cycle effect has been clearly demonstrated by Pint et al. (2002) for a variety of Ni base superalloys, NiAl and its variants, alloys containing Pt, and Fe based alloys. Outside of the actual operating environment, generally two distinct types of oxidation tests are used: (1) isothermal tests and (2) cyclic tests.

1. Oxidation kinetics is studied in isothermal tests by measuring growth of oxide scale thickness or change in weight as a function of time of exposure at various temperatures (Fig. 4.2). Such tests eliminate the effect of cycling of temperature. The isothermal exposure results in the formation and growth of oxide scale on the surface of exposed alloy. Posttest oxide scale thickness is measured optically using metallographically prepared samples. Weight changes are generally recorded in situ using microbalances (Birks and Meier, 1983). The isothermal test is conducted in a furnace. Button-sized samples (typically 0.5 in. diameter × 0.125 in. thickness) with surfaces polished with 600 grit abrasives are suspended in a tube furnace. The temperature and gas conditions are maintained close to those observed in the application environment, such as in the high-pressure turbine section of a gas turbine engine. Gas velocities, however, are seldom commensurate with the real environment. For example, gas turbine engines generate gas velocities of the order of the velocity of sound ($\sim$330 m/s at room temperature, increasing with temperature to 1000 m/s at 1093°C, 2000°F), which is not possible in a furnace. There are direct and indirect influences of gas velocity on oxidation. The heat transfer coefficient and, therefore, the temperature increase with higher gas velocity. Higher velocity may also induce erosion of protective oxide scales if the gas has entrapped particles, as happens in gas turbine engines. The change in weight due to oxidation over time at selected temperatures is automatically recorded in furnace tests. A typical isothermal time–temperature regime consists of heat-up times of 10 to 15 minutes to the desired temperature, hold time of 20 hours or longer, and a cool-down rate in the range of 10 to 15°C (18 to 27°F) per minute. While the furnace test provides a standard and well-controlled condition to study

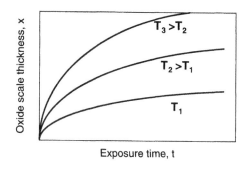

Figure 4.2 Schematic of oxidation scale growth.

fundamental processes involved in oxidation, it does not simulate actual operating conditions or provide oxidation life of components in actual use. This is due to the lack of fast transients and associated damage to oxide scales, and to brittle coatings for coated test samples. Damages resulting in loss of oxide scale enhance oxidation rates by exposing fresh metal surfaces to the oxidizing environment.

2. The cyclic test is conducted either by cycling in and out of a furnace or in a burner rig. Guidelines for test standardization of cyclic oxidation including apparatus for testing have been discussed by Nicholls and Bennett (2000). In the furnace cyclic test, in one of several possible automated versions, the sample in the form similar to that in the isothermal test is moved in and out of the hot zone of an electrically heated furnace by a chain attached to a motor. A typical cycle may consist of 30 minutes in the hot zone followed by 10 minutes out of the furnace at temperature at or below 100°C (212°F).

The burner rig test, described in Fig. 4.3 (Vargas et al., 1980), is more versatile in introducing effects of temperature transients experienced in actual use, although it does not lend itself well to weight loss monitoring. In this test, a combustion flame is created in a ducted burner-can by injecting jet fuel and air and igniting the mixture. Burner rigs can also be run without ducts. The advantage of ducted rig is the ability to control partial pressures of gases such as in hot corrosion, discussed in Section 5.1. Coated samples in the form of cylindrical bars are mounted on a rotating carousel and exposed to the flame. The rotation of the carousel holding multiple samples ensures uniform exposure of all samples. The test temperature is usually measured by an optical pyrometer directed at the hottest zone of a sample in the flame. After exposure to the flame for a fixed period, the carousel is moved so that the samples are out of the flame. The samples are cooled now by forced ambient air for a fixed duration. At the end of the cooling step, the samples are reintroduced into the flame by moving the carousel appropriately. This sequence defines one cycle equivalent of exposure. The samples are visually monitored and their weights are routinely measured at specific intervals.

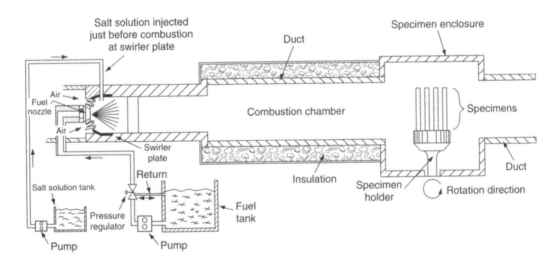

Figure 4.3 Burner rig testing of alloys and coatings by exposing samples to a combustion flame. The rig is also used to conduct a hot corrosion test, described in Chapter 5, by injecting salt solution into the flame (J. R. Vargas, N. E. Ulion, and J. A. Goebel, Advanced coating development for industrial utility gas turbine engines, *Thin Solid Films*, 1980, 73(2), 407–414). Reprinted with permission from Elsevier.

The oxide scale growth (or weight gain) in isothermal oxidation follows one of the relationships:

$$\text{Parabolic growth: } x^2 = kt,$$

$$\text{Linear growth: } x = x_0 + kt,$$

$$\text{Logarithmic growth: } x = k_0 + \log(kt + 1),$$

where x is the oxide scale thickness after exposure for time t; k is the rate constant, a function of temperature T, and x_0 and k_0 are constants.

Parabolic Growth

Many systems oxidize such that the growth of oxide scale, when exposed isothermally, follows the parabolic equation. Oxidation in such cases is thermally activated and controlled by diffusion of species through the oxide scale. The scale growth rate can be easily deduced from Fick's law of diffusion. For example, if oxygen is the diffusion-controlling species, for the geometry shown in Fig. 4.4, M is a metal atom, O is an oxygen atom, and e^- is an electron, Fick's law of diffusion provides $J = -D\ dc/dx$, where D is the diffusivity and dc/dx is the concentration gradient for the diffusing species.

This equation can be rewritten in terms of weight gain w and thickness x as

$$dw/dt = -D\ dc/dx,$$

where $w = r'x$ for unit area of the cross section, and r' is the oxide density. Rearranging the equation, we have

$$r'\ dx/dt = -D\ dc/dx = D\,\Delta c/x \text{ or } x\ dx = D\,\Delta c\ dt/r.$$

Here we have assumed a linear gradient so that $dc/dx = \Delta c/x$, where Δc is the concentration change over distance x. Integrating $x\ dx = D\ dc\ dt/r$ over x, increasing from 0 to x when time goes from 0 to t, we get

$$x^2 = 2D\,\Delta c\ t/r'.$$

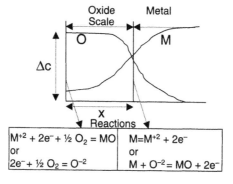

Figure 4.4 Diffusion profile for oxide scale growth.

Substituting $k = 2D\Delta c/r'$, we get

$$x^2 = kt.$$

The oxide scale growth depends on temperature and oxygen partial pressure. The temperature dependency appears through the diffusivity factor, which follows the relationship

$$D = D_0 \exp(-\Delta H/RT),$$

where D_0 is a constant, R is the gas constant, ΔH is the activation energy of diffusion, and T is the temperature in kelvins.

An example of parabolic oxidation is shown in Fig. 4.5 for CoNiCrAlY and NiCoCrAlY coatings isothermally oxidized at 1050°C (1922°F) (Brandl et al., 1996). The rate constant k is on the order of $4 \times 10^{-10}\,\mathrm{g^2/cm^4\,s}$.

The parabolic rate constant k for the growth of Cr_2O_3, Al_2O_3, and SiO_2 scales is summarized in Fig. 4.6 (Goward, 1970). Saunders and Nicholls (1989) provide additional data including growth of SiO_2.

Linear Growth

Linear growth kinetics results if the rate-controlling step is a metal–oxygen reaction at the interface between the oxide scale and the metal. In such cases, the growth rate is independent of thickness,

$$dx/dt = k_0,$$

with k_0 being a constant. Integration gives

$$x = x_0 + k_0 t,$$

where x_0 is the initial scale thickness at $t = 0$. Linear rate laws have been observed when the oxide scale is not protective, as in the case of alkali metals, and in oxidation in weak oxidation environments such as dilute oxygen or mixtures of CO and CO_2.

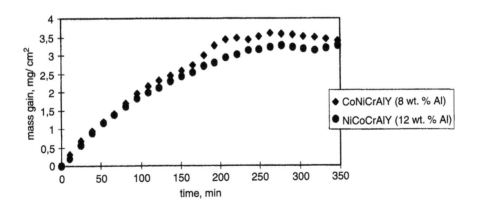

Figure 4.5 Isothermal oxidation of MCrAlY coatings (F. W. Brandl, H. J. Grabke, D. Toma, and J. Kruger, The oxidation behavior of sprayed MCrAlY coatings, *Surf. Coat. Technol.*, 1996, 86–87, 41–47). Reprinted with permission from Elsevier.

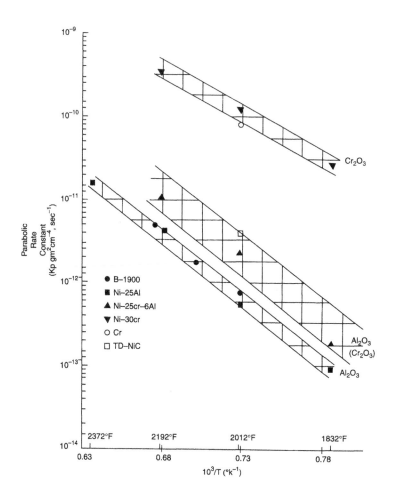

Figure 4.6 Range of parabolic growth rates for selected oxides on several alloys. Note that alumina has a lower growth rate compared with chromia. (G. W. Goward, Current research on the surface protection of superalloys for gas turbine engines, *JOM,* October 1970, pp. 31–39). Reprinted with permission from The Minerals, Metals & Materials Society.

Logarithmic Growth

Logarithmic rate laws are observed for thin films growing at low temperatures such as nickel oxidizing below 200°C (392°F). The mechanism of growth involves electric fields localized very close to the metal surface. Adsorbed oxygen atoms acquire electrons from the metal atoms, creating large electric fields across the thin film between positive metal ions and negative oxygen ions. Metal atoms are pulled by the electric field through the oxide film. In such cases the growth rate varies inversely with time,

$$dx/dt = k/t,$$

which on integration reduces to

$$x = x_0 + k \ln t.$$

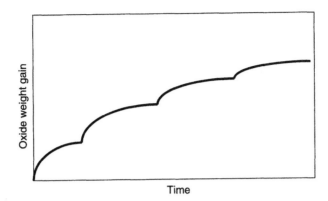

Figure 4.7 Schematic of breakaway oxidation showing growths followed by spalls of the oxide scales.

Breakaway Oxidation

Large stresses build up as oxide scales grow on continued exposure in the oxidizing environment. These stresses could be either compressive or tensile, and when they exceed the cohesive or adhesive strengths of the scales, repeated periodic breaks occur. After each breakaway, parabolic growth continues, eventually leading to complete spall or a linear growth profile (Fig. 4.7).

Influence of Thermal Cycling on Oxidation

Oxide scales have lower coefficients of thermal expansion (CTEs) than those of the underlying metals. The CTE differences generate stresses ($E \Delta \alpha \Delta T$, where $\Delta \alpha$ is the CTE difference between the oxide and the underlying metal, and ΔT is the temperature range of thermal cycling). Stresses are low at the elevated temperatures ($\Delta T \sim 0$) at which oxide scales form but increase to large values at lower temperatures ($\Delta T \sim$ large). The large stresses associated with thermal cycling result in sporadic spalls of the oxide scale, leading an initial parabolic growth into loss of thickness or weight.

Figure 4.8 (Levy et al., 1986) compares the isothermal and cyclic oxidation behavior of the same alloys, which include polycrystalline, single-crystal, and directionally solidified microstructures. The polycrystalline alloy is MAR-M 200. It is clear that the oxidation curve in the cyclic test is the result of a combination of scale growth or weight gain during the heating part of the cycle and a loss of scale or weight loss associated with scale spall during the cooling part of the cycle.

The weight gain curves observed in isothermal oxidation can also be constructed for controlled cyclic tests by capturing and weighing the spalled oxide chips and adding to the weight of the sample. Figure 4.9 is the oxidation kinetics reconstructed this way for MA956HT in cyclic testing (Wright et al., 2001). For many alloys, the extent of oxide scale spalled is found to be a function of $x^2 q_0$, where x is the scale thickness and q_0 is a spalling constant (Lowell et al., 1991). The oxidation kinetics can be empirically expressed as

$$x = k^{1/2} t^{1/2} - k^{\circ} t.$$

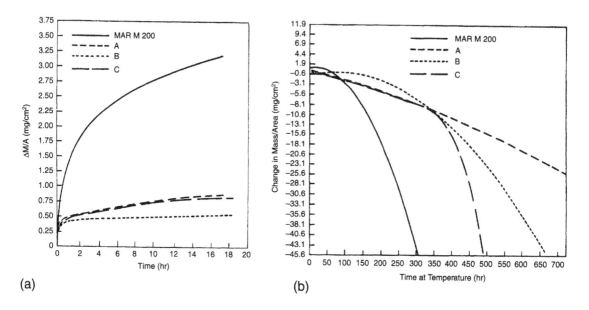

(a)   (b)

Figure 4.8 Oxidation weight change at 2000°F (1093°C) of single-crystal nickel base alloys A, B, C, and directionally solidified Mar M 200 in (a) isothermal and (b) cyclic test (M. Levy, P. Farrell, and F. Pettit, Oxidation of some advanced single-crystal nickel–base superalloys in air at 2000°F [1093°C], *Corrosion*, 1986, 42(12), 708–717). Reprinted with permission from NACE International.

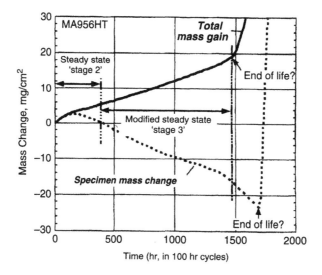

Figure 4.9 Reconstructed oxidation kinetics of MA956HT exposed to 1300°C (2372°F) in 100-hour cycles (I. G. Wright, B. A. Pint, L. M. Hall, and P. F. Tortorelli, Oxidation lifetimes: Experimental results and modeling, in *Lifetime Modeling of High-Temperature Corrosion Process, Proc. European Federation of Corrosion Workshop,* Eds. M. Schutze, W. J. Quadakkers, and J. R. Nicholls, EFC Publications No. 34, Maney Publishing, 2001, pp. 339–358). Reproduced with permission from Maney Publishing.

4.3 OXIDATION OF ALLOYS

Metals are seldom used in pure form in load-bearing applications, particularly at elevated temperatures because of low strength and environmental resistance. For engineering applications, metals are strengthened and their environmental resistance improved by appropriate alloying. The basic mechanisms operating in pure metal oxidation are also operative in the oxidation of alloys with added complications. These complications include the formation of multiple oxides, mixed oxides, internal oxides, and diffusion interactions within the metals. The effect of alloying on oxidation behavior can be understood first by considering binary alloy model AB consisting of element A, the major component, and B, the minor component (Wagner, 1959). We have two distinct possibilities depicted in cases 1 and 2 in Figs. 4.10 and 4.11 (Smialek and Meier, 1987).

Binary Alloy Systems

Case 1

Case 1, shown in Fig. 4.10, considers element A as more noble and B as more reactive. Thus, at atmospheric pressure of oxygen, A does not form AO (oxygen pressure is too low), whereas B converts to BO. We have two distinct situations depending on the concentration of B in A. In case (a) of Fig. 4.10, the alloy is dilute in B. Assuming oxygen has a finite solubility in the alloy, it diffuses to internally oxidize and form dispersed precipitates of BO in A. Because there is not enough B available, a continuous BO scale does not form. In case (b) of Fig. 4.10, the alloy is concentrated in B. In this case sufficient B is available to form a continuous BO

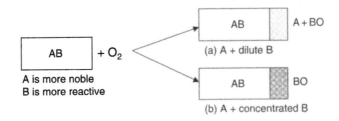

Figure 4.10 Binary alloy oxidation: Component A is more noble, B more reactive.

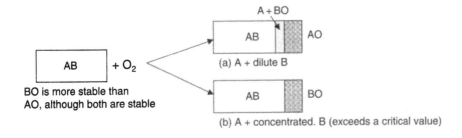

Figure 4.11 Binary alloy oxidation: Component A and B form stable oxides but BO is more stable than AO.

scale. The critical concentration of B required to form a continuous scale is given by (Smialek and Meier, 1987):

$$B_{critical} \geq [0.3N_O(D_O/D_B)(V_m/V_{mo})]^{1/2},$$

where D_O, D_B are the respective diffusivities of oxygen and element B in the alloy, V_m and V_{mo} are the molar volumes of metal and oxide formed, and N_O is the solubility of oxygen in the alloy. The foregoing relationship shows that an alloy with lower concentration of B will form continuous oxide scale provided N_O, D_O, and V_m are low and D_B and V_{mo} are high.

Case 2

In this case A and B are both reactive to oxygen with BO more stable than AO. The concentration of B again dictates the oxide morphologies. If the alloy is dilute in B, case (a) in Fig. 4.11, a stable oxide AO forms as the outer scale. Below the oxide scale, at the AO–alloy interface, the O_2 activity is high enough to oxidize B into BO precipitates. In case (b) of Fig. 4.11, the concentration of B exceeds the critical level required to form a continuous BO scale. Case (b) is the basic model for creating oxidation-resistant alloys and coatings. The oxide growth rate is parabolic, with activation energy characteristic of the growth of BO. The actual rate depends on how protective the BO scale is and on the presence of additional alloying elements.

Binary alloys of Ni with Cr or Al illustrate Fig. 4.11(b). They form the basis of many oxidation-resistant commercial alloys and coatings and have been studied to understand oxidation mechanisms at high temperature.

Ni–Cr Alloys: There are three composition (in wt %) regimes with distinct oxidation characteristics:

   Cr < 10%: Such alloys form an NiO external scale and internal Cr_2O_3. This follows from mechanisms shown in case 4.11(a).

   30% > Cr > 10%: In such alloys the outer scale consists of NiO on grains and Cr_2O_3 in grain boundaries.

   Cr > 30%: For such alloys the external scale consists of Cr_2O_3, which is maintained because of the large Cr reservoir.

Ni–Al Alloys: There are three composition regimes also for this alloy system with distinct oxidation characteristics:

   Al < 6%: Such alloys form an NiO external scale and internal Al_2O_3 and $NiAl_2O_4$. This follows from mechanisms shown in case Fig. 4.11(a).

   17% > Al > 6%: In such alloys the outer scale initially consists of Al_2O_3. However, on continuous exposure, Al depletion occurs in the alloy adjacent to the oxide scale. In the depleted zone, the Al activity is well below the requirement to form continuous Al_2O_3. NiO, therefore, overtakes Al_2O_3. The overall result is the formation of a mixture of NiO, $NiAl_2O_4$ spinel formed by the combination of NiO and Al_2O_3.

   Al > 17%: For such alloys, following Fig. 4.11(b), the external scale of Al_2O_3 is maintained because of the large Al reservoir.

Addition of Cr, as low as 5%, to the Ni–Al system reduces the Al requirement from 17% to as low as 5% to form a continuous Al_2O_3 scale. Cr is thought to act as an oxygen getter initially to form a Cr_2O_3 subscale, which converts to Al_2O_3 by Al substitution.

Ternary and Multicomponent Alloy Systems

Practical alloys and metallic coatings for high-temperature application are seldom binary. These alloys are typically Cr_2O_3 and Al_2O_3 formers. Silica (SiO_2) also forms a protective scale, particularly for refractory metals with which it has better thermal expansion matching. However, SiO_2 is not stable at low pressures. It decomposes to gaseous species such as SiO. It also reacts with water vapor at high temperatures, forming $Si(OH)_4$ gas. The use of Cr_2O_3 scale-forming alloys is limited to temperatures below 1000°C (1832°F).

Volatile CrO_3 forms above this temperature in the presence of oxygen due to the reaction

$$\tfrac{1}{2}\,Cr_2O_3(s) + \tfrac{3}{4}O_2\,(g) = CrO_3\,(g).$$

Thermodynamic analysis of the foregoing reaction shows oxygen partial pressure dependence to be $P_{O_2}^{3/4}$. Thus, the volatilization becomes important at high oxygen partial pressures. Volatilization of Cr_2O_3 has actually been observed even at lower temperatures between 850°C (1562°F) and 900°C (1652°F) (Radcliff, 1987). Also, in the presence of water vapor, $CrO_2(OH)_2$ forms above the Cr_2O_3 scale. The scale rapidly vaporizes, resulting in continuing metal recession. SiO_2 scales are also subject to water vapor-enhanced volatility. For high-temperature application above 1000°C (1832°F), useful alloys are, therefore, designed to be Al_2O_3 formers. In such alloys, the alumina scale can have one of several allotropic forms, depending on the alloy composition and temperature of oxidation. The allotropes include the transient phases γ, δ, and θ which, on thermal exposure, convert into the stable phase α (Prasanna et al., 1996). The composition of the scale formed and its stability depend on the alloy composition, the temperature, and the cyclic nature of thermal exposure, as is evident from the ternary diagrams in Fig. 4.12 (Smialek and Meier, 1987; Giggins and Pettit, 1971; Wallwork and Hed, 1971; Barrett and Lowell, 1977; Tumarev and Panyushin, 1970/1959). The ternary shows the three primary regions, (I) the external NiO scale and a subscale composed of Cr_2O_3 and Al_2O_3, (II) external scale of Cr_2O_3 and subscale of Al_2O_3, and (III) the external scale of Al_2O_3. Cyclic oxidation for an extended period of time results in oxide scale spalling. This effectively contracts the three zones to higher Al and Cr contents, and transition zone T opens up in the zone left behind. This transition zone consists of a composite of NiO, $NiCr_2O_4$, $NiAl_2O_4$, and Al_2O_3. The alloy composition corresponding to the location of the triangle inset is 10% Al, balance Ni. The oxide scale formed on this alloy is NiO. The alloy composition corresponding to the location of the diamond inset in Fig. 4.12 is 5.3% Cr, 3.3% Al, with the balance being Ni. Although this alloy has less aluminum than the binary, it forms Al_2O_3 and not NiO. This occurs because of the gettering effect of Cr.

The rates of growth of the oxide scales in the three zones differ by about an order of magnitude for each grouping in the following order: $k(I) > k(II) > k(III)$. The types of the oxide scales formed on a number of known alloys on exposure to oxygen at 1100°C (2012°F) are shown in Fig. 4.13 (Hindam and Whittle, 1982; Wallwork and Hed, 1971).

Commercial superalloys contain many alloying elements of significance over and above chromium and aluminum. The oxidation behavior of these alloys is very complex and oxidation resistance varies widely, although the general mechanisms described earlier still apply. The complexities arise from significant influence of the individual elemental constituents. Nickel

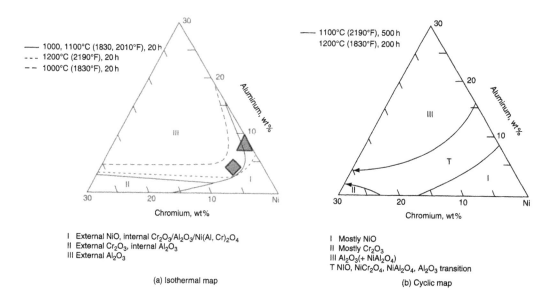

I   External NiO, internal Cr₂O₃/Al₂O₃/Ni(Al, Cr)₂O₄
II External Cr₂O₃, internal Al₂O₃
III External Al₂O₃

I   Mostly NiO
II Mostly Cr₂O₃
III Al₂O₃(+ NiAl₂O₄)
T   NIO, NiCr₂O₄, NiAl₂O₄, Al₂O₃ transition

(a) Isothermal map

(b) Cyclic map

Figure 4.12 Oxidation map of Ni–Cr–Al ternary alloys in isothermal as well as cyclic thermal exposure (J. L. Smialek, G. M. Meier, High temperature oxidation, in *Superalloy II*, Eds. C. T. Sims, N. S. Stoloff, and W. C. Hagel, Wiley, 1987, pp. 293–323. Reprinted with permission from John Wiley & Sons, Inc.; C. S. Giggins and F. S. Pettit, Oxidation of Ni–Cr–Al Alloys Between 1000° and 1200°C, *J. Electrochem. Soc.*, 1971, 118, 1782. Used with permission from The Electrochemical Society; G. R. Wallwork and A. Z. Hed, *Oxid. Met.*, 1971, 3, 171; C. A. Barrett and C. E. Lowell, Resistance of Ni–Cr–Al Alloys to Cyclic Oxidation at 1000° and 1200°C, *Oxid. Met.*, 1977, 11. Used with permission from Springer. Additional reference: A. S. Tumarev and L. A. Panyushin, NASA TT F-13221, 1970; *Izv. Vyssh. Uchebn. Zaved. Chern. Metall.*, 1959, 9, 125).

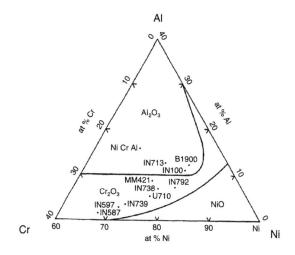

Figure 4.13 The type of oxide scales formed on several alloys at 1100°C (1832°F) (H. Hindam and D. P. Whittle, Microstructure, adhesion and growth kinetics of protective scales on metals and alloys, *Oxid. Met.*, 1982, 18(5/6), 245–284; G. R. Wallwork and A. Z. Hed, Some Limiting Factors in the Use of Alloys at High Temperatures, *Oxid. Met.*, 1971, 3(2), 171–184. Used with permission from Springer; Data sources include R. Kosak and R. Rapp, PhD Thesis, The Ohio State University, 1969; also F. S. Pettit, *Trans. AIME*, 1967, 239, 1296. Reprinted with permission from The Minerals, Metals & Materials Society).

base superalloys containing elements such as Co, Cr, Al, Ti, W, and Ta exhibit general behavior similar to simple NiCrAl alloys.

The cyclic oxidation of superalloys consists of several steps. The process involves an initial transient period during which the oxides of the individual constituents form. These include NiO, CoO, Cr_2O_3, TiO_2, and Ta_2O_5. Continued oxidation leads to the formation of the most stable oxide, which is Al_2O_3 for alumina-forming alloys. The parabolic rate constants of the transient oxides are larger than that of the stable oxide. In a cyclic environment the oxide scale cracks and spalls. Aluminum in the alloy diffuses to the oxide–alloy interface to reform the scale. The oxide scale spalling followed by the reformation process continues. An aluminum-depleted zone forms in the alloy below the oxide scale. The thickness of the depleted zone is dependent on the aluminum content of the alloy. Isothermal oxidation of NiCoCrAlY with 12 wt % Al, for example, exhibits depleted zone thickness 3 times the thickness of the alumina scale (Brandl et al., 1996). Because of the loss of aluminum, the depleted zone becomes enriched with the other alloying constituents, some of which, having low solubility in the depleted zone, precipitate out in the form of acicular phases. Additionally, nitrogen, which diffuses into the alloy during oxidation, now exceeds its solubility limit and precipitates out as nitrides of such elements as Ti. The new precipitate phases penetrate into the alloy. The alloy is finally depleted of aluminum to below such a critical level that the regeneration of a continuous scale of Al_2O_3 is no longer feasible. At this stage for nickel base alloys, breakaway oxidation starts with the formation of continuous NiO, which is not protective. As a result, oxygen now diffuses into the alloy, forming internal oxides of the remaining aluminum, chromium, and other reactive constituents of the alloy. The alloy substrate now loses wall thickness and load-bearing capability. The sequence of oxidation steps that occur on cyclic oxidation of a nickel base superalloy is illustrated in Fig. 4.14. In order to alleviate this situation of loss of oxidation resistance, the following conditions need to be satisfied: (1) increase aluminum activity as much as possible; (2) increase aluminum diffusivity to the alloy surface; (3) inhibit diffusivity of oxygen into the alloy; and (4) improve the adherence of the alumina scale to the substrate.

Similar processes and arguments also hold true for cobalt base alloys, although some differences exist in the detail. For comparable oxidation resistance, cobalt base alloys need higher combined aluminum and chromium than do nickel base alloys. Also, once the protective alumina scale fails, CoO forms on cobalt base alloys, which spalls off catastrophically as opposed to the gradual failure of NiO formed on nickel base alloys.

The substrate alloy compositions are generally a result of an effort to maximize structural properties such as creep, tensile, and fatigue strength. The alloying constituents, however, affect the oxidation resistance significantly.

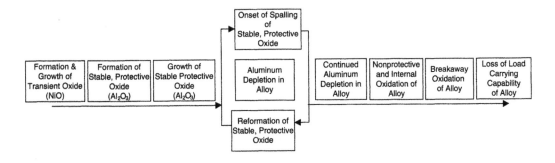

Figure 4.14   General sequence of steps in superalloy cyclic oxidation.

4.4 ROLES OF SPECIFIC ALLOYING CONSTITUENTS

Aluminum

Aluminum obviously has the greatest effect in oxidation resistance, which generally increases with its concentration (Barrett et al., 1983).

Chromium

In addition to improving hot corrosion resistance and increasing the activity of aluminum, chromium effectively reduces oxygen diffusion into the alloy by lowering the oxygen activity at the oxide scale–alloy interface.

Cobalt

Generally, the oxidation resistance increases as the cobalt concentration decreases. The rate of growth of CoO is usually greater than that of NiO. Thus, once the formation of protective Al_2O_3 is no longer possible due to aluminum depletion, high Co concentration may lead to faster growing CoO scale.

Silicon

Silicon has a beneficial effect. The effect is derived from the formation of SiO_2 subscale. B-1900 alloy with 0.5 to 1.5% Si exhibits oxidation resistance comparable to that of aluminide-coated B-1900 without Si. Improvement, although not to the same extent, is also seen for MAR-M 200 and IN-713. However, Si addition tends to make the alloys brittle. Coatings containing Si have similar beneficial effects with some exceptions. Examples of Si-containing coatings include diffusion aluminides (discussed in Section 6.3) Sermaloy J (3 to 5% Si), and JoCoat.

Boron

Boron has a strong detrimental effect on oxidation resistance. It encourages NiO scale formation rather than Al_2O_3. Boron in substrate alloys also negatively affects coatings formed or deposited on them.

Titanium

Titanium typically improves oxidation resistance of FeCrAl as shown in Fig. 4.15 (Sarioglu et al., 2000). However, addition of a similar amount of Ti to Ni base alloy, such as PWA 1480, does not have a comparable effect. Ti is known to increase the growth rate of Al_2O_3 and has a detrimental effect on scale adherence on Ni base alloys. For Ni–Cr alloys, on the other hand, Ti promotes Cr_2O_3 formation.

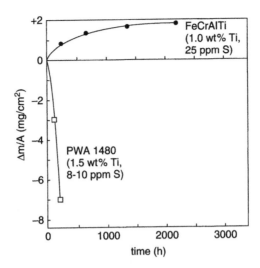

Figure 4.15 Effect of Titanium addition on the oxidation resistance of some alloys (C. Sarioglu, M. J. Stiger, J. R. Blachere, R. Janakiram, E. Schumann, A. Ashary, F. S. Pettit, and G. H. Meier, The adhesion of alumina films in metallic alloys and coatings, in *Materials and Corrosion 51*, pp. 358–372, Wiley-VCH Verlag GmbH, Weinheim, Germany, 2000). Reprinted with permission from John Wiley & Sons, Inc.

Manganese

Manganese promotes the formation of Cr_2O_3 on Ni–Cr alloys. Additionally, manganese has been known to promote formation of Al_2O_3 scale on binary Fe–Al alloys.

Tantalum, Molybdenum, Tungsten

Tantalum in small concentration improves oxidation resistance (Barrett et al., 1983). Refractory elements in general, however, reduce the diffusivity of aluminum to the surface.

Oxygen Reactive Elements (REs)

Elements such as Y, Hf, La, Sc, Ce, and Zr are known as oxygen reactive elements or simply reactive elements (REs) because their oxides are generally more stable than the oxide scales formed on most alloys when exposed to a high-temperature oxidizing environment. The presence of an RE improves the adherence of Al_2O_3 as well as Cr_2O_3 scales to alloy substrates, although the growth rates of the two oxides are influenced differently. The growth rate of Cr_2O_3 is reduced, whereas that of Al_2O_3 does not change significantly. Even when added in their oxide form, some of the REs improve oxide scale adherence. The beneficial effect of a few of these element additions on the oxidation kinetics is shown in Fig. 4.16 (Sarioglu et al., 2000). All of the RE-containing alloys exhibit better oxidation resistance than the undoped NiCrAl. A number of mechanisms have been proposed to explain the beneficial effects of oxygen reactive elements:

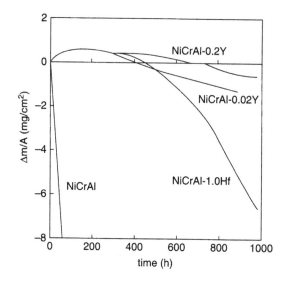

Figure 4.16 Effect of reactive element addition on the oxidation resistance of Ni–Cr–Al alloys (C. Sarioglu, M. J. Stiger, J. R. Blachere, R. Janakiram, E. Schumann, A. Ashary, F. S. Pettit, and G. H. Meier, The adhesion of alumina films in metallic alloys and coatings, in *Materials and Corrosion 51*, 358–372, 2000, Wiley-VCH Verlag GmbH, Weinheim, Germany, 2000). Reprinted with permission from John Wiley & Sons, Inc.

- REs mechanically key the oxide scale to the alloy by forming oxide pegs at the oxide–alloy interface.

- REs segregate at the interface and increase the bond strength of the oxide scale to the alloy. A study by Jedlinski and Mowec (1987) on yttrium ion–implanted NiAl showed that the Al_2O_3 oxide scale loss on furnace cycling decreased significantly with increased dose of yttrium. This is a clear indication of the influence of yttrium on oxide scale adherence.

- REs act as vacancy sinks to suppress the formation of voids.

- REs tie up sulfur, which otherwise would segregate at the oxide–alloy interface and reduce the bond strength of the oxide scale to the alloy (Smeggil et al., 1986). Yttrium, for example, ties up sulfur in the form of yttrium sulfides or complexes. This mechanism gets credence from the fact that the level of sulfur that can be tolerated in the alloy correlates with the type and amount of REs added.

- REs alter the growth mechanism of alumina and chromia from a combination of outward diffusion of aluminum or chromium and inward diffusion of oxygen, to predominantly inward diffusion of oxygen. This change reduces lateral growth of the oxides and the associated residual stresses. Confirmation of the change of the diffusion mechanism has come from ^{18}O tracer experiments (Reddy et al., 1982; Pint et al., 1993).

All of these mechanisms may not be operating for all of the REs or when they are added as oxides as opposed to the elemental form. For example, the sulfur tie-up mechanism may work for elemental addition of Y, whereas the mechanism involving reduction in residual stress may work for addition of Y_2O_3.

Rhenium/Ruthenium

Rhenium (Re) and ruthenium (Ru) additions to superalloys, particularly single crystals, improve mechanical properties such as creep resistance by reducing the rate of coarsening of γ' precipitates. The growth of the precipitates is a diffusion-controlled process. Re, which goes into solid solution in the γ phase, reduces diffusivity of alloying elements. Because oxidation also involves diffusion in the matrix, effects of Re on oxidation of MCrAlY alloys, particularly coatings, have been investigated. The results confirm improved oxidation capability of MCrAlY coatings in the presence of Re (Czech et al., 1994). In the case of superalloys, however, recent data (Kawagishi et al., 2006) show that Re as well as Ru degrade oxidation resistance. Their oxides have high vapor pressures at high temperature and they render alumina scales to have uneven structures.

Reduction of Sulfur Level

Coatings and particularly alloys always contain tens of parts per million of sulfur. During oxidation, sulfur segregates at the oxide–alloy interface, which results in the reduction of the adhesive strength of the oxide scale to the alloy. Removal of sulfur through repeated oxidation and polishing, high-temperature annealing in vacuum, or desulfurization in hydrogen at high temperature improves oxidation resistance. Figure 4.17 (Sarioglu et al., 2000) shows the beneficial effect of desulfurization as well as that of addition of yttrium, compared with oxidation in the presence of platinum aluminide coating and with bare single-crystal nickel base superalloys.

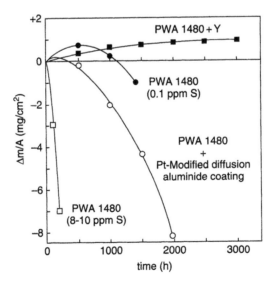

Figure 4.17 Effects of desulfurization, yttrium addition, and diffusion aluminide coating on oxidation resistance of single crystal nickel base alloys (C. Sarioglu, M. J. Stiger, J. R. Blachere, R. Janakiraman, E. Schumann, A. Ashary, F. S. Pettit, and G. H. Meier, The adhesion of alumina films to metallic alloys and coatings, in *Materials and Corrosion, 51*, pp. 358–372, Wiley-VCH Verlag GmbH, Weinheim, Germany, 2000). Reprinted with permission from John Wiley & Sons, Inc.

4.5 OXIDATION IN THE PRESENCE OF WATER VAPOR

The presence of water vapor in the environment is known to accelerate the spalling of the oxide scales formed on many superalloys such as CMSX 4, PWA 1480, and PWA 1484. As a result, the oxidation rate increases (Sarioglu et al., 2000). The aluminum-depleted zones are larger when oxidized in wet air than in dry air. This observation is consistent with increased oxidation rate. Alloys containing reactive elements such as PWA 1487 and yttrium-containing Rene N5 are less susceptible to this phenomenon, principally because of the better adherence of the alumina scale formed on these alloys.

4.6 OXIDATION OF POLYCRYSTALLINE VERSUS SINGLE-CRYSTAL ALLOYS

Oxidation resistance of alloys is primarily controlled by the composition and does not conform to the anisotropy of single-crystal alloys. In order to improve creep and thermal fatigue resistance, nickel base superalloys, strengthened with coherent γ′ precipitates, have been developed without grain boundaries. The grain boundaries act as failure sites when alloys are subjected to stress. The "single crystal" alloys so developed (see Fig. 3.10), therefore, do not need the grain boundary strengthening elements such as carbon, boron, and zirconium. However, the alloys

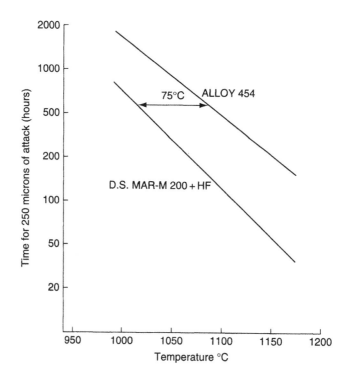

Figure 4.18 Comparison of oxidation behavior between single-crystal (Alloy 454, same as PWA 1480) and directionally solidified (DS) superalloys as a function of temperature (M. Gell, D. N. Duhl, and A. F. Giamei, The development of single crystal superalloy turbine blades, in *Superalloys 1980*, Eds. John K. Tien, Stanley T. Wlodek, Hugh Morrow III, Maurice Gell, and Gernant E. Maurer, American Society of Metals, 1980, pp.205–14). Reprinted with permission from ASM International.

do have the γ–γ' phase boundaries. The elimination of the grain boundary removes the need for boundary-strengthening elements and also the tramp impurity elements, that segregate there. These alterations in composition profoundly affect the oxidation behavior, as is demonstrated in Fig. 4.18 (Gell et al. 1980), which compares a directionally solidified alloy with columnar grain boundaries with a single crystal having no grain boundaries, the compositions of the two being close. Absence of the grain boundary elements improves the alumina scale adherence.

REFERENCES

Alcock, J. C. B., and E. Easterbrook, Thermodynamics and kinetics of gas–metal systems, in *Corrosion*, 3rd ed., Vol. 1, Ed. L. L. Shreir, R. A. Jarman, and G. T. Burstein, Butterworth-Heinemann, 1994.

Barrett, C. A., and C. E. Lowell, Resistance of Ni–Cr–Al alloys to cyclic oxidation at 1000° and 1200°C, *Oxid. Met.*, 1977, 11, 199–223.

Barrett, C. A., R. V. Miner, and D. R. Hull, The effects of Cr, Al, Ti, Mo, W, Ta, and Cb on the cyclic oxidation behavior of cast Ni-base superalloys at 1100 and 1150°C, *Oxid. Met.*, 1983, 20(5/6), 255–278.

Birks, N. G., and G. H. Meier, *Introduction to High Temperature Oxidation of Metals*, p. 6, Edward Arnold, London, 1983.

Brandl, W., H. J. Grabke, D. Toma, and J. Kruger, The oxidation behavior of sprayed MCrAlY coatings, *Surf. Coat. Technol.*, 1996, 86–87, 41–47.

Czech, N., F. Schmitz, and W. Stamm, Improvement of MCrAlY coatings by addition of rhenium, *Surf. Coat. Technol.*, 1994, 68/69, 17–21.

Gell, M., D. N. Duhl, and A. F. Giamei, The development of single crystal superalloy turbine blades, in *Superalloys 1980*, pp. 205–214, Eds. John K. Tien, Stanley T. Wlodek, Hugh Morrow III, Maurice Gell, and Gernant E. Maurer, American Society of Metals, 1980.

Giggins, C. S., and F. S. Pettit, Oxidation of Ni–Cr–Al alloys between 1000° and 1200°C, *J. Electrochem. Soc.*, 1971, 118, 1782–1789.

Goward, G. W., Current research on the surface protection of superalloys for gas turbine engines, *JOM*, October 1970, pp. 31–39.

Hindam, H., and D. P. Whittle, Microstructure, adhesion and growth kinetics of protective scales on metals and alloys, *Oxid. Met.*, 1982, 18(5/6), 245–284.

Jedlinski, J., and S. Mowec, The influence of implanted yttrium on the oxidation behavior of β-NiAl, *Mater. Sci. Eng.*, 1987, 87, 281–287.

Kawagishi, K., H. Harada, Akihiro Sato, Atsushi Sato, and T. Kobayashi, The oxidation properties of fourth generation single-crystal nickel-based superalloys, *JOM*, January 2006, pp. 43–46.

Levy, M., P. Farrell, and F. Pettit, Oxidation of some advanced single crystal nickel-base superalloys in air at 2000°F (1093°C), *Corrosion*, 1986, 42(12), 708–717.

Lowell, C. E., C. A. Barrett, R. W. Palmer, J. V. Auping, and H. B. Probst, COSP: A computer model of cyclic oxidation, *Oxid. Met.*, 1991, 36(1/2), 81–112.

Nicholls, J. R., and M. J. Bennett, Cyclic oxidation—guideline for test standardization aimed at service behavior, *Mater. High Temp.*, 2000, 17(3), 413–428.

Pint, B. A., J. R. Martin, and L. W. Hobbs, ^{18}O/SIMS characterization of the growth mechanism of doped and undoped α-Al_2O_3, *Oxid. Met*, 1993, 39, 167–195.

Pint, B. A., P. F. Tortorelli, and I. G. Wright, Effect of cycle frequency on high-temperature oxidation behavior of alumina-forming alloys, *Oxid. Met.*, 2000, 58(1/2), 73–101.

Prasanna, K. M. N., A. S. Khanna, Ramesh Chandra, and W. J. Quadakkers, Effect of θ-alumina formation on the growth kinetics of alumina-forming superalloys, *Oxid. Met.*, 1996, 46(5/6), 465–480.

Radcliff, S., Factors influencing gas turbine use and performance, *Mater. Sci. Technol.*, 1987, 3, 554–561.

Reddy, K. P. R., J. L. Smialek, and A. R. Cooper, [18]O tracer studies of Al_2O_3 scale formation on NiCrAl alloys, *Oxid. Met.*, 1982, 17, 429–449.

Richardson, F. D., and J. H. E. Jeffes, *J. Iron Steel Inst.*, The Metals Society, 1948, 160, 261.

Sarioglu, C., M. J. Stiger, J. R. Blachere, R. Janakiram, E. Schumann, A. Ashary, F. S. Pettit, and G. H. Meier, The adhesion of alumina films in metallic alloys and coatings, *Mater. Corrosion* 2000, 51, 358–372.

Saunders, S. R. J., and J. R. Nicholls, Coatings and surface treatments for high temperature oxidation resistance, *Mater. Sci. Technol.*, 1989, 5(8), 780–798.

Smeggil, J. G., A. W. Funkenbusch, and N. S. Bornstein, A relationship between indigenous impurity elements and protective oxide scales adherence characteristics, *Met. Trans.*, 1986, 17A, 923–932.

Smialek, J. L., and G. M. Meier, High temperature oxidation, in *Superalloy II*, pp. 293–323, Eds. C. T. Sims, N. S. Stoloff, and W. C. Hagel, Wiley, New York, 1987.

Tumarev, A. S., and L. A. Panyushin, NASA TT F-13221, 1970; *Izv. Vyssh. Uchebn. Zaved. Chern. Metall.*, 1959, 9, 125.

Vargas, J. R., N. E. Ulion, and J. A. Goebel, Advanced coating development for industrial utility gas turbine engines, *Thin Solid Films*, 1980, 73(2), 407–414.

Wagner, C., *Bur. Bunsenges. Phys. Chem.*, 1959, 63, 772.

Wallwork, G. R., and A. Z. Hed, Some limiting factors in the use of alloys at high temperatures, *Oxid. Met.*, 1971, 3(2), 171–184.

Wright, I. G., B. A. Pint, L. M. Hall, and P. F. Tortorelli, Oxidation lifetimes: Experimental results and modeling, in *Lifetime Modeling of High-Temperature Corrosion Process, Proc. European Federation of Corrosion Workshop*, pp. 339-358, Eds. M. Schutze, W. J. Quadakkers, and J. R. Nicholls, EFC Publications No. 34, Maney Publishing, 2001.

Chapter 5

HIGH-TEMPERATURE CORROSION

Oxidation of metals and alloys has been discussed in detail in Chapter 4. In industrial processes, materials are exposed not only to oxygen but also to other environmental constituents in the form of gases such as CO_2 and SO_2, fused or molten salts like alkali and alkaline earth sulfates, chlorides, and solid particles in the form of sand and fly ashes. Some of the constituents are the exhaust products of the industrial processes, whereas others are ingested from extraneous sources. The solid particles may melt during their transit through the system or remain in particulate form before exiting the system. The interaction of these environmental constituents with the materials of construction results in corrosion and erosion. Corrosion consists of parasitic chemical conversion reactions, whereas erosion involves mechanical impact and associated material loss. The latter may, however, accelerate environmental interactions by removing protective oxide scales.

5.1 HOT CORROSION PROCESSES

Corrosion of materials including Ni and Co base superalloys induced by molten salts in an oxidizing gas at elevated temperatures is termed "hot corrosion" (Pettit and Giggins, 1987; Goward, 1986; Rapp and Zhang, 1994; Sidhu et al., 2006) to distinguish it from the traditional low-temperature corrosion. It is the result of accelerated oxidation at temperatures typically between 700°C (1300°F) and 925°C (1700°F) when metals and alloys become covered with contaminant salt films. The salts in the vapor phase are generally benign. The temperature range within which hot corrosion occurs strongly depends on the salt chemistry and gas constituents, as well as the alloy composition. This type of corrosion has been observed in coal gasifiers, petrochemical process equipment, internal combustion engine exhaust systems, boilers, gas turbine engines on aircraft, and particularly in marine and industrial applications (Fig. 5.1) (Keienburg et al., 1985), and in waste incinerators. The temperatures and partial pressures of oxygen and sulfur in some of the industrial processes is shown in Fig. 5.2 (Natesan, 1985). Although the gases by themselves do not have significant corrosive effects, the overall environment in these processes leads to the formation of solid and molten salts that take part in corrosion.

It is now customary to call corrosion that occurs above the melting point of the salt "type I hot corrosion." Typically, this type of corrosion occurs at the upper end of the temperature range indicated earlier. The corrosion at the lower end of the temperature range is called "type II hot corrosion". Type II may also occur above the salt melt temperature if the

Figure 5.1 Solid turbine blade of Ni base alloy Udimet 520 after 27,293 hours service at 750°C (1382°F) in a natural gas burning industrial gas turbine engine at Kraftwerk Union. Notice extensive hot corrosion, loss of cross section, and corrosion products (K.-H. Keienburg, W. Esser, and B. Deblon, Refurbishing procedures for blades of large stationary gas turbines, *Mater. Sci. Technol.*, 1985, 1, 620–628). Reproduced with permission from Maney Publishing.

deposited salts form a eutectic mixture with the melting point significantly lower than that of the individual constituents. These constituents include the product of reaction of the salts with the oxides formed on corroding metals and alloys.

Figure 5.3 schematically illustrates the range of the two types of hot corrosion measured in terms of metal loss as a function of temperature, superimposed on rate of traditional oxidation in absence of corrosion. In both types of hot corrosion, fluxing with corroding salts defeats the protective oxide scale that forms on superalloys and coatings. Once the protective scale is rendered ineffective, the substrate alloy or coating becomes vulnerable to internal oxidation and sulfidation, the latter occurring particularly in a type I situation. There are generally two stages to hot corrosion. The first stage is known as the initiation stage and involves the breakdown of the protective oxide scale. The second stage is called the propagation stage, in which the salts have access to the now unprotected metal and corrosion continues unabated at exceedingly high rates. Figure 5.4 (Birks et al., 1989) shows typical microstructures of type I and type II hot corrosion. As shown in Fig. 5.4, type I is characterized by an outer porous oxide layer over an intermediate region consisting of a mix of alloy and oxides and an inner region with extensive sulfidation in the zone depleted of γ' phases of the aluminide-coated Ni base alloy Nimonic 105. Type II hot corrosion, first observed in a marine application, in a gas turbine engine LM2500 powering the ship *William Callahan* (Wortman et al., 1976), typically involves nonuniform attack in the form of pitting with minimal to no sulfidation.

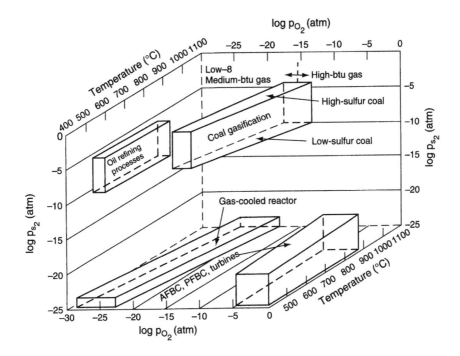

Figure 5.2 Operating conditions temperature–partial pressure of oxygen and sulfur in some industrial processes (K. Natesan, High-temperature corrosion in coal gasification systems, *Corrosion*, 1985, 41, 646–655. Reprinted with permission from NACE International.

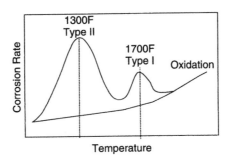

Figure 5.3 Schematic illustration of temperature effect on rate of damage to superalloys due to types I and II hot corrosion superimposed on contribution due to oxidation.

The Corroding Salts

The salts involved in hot corrosion are typically alkali and alkaline earth sulfates. The exact composition of the salts depends on the particular industrial process, impurities involved, and the fuel, air, and coolant compositions. For example, in gas turbine engines used in marine and industrial power generation, salts deposited on turbine blades in stages 1 and 2 have been identified and quantitatively measured (Bornstein, 1996) as shown in Table 5.1

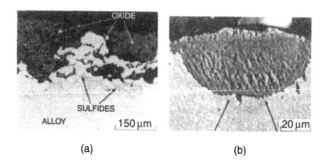

(a)   (b)

Figure 5.4 Hot corrosion morphology. (a) Type I developed on Ni–25Cr–6Al alloy coated with sodium sulfate. (b) Type II developed on CoCrAlY coating on vane in marine service (N. Birks, G. H. Meier, and F. S. Pettit, High temperature corrosion; in *Superalloys, Supercomposites and Superceramics,* Eds. John K. Tien and Thomas Caulfield, Academic Press, 1989; pp. 439–489). Reproduced with permission from Elsevier.

Table 5.1   Salts Recovered from Marine and Industrial Gas Turbines

Salt Constituent	Deposit Chemistry (mole %)	
External Airfoil Surface	First Stage	Second Stage
Na_2SO_4	40	28
K_2SO_4	4	3
$CaSO_4$	40	59
$MgSO_4$	13	8
Internal Cooling Passages		
Na_2SO_4	45	37
K_2SO_4	3.2	4.4
$CaSO_4$	41	46
$MgSO_4$	9.5	11.5

The data show that neither the external surfaces nor the internal cooling passages of air-cooled turbine blades contain any deposits of NaCl, the predominant salt constituent of air. Thus, NaCl does not play any *direct* role in hot corrosion. The effect is more through influencing the melting point of the salt and spallation of oxide scale, as we will see later in this chapter. Being the most water-soluble constituent of the salt content, NaCl in the air continues to remain soluble until it reaches the combustion chamber of turbines. A more important observation from the data in Table 5.1 and from other studies is that the predominant constituent of the salt deposit is sodium sulfate. The source of the sulfate salts is generally twofold. The process air and water may contain minor amounts of sulfates. Additionally, the fuel used in many of the industrial processes, including gas turbine engines, contains sulfur. Even aviation fuel, which is designated as "clean," contains sulfur levels of 0.1% or less. The heavy oils used in many industrial gas turbine engines to generate electricity may contain sulfur levels as high as 4%. Combustion of the fuel in the presence of air or oxygen generates SO_2. As discussed earlier, the intake air contains sea salts, particularly for plants operating within several hundred miles of the seacoasts. Seawater salt composition (Jaffee and Stringer, 1971) is given in Table 5.2. More than 10% of the salt content is sodium sulfate.

Table 5.2 Seawater Salt Composition

Salt Constituents	Concentration, g/L (%)
NaCl	23 (52)
$MgCl_2\, 6H_2O$	11 (25)
KCl, KBr	1.1 (2)
$CaCl_2\, 2H_2O$	1.4 (3)
$Na_2SO_4\, 10H_2O$	8 (18)
Total	44.5 (100)

Sulfates are also the products of the reactions between NaCl in the air that made it to the combustion chamber (Gupta et al., 1989) and the sulfur-containing product gas:

$$2NaCl + SO_2 + O_2 = Na_2SO_4 + Cl_2,$$

$$4NaCl + 2SO_2 + 2H_2O + O_2 = 2Na_2SO_4 + 4HCl.$$

The other alkali and alkaline earth sulfates are the products of similar reactions between sulfur dioxide and the remaining constituents of sea salt, such as KCl, $CaCl_2$, and $MgCl_2$. In hot corrosion, it is essential that the components be in contact with these fused salts because it has been found that in the vapor phase the sulfates are innocuous (DeCrescente and Bornstein, 1968). Droplets carrying the salts enter the system, such as a gas turbine engine, through its various modules, such as the compressor, continuing to evaporate the solvent water. In this journey, the first to precipitate are the least soluble salt components, whereas the most soluble salt, NaCl, continues past the compressor to the combustor. The residence times of the evaporating droplets in the hot zone, however, are insufficient to complete the evaporation of the solvent (Stearns et al., 1979; Rosner et al., 1979). The mechanism of deposition of salt is, therefore, thought to be predominantly impaction rather than equilibrium condensation. The impactions result in precipitates depositing as solids on compressor airfoils. Having the highest solubility, the NaCl in the droplets continues into the combustor, wherein it is converted to sulfates by the reaction with sulfur oxides. The sulfate particles deposited on the compressor airfoils frequently shed into the gas stream. Particles of sizes less than about $10\,\mu m$ follow the aerodynamic contour of the gas and exit the engine, while larger particles with higher momentum deviate and strike the turbine airfoils (McCreath, 1983). Depending on the temperature, the sulfate particles stick to the surface of downstream components.

Formation of a liquid film of sodium sulfate on an alloy substrate, a requirement for type I hot corrosion, needs exposure to a temperature above its melting point, which is about 1623°F (884°C). However, in the presence of the other sulfates, a eutectic forms, as indicated by the ternary diagram of Fig. 5.5, decreasing the melting point to as low as 650°C (1200°F). Hot corrosion may now occur far below the melting point of pure sodium sulfate.

Acid and Base Characteristics of the Salts

For molten sodium sulfate at 1700°F (927°C), equilibrium is represented by the reaction

$$Na_2SO_4 = Na_2O + SO_3.$$

What this equilibrium conveys is that the rate of formation (reaction from right to left) of Na_2SO_4 is equal to the rate of decomposition (reaction from left to right). Using free energies

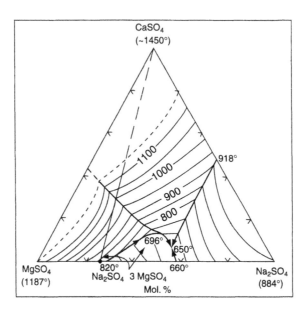

Figure 5.5 Na–Mg–Ca Sulfate ternary diagram showing the low-melting eutectic (*Phase Diagram for Ceramists*, Vol. II, American Ceramic Society, 1969). Reprinted with permission from Blackwell Publishing.

of formation, one can easily see that at equilibrium, partial pressure P of SO_3 is related to the activity a of Na_2O through the relation

$$\log P_{SO3} + \log a_{Na2O} = -16.7.$$

The stability ranges of various equilibrium phases in the Na–O–S system as a function of $-\log a_{Na2O}$ (the basicity of the melt) or partial pressure of SO_3 (acidity of the melt, given by $+\log P_{SO3}$) and the oxygen pressure is shown in Figure 5.6 (Park and Rapp, 1986).

What the phase stability diagram shows is that Na_2SO_4 is stable over a wide range of acidity (rich in SO_3) or basicity (rich in Na_2O).

For acidic conditions, which generally occur when in addition to the presence of the sulfate, SO_3 gas (in presence of oxygen, SO_2 is catalytically converted to SO_3) is available, the sulfate salt reacts with the protective oxide scale, such as Al_2O_3, on the alloy:

$$\text{(Acidic fluxing)} \quad Al_2O_3 + 3Na_2SO_4 = \mathbf{Al_2(SO_4)_3} + 3Na_2O,$$

$$3Na_2O + 3SO_3 = 3Na_2SO_4.$$

The last reaction can alternately be written as

$$3Na_2O + 3SO_2 + \tfrac{3}{2}O_2 = 3Na_2SO_4.$$

This type of acidic fluxing, first proposed by Goebel and Pettit in the early 1970s (Goebel and Pettit, 1970; Goebel et al., 1973), takes place in type II hot corrosion, which typically occurs well below the melting point of pure Na_2SO_4. Around the same time Cutler (1971) quantified the role of partial pressures of SO_3 and O_2 by utilizing an electrochemical "potentiometric" technique for hot corrosion of 18Cr8Ni steel in a low-melting eutectic of sulfate salts of Li, Na, and K.

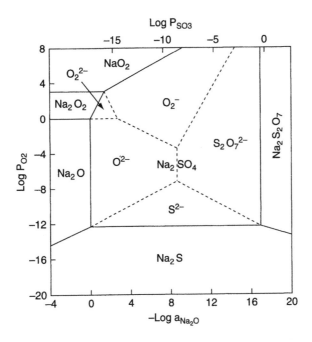

Figure 5.6 The Na–O–S phase stability diagram at 1173 K (C. O. Park and R. A. Rapp, *J. Electrochem. Soc.*, 1986, 133, 1636–1641). Reprinted with permission from the Electrochemical Society.

The melting point of Na_2SO_4 is significantly higher than the type II temperature regime. However, other sulfates in the deposit form low-melting eutectics, or reactions with oxides of the alloy constituents depress the melting point of Na_2SO_4. For example, $CoSO_4$ is the product of oxidation of CoCrAlY in oxygen containing SO_3. In the presence of Na_2SO_4, $CoSO_4$ forms a eutectic with a melting point around 565°C (1049°F), a depression of the melting point of the salt by as much as 319°C (574°F). This eutectic contains the principal corrosive species that were first found in the field on turbine blades in General Electric marine engine LM2500. The turbine blades were made of coated Ni base superalloy Rene 80 with a coating of composition given by CoCrAlY (Wortman et al., 1979).

Basic conditions, on the other hand, occur at temperatures where pure Na_2SO_4 is molten and the melt is richer in Na_2O. The sulfate salt reacts with the protective scale, forming aluminate (or chromate for reaction with Cr_2O_3):

$$\text{(Basic fluxing)} \quad 2Al_2O_3 + 2Na_2O + O_2 \text{ (gas)} = 4NaAlO_2.$$

The basic fluxing, first proposed by Bornstein and DeCrescente (1969), occurs in type I hot corrosion.

In many instances, the combustion environment sets the acidic or basic condition of the salts as well as the temperature in the industrial processes. For example, calculations by Tschinkel (Tschinkel cited in Goebel and Pettit, 1970) show that at 1000°C (1832°F), in the turbine section of a gas turbine engine, $P_{O2} > 0.2$ atm and $P_{SO3} \sim 10^{-4}$ to 10^{-5} atm ($\log P_{SO3} + \log a_{Na2O} = 16.7$ at 927°C, 1700°F). Thus, $a_{Na2O} \sim 10^{-12.7}$ to $10^{-11.7}$, which indicates acid condition in the salt. These estimates agree with Cutler's (1987) estimates of P_{O2} and $P_{SO3} + P_{SO2}$ of 1 and 5×10^{-4} atm, respectively.

5.2   HOT CORROSION OF METALS AND ALLOYS

Solubility of Oxides in Molten Salts

Materials including superalloys derive their stability at high temperature in oxidizing atmospheres by the formation of slow-growing protective oxide scales. Hot corrosion resistance of these materials depends on the solubility of these oxide scales in molten salts. The solubility of several of the oxides in pure molten Na_2SO_4 at 1200 K (1700°F) and 1 atm O_2 has been measured electrochemically by Rapp and colleagues. Figure 5.7 (Rapp, 2002) shows a compilation of some of the solubility data. A number of features are noteworthy:

- Each oxide, with the exception of SiO_2, exhibits a solubility minimum.

- There is a difference in basicity of about six orders of magnitude between the solubility minimums for the basic oxides (Ni, Co) and those of the protective acidic oxides (Al, Cr, Si). Many of the coatings, which form these oxide scales, are the subjects of the next chapter.

- The solubility minimum for each oxide corresponds to a particular basicity, $-\log a_{Na2O}$ (or acidity P_{SO3}).

Mechanism of Sustained Hot Corrosion

Detailed understanding of the mechanism of hot corrosion is not complete yet. However, based on the data available to date, a model has been proposed by Rapp and

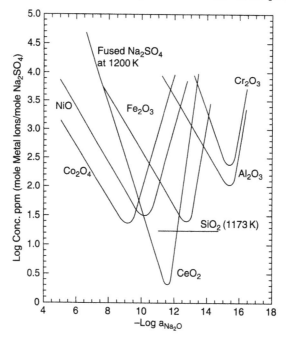

Figure 5.7   Solubilities of oxides in molten pure Na_2SO_4 at 1700°F (1200 K) (Robert A. Rapp, Hot corrosion of materials: A fluxing mechanism, *Corros. Sci.*, 2002, 22, 209–221). Reprinted with permission from Elsevier.

Goto (1981). For sustained hot corrosion, the model requires a negative solubility gradient (Fig. 5.8), i.e.,

$$\{d(\text{oxide solubility})/dx\}_{\text{at the oxide–salt interface}} < 0,$$

where x is the distance from the oxide–salt interface. The molten salt film in contact with the oxide scale is more basic and, therefore, has higher solubility of the oxide, which will dissolve to the saturation level. The dissolved solute will follow the negative concentration gradient to diffuse to the salt–gas interface. Because the solubility is low at this interface and the concentration exceeds saturation, the solute will reprecipitate as a nonprotective oxide. This mechanism does not consume the molten salt and therefore will be sustained until either the melt becomes more basic and oxide ions are not produced or the oxide scale is completely removed and the metal becomes accessible to the salt.

Role of Vanadium

Vanadium, nickel, and iron are constituents of the "porphyrins" that form the complex molecular structure of petroleum-based fuel oils. The concentration of vanadium varies widely in crude oils from a high of 1200 ppm in Venezuelan crude to a low of a few parts per million in Arabian light (Moliere and Sire, 1993). The oil refining process reduces vanadium and other heavy metals. However, many fuels, particularly residual oils, still contain vanadium in the

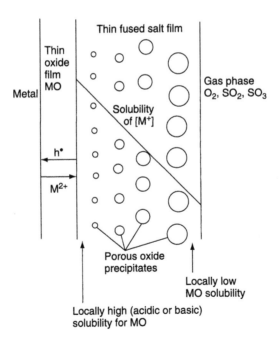

Figure 5.8 Negative solubility gradient in molten film of salts sustains oxide dissolution at the oxide–salt interface and oxide precipitation at the salt–gas interface (R. A. Rapp and K. S. Goto, Hot corrosion of metals by molten salts, *Proc. 2nd Int. Symp. on Molten Salts*, Eds. J. Braunstein and J. R. Selman, Electrochemical Society, Pennington, NJ, 1981, pp. 159–177). Reproduced with permission from the Electrochemical Society.

form of metal–organic complexes. When these fuels combust, vanadium oxides such as V_2O_4 and V_2O_5 are produced. These oxides react with sodium sulfate in the environment, forming low-melting sodium vanadates through the reactions

$$V_2O_5 + Na_2SO_4 = 2NaVO_3 \text{ (melting point } \sim 630°C, 1166°F) + SO_3,$$

$$V_2O_5 + 2Na_2SO_4 = Na_4V_2O_7 \text{ (melting point } \sim 635°C, 1175°F) + 2SO_3,$$

$$V_2O_5 + 3Na_2SO_4 = 2Na_3VO_4 \text{ (melting point } 850°C, 1562°F) + 3SO_3.$$

These vanadates are very acidic and aggressively attack the oxide scales by the process of fluxing. The solubilities of the oxides are higher and the minima are shifted to more basic regimes of the molten sulfate salts. One method used to alleviate the situation is to tie up vanadium with additives such as MgO. MgO reacts with V_2O_5, forming higher melting magnesium vanadates (for a version of binary phase diagram, see Speranskaya, 1971; Rhys-Jones et al., 1987; Moliere and Sire, 1993) through the reactions

$$3MgO \text{ (melts at } 2695°C) + V_2O_5 \text{ (melts at } 675°C, 1247°F)$$

$$= Mg_3V_2O_8 \text{ (melts at } 1070°C, 1958°F),$$

$$MgO + V_2O_5 = MgV_2O_6 \text{ (melts at } 740°C, 1364°F),$$

$$2MgO + V_2O_5 = Mg_2V_2O_7 \text{ (melts at } 930°C, 1706°F).$$

The product of the reactions is one of the high-melting magnesium vanadates, which deposits as a benign solid with minimal corrosion. As some of the MgO is also converted to sulfate, excess MgO is required to be added to the fuel than is called for based on the stoichiometry of the reactions. The typical additive ratio used to neutralize vanadium oxides by weight is Mg/V ~ 3.

5.3 ROLE OF SPECIFIC ALLOYING ELEMENTS IN HOT CORROSION OF NI AND CO BASED ALLOYS AND COATINGS

The role of various alloying elements in alloys and coatings depends on the details of the hot corrosion mechanism. In general, however, the following analysis can be made.

Chromium

Of all the constituents in alloys as well as coatings, chromium is most effective in imparting resistance to hot corrosion, particularly type II. The beneficial effect of increased chromium content in combating hot corrosion is well illustrated by data in Fig. 5.9 (Goward, 1983).

The key to the success of Cr in inhibiting corrosion lies in the fact that Cr_2O_3 can dissolve to form several valence states and that the solubility in basic melt is oxygen pressure dependent:

$$Cr_2O_3 + 3SO_3 \text{ (in } Na_2SO_4) = Cr_2(SO_4)_3 \text{ in acidic melt,}$$

$$Cr_2O_3 + Na_2O \text{ (in } Na_2SO_4) = 2NaCrO_2 \text{ in basic melt of low oxygen activity,}$$

$$2Cr_2O_3 + 4Na_2O \text{ (in } Na_2SO_4) + 3O_2 = 4Na_2CrO_4 \text{ in basic melt of high oxygen activity,}$$

$$2Cr_2O_3 + 2Na_2O \text{ (in } Na_2SO_4) + 3O_2 = 2Na_2Cr_2O_7 \text{ in basic melt of high oxygen activity.}$$

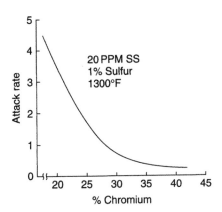

Figure 5.9 Effect of chromium on hot corrosion of MCrAlY coatings, M being Ni and Co (G. W. Goward, Recent developments in high temperature coatings for gas turbine airfoils, in *High Temperature Corrosion*, Ed. Robert A. Rapp, National Association of Corrosion Engineers, Houston, TX, 1983, pp. 553–560). Reproduced with permission from NACE International.

What all these reactions lead to are the following:

- In the presence of a thin sulfate film, the oxide–sulfate interface has low oxygen activity. Cr_2O_3 dissolves as $NaCrO_2$, which has a lower solubility than Na_2CrO_4 has at the salt–gas interface (Fig. 5.10) (Rapp, 1987).

- At the salt–gas interface the oxygen activity is higher. Therefore, Cr_2O_3 will dissolve as Na_2CrO_4, which has higher solubility than $NaCrO_2$ has at the oxide–salt interface.

- The result is a positive solubility gradient, lower at the oxide–salt interface and higher at the salt–gas interface. The positive gradient does not sustain the dissolution (at the salt–gas interface) and reprecipitation (at the oxide–salt interface) of Fig. 5.8. Instead, Cr_2O_3 will dissolve to its saturation level at the oxide–salt interface and cease to dissolve any further.

- This mechanism does not hold good for Al_2O_3 because it lacks multiple valence states. The solubility gradient is negative (Fig. 5.8). Solution at the oxide–gas interface and precipitation at the oxide–salt interface promote the dissolution precipitation mechanism, defeating the oxide scale. The model also does not hold good for Cr_2O_3 in an acid melt, in which case a negative solubility gradient exists, leading to dissolution at the oxide–salt interface and reprecipitation at the salt–gas interface.

An additional advantage of Cr relative to Al is that Cr_2O_3 is a faster growing oxide and, therefore, is able to reform sooner in case of loss of scale due to hot corrosion.

Nickel and Cobalt

As the solubility curves in Fig. 5.7 indicate, both nickel and cobalt are susceptible to hot corrosion by basic fluxing. In the type II environment, Ni base alloys exhibit somewhat better corrosion resistance than Co base alloys of otherwise equivalent composition. One of the contributing factors includes the higher eutectic temperature (by about 100°C, 180°F) Na_2SO_4

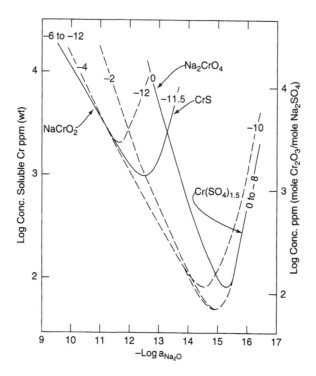

Figure 5.10 Solubilities of Cr_2O_3 in molten Na_2SO_4 at 1700°F (1200 K) (Robert A. Rapp, Chemistry and electrochemistry of hot corrosion of metals, *Mater. Sci. Eng.*, 1987, 87, 319–327). Reprinted with permission from Elsevier.

has with $NiSO_4$ than with $CoSO_4$, giving Ni higher stability toward hot corrosion. Additionally, the sulfides of Co are thermally more stable than the sulfide of Ni. This helps formation of protective scales on Ni alloys at lower temperatures than on equivalent Co alloys.

Silicon

The oxide of Si has a low solubility in the acidic range of the sulfate salt. As a result, SiO_2 scale-forming alloys and coatings exhibit superior hot corrosion resistance in a type II hot corrosion environment.

Tungsten and Molybdenum

The oxides WO_3 and MoO_3 react with Na_2SO_4, forming low-melting reaction products and resulting in increased P_{SO3}:

$$WO_3 + Na_2SO_4 = Na_2WO_4 \text{ (melts at 696°C, 1285°F)} + SO_3.$$

Initially the local partial pressure of SO_3 increases, causing acidic fluxing of oxide scale and sulfidation of underlying metal. The oxide MoO_3 behaves similarly. The effects of the reactions increase the susceptibility of W- and Mo-containing alloys (B1900, IN 738, Rene 80) to hot corrosion. Some alloys exhibit an incubation period during which hot corrosion is very slow or nonexistent. Beyond this period the rate of attack increases significantly. The concentration of Mo and W in the alloy appears to determine the length of the incubation period. The SO_3 pressure in the gas obviously influences the overall effects.

Vanadium

The effect of vanadium as a contaminant of fuel used in some industrial processes has already been discussed. Vanadium as an alloy constituent has a similar detrimental effect.

Titanium

Titanium as an alloy constituent for alumina formers has a detrimental effect on the growth rate of the alumina scale and therefore reduces oxidation resistance. However, when present in minor levels, particularly with increasing Ti/Al ratio, Ti (Felix, 1973) generally reduces the rate of hot corrosion.

Rare Earth Elements

When present as oxides such as within the surface scale of superalloys or coatings, the elements Ce, La, and Gd are known to reduce the extent of hot corrosion. In metallic form, however, they are not as beneficial. As oxides, these elements act as getters for sulfur, forming oxysulfides of the form M_2O_2S, M being the rare earth elements (Seybolt, 1971). Although thermodynamically the oxides of these elements are more stable in an oxidizing environment, with higher partial pressures of S, the oxysulfides and sulfides gain increased stability.

Platinum

Diffusion aluminide coatings in which platinum has been incorporated exhibit better resistance to type I hot corrosion. This is clearly demonstrated by cyclic hot corrosion test data of several coatings on three different nickel base superalloy substrates, polycrystalline Rene 80, single crystal PWA 1480, and CMSX-3 in Fig. 5.11 (Meier and Pettit, 1989). The extent of corrosion is significantly higher for coatings without platinum than those with platinum, the latter shown as the band of data at the top of Fig. 5.11.

The benefit of platinum is much less prominent in type II hot corrosion, as depicted in Fig. 5.12 (Meier and Pettit, 1989). Platinum becomes more effective against type II hot corrosion if it is present as a continuous $PtAl_2$ phase at the coating surface.

5.4 INFLUENCE OF OTHER CONTAMINANTS

Presence of Carbon

Carbon deposits on hot components generally occur because of incomplete combustion of fuel. Such deposits have been found to enhance hot corrosion. The dissolution of oxide scales is accelerated, and sulfur penetration into the base alloy is promoted by the presence of carbon.

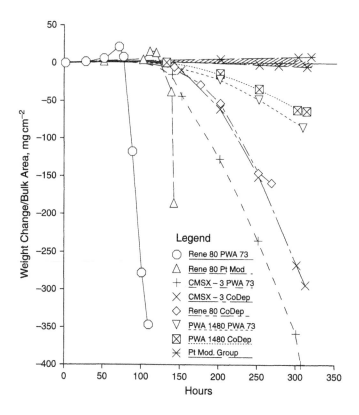

Figure 5.11 Type I cyclic hot corrosion data for a number of coating/alloy systems at 1000°C (1830°F) (G. H. Meier and F. S. Pettit, High-temperature corrosion of alumina-forming coatings for superalloys, *Surf. Coat. Technol.*, 1989, 39/40, 1–17). Reprinted with permission from Elsevier.

The enhanced sulfidation is thought to result from breakdown of sulfate salt in the presence of carbon (McKee and Romeo, 1974, 1975):

$$2Na_2SO_4 + 6C + 3O_2 = 2Na_2O + 2S + 6CO_2.$$

The released sulfur forms sulfides with the alloying elements. The physical form of the carbon, such as crystal structure and particle size, has a strong effect on the corrosion rate (Huang et al., 1979).

Presence of Chlorides

The presence of chlorides in the form of NaCl and HCl is known to disrupt the protective oxide scales on alloys and coatings. The sulfate salts containing chlorides now find access paths to the metallic coating and the substrate, which are susceptible to hot corrosion (Conde and McCreath 1980; Hancock et al., 1973). The chlorides accumulate at the metal oxide scale interface. The concentration at the interface often exceeds that at the surface, indicating that the migration is not concentration gradient dependent and does not involve ionic species (Kofstad, 1988). NaCl (melting point 742°C, 1368°F) also forms low melting eutectic with Na_2SO_4, reducing the melting temperature from 884°C to 620°C (from 1623°F to 1148°F) (Hancock, 1987). Large concentrations of chlorides also depletes Al and Cr from the coating or the exposed alloy, rendering them unable to form protective oxide scales (Nagarajan et al., 1982).

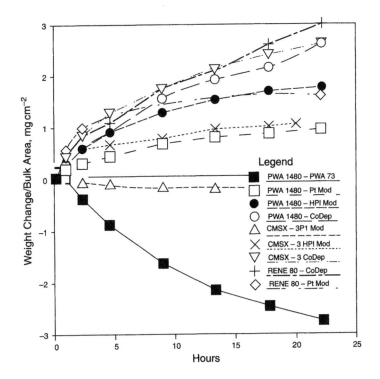

Figure 5.12 Type II cyclic hot corrosion data for a number of coating/alloy systems at 700°C (1290°F) (G. H. Meier and F. S. Pettit, High-temperature corrosion of alumina-forming coatings for superalloys, *Surf. Coat. Technol.*, 1989, 39/40, 1–17). Reprinted with permission from Elsevier.

5.5 HOT CORROSION OF TBCs

Thermal barrier coatings (TBCs), which are discussed in detail in Chapter 7, are used to reduce temperatures of metallic components by the application of coating of ceramics of low thermal conductivity. Zirconia, ZrO_2, is an industry standard for TBC. In order to avoid a phase change of zirconia during thermal cycling, it is stabilized by alloying of the ceramic with oxides such as MgO, CaO, and Y_2O_3, the last providing the best stability and performance.

Molten sulfates in the presence of sulfur dioxide and trioxide are known to react with the stabilizers, MgO being the most reactive. The reactions result in the tie-up of the stabilizers as sulfates, as illustrated for magnesia and yttria stabilizers:

$$MgO + Na_2SO_4 = MgSO_4 + Na_2O,$$

$$SO_2 + 1/2O_2 = SO_3,$$

$$Na_2O + SO_3 = Na_2SO_4,$$

$$Y_2O_3 + 3Na_2SO_4 = Y_2(SO_4)_3 + 3Na_2O,$$

$$3SO_2 + 3/2O_2 = 3SO_3,$$

$$3\,Na_2O + 3SO_3 = 3Na_2SO_4.$$

The presence of $MgSO_4$ has been known to occur in gas turbine engine combustion chambers utilizing magnesia stabilized zirconia TBC (Goward, personal communication, 2005). Yttria-stabilized zirconia, on the other hand, is relatively more stable, and as the foregoing reactions show, requires higher SO_3 partial pressure. The leaching of the stabilizers makes the ceramic prone to the detrimental phase transformation, resulting in failure of TBCs on thermal cycling (Lau and Bratton, 1982).

Vanadium compounds formed as a result of burning of vanadium-containing fuels also react with stabilizers of TBC in a similar fashion. The reaction forms vanadates,

$$Y_2O_3 + V_2O_5 = 2YVO_4,$$

resulting in destabilization and ultimate failure of the TBC.

5.6 HOT CORROSION-LIKE DEGRADATION

A type of degradation of coatings and superalloy substrates has been observed in the early 1990s in helicopter engines conducting low-altitude missions in the Persian Gulf region (Fig. 5.13) (Smialek et al., 1994). The degradation has some semblance to hot corrosion, including the occasional presence of sulfur. Postexposure analysis of T700 and T55 engine turbine vanes of Co base alloy X-40, coated with diffusion aluminide, revealed cooling hole blockages by fine sandy materials, Ca–Al–Fe silicate deposits containing CO_2- and SO_2-filled gas bubbles, worn-out coating in some areas, internal oxidation, and substrate attack. The fine desert sand of the region is involved in the degradation process. Blockage of cooling holes resulted in overheating to temperatures at which the sands melt. Sands collected over a vast region in the Persian Gulf area consisted of various concentrations of Ca–Al–Fe silicate, Ca–Mg carbonates, feldspars $(Na–K)AlSi_3O_8$, NaCl, and gypsum $CaSO_4 2H_2O$. Although the design temperature of the vanes was limited to 815°C (1500°F), the deposit and the profile of the reaction product suggest significant overheating.

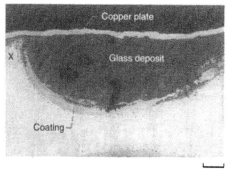

100 μm

Figure 5.13 Leading edge cooling hole of T700 helicopter turbine vane (X-40 alloy): hole plugged with porous deposit and diffusion aluminide coating degraded (James L. Smialek, Frances A. Archer, and Ralph G. Garlick, Turbine airfoil degradation in the Persian Gulf War, *JOM*, 1994, 12, 39–41). Reprinted with permission from The Minerals, Metals & Materials Society.

REFERENCES

Birks, N., G. H. Meier, and F. S. Pettit, High temperature corrosion, in *Superalloys, Supercomposites and Superceramics*, pp. 439–489, Eds. John K. Tien and Thomas Caulfield, Academic Press, New York, 1989.

Bornstein, N., Review of sulfidation corrosion yesterday and today, *JOM*, 1996, 48(11), 38.

Bornstein, N. S., and M. A. DeCrescente, *Trans. Met. Soc. AIME*, 1969, 245, 1947.

Conde, J. F., and C. G. McCreath, in *Proc. Conf. on Behavior of High Temperature Materials in Aggressive Environments*, pp. 497–512, The Metals Society, London, 1980.

Cutler, A. J. B., The effect of oxygen and SO_3 on corrosion of steels in molten sulphates, *J. Appl. Electrochem.*, 1971, 1, 19–26.

Cutler, A. J. B., Corrosion reaction in molten sulfates, *Mater. Sci. Technol.*, 1987, 3, 512.

DeCrescente, M. A., and N. S. Bornstein, Formation and reactivity thermodynamics of sodium sulfate with gas turbine alloys, *Corrosion*, 1968, 24, 127.

Felix, P. C., Evaluation of gas turbine materials by corrosion rig tests, in *Deposition and Corrosion in Gas Turbines*, p. 331, Eds. A. B. Hart and A. J. B. Cutler, Wiley, New York, 1973.

Goebel, J. A., and F. S. Pettit, Na_2SO_4-induced accelerated oxidation (hot corrosion) of nickel, *Metall. Trans*, 1970, 1, 1943–1954.

Goebel, J. A., F. S. Pettit, and G. W. Goward, Mechanisms for the hot corrosion of nickel-base alloys, *Metall. Trans*, 1973, 4, 261–278.

Goward, G. W., Recent developments in high temperature coatings for gas turbine airfoils, in *High Temperature Corrosion*, pp. 553–560, Ed. R. A. Rapp, National Association of Corrosion Engineers, Houston, TX, 1983.

Goward, G. W., Low-temperature hot corrosion in gas turbines: A review of causes and coatings therefor, *J. Eng. Gas Turbine Power, Trans ASME*, 1986, 108, 421–425.

Gupta, A. K., J. P. Immarigeon, and P. C. Patnaik, A review of factors controlling the gas turbine environment and their influence on hot salt corrosion test methods, *High Temp. Technol.*, 1989, 7(4), 173–186.

Hancock, P., Vanadic and chloride attack of superalloys, *Mater. Sci. Technol.*, 1987, 3(7), 536–544.

Huang, T., E. A. Gulbransen, and G. H. Meier, Hot corrosion of Ni-base turbine alloys in atmospheres in coal-conversion systems, *JOM*, 1979, 3, 28–35.

Hurst, R. C., J. Johnson, M. Davies, and P. Hancock, in *Deposition and Corrosion in Gas Turbines*, pp. 153–157, Ed. A. B. Hart and A. J. B. Cutler, Barking, Applied Science, 1973.

Jaffee, R. I., and J. Stringer, High-temperature oxidation and corrosion of superalloys in the gas turbine (A review), in *Source Book on Materials for Elevated-Temperature Applications*, pp. 19–33, Ed. Elihu F. Bradley, American Society of Metals, 1997; Original Source: *High Temperatures—High Pressures*, 3, 1971.

Keienburg, K.-H., W. Eßer, and B. Deblon, Refurbishing procedures for blades of large stationary gas turbines, *Mater. Sci. Technol.*, 1985, 1, 620–628.

Kofstad, P., *High Temperature Corrosion*, Elsevier Applied Science, New York, 1988.

Lau, S. K., and R. J. Bratton, Degradation mechanisms of ceramic thermal barrier coatings in corrosive environments, in *High Temperature Protective Coatings*, pp. 305–317, Ed. Subhash C. Singhal, American Institute of Mining, Metallurgical, and Petroleum Engineers, New York, 1982.

McCreath, C. G., Environmental factors that determine hot corrosion in marine gas turbine rigs and engines, *Corros. Sci.*, 1983, 23, 1017–1023.

McKee, D. W., and G. Romeo, *Second US/UK Navy Conf. on Gas Turbine Materials in Marine Environment, MCIC Report 75-27*, Castine, ME, July 1974.

McKee, D. W., and G. Romeo, Effects of transient carbon deposition on the sodium sulfate induced corrosion of nickel base alloys, *Metall. Trans.*, 1975, 6A, 101–108.

Meier, G. H., and F. S. Pettit, High-temperature corrosion of alumina-forming coatings for superalloys, *Surf. Coat. Technol.*, 1989, 39/40, 1–17.

Moliere, M., and J. Sire, Heavy duty gas turbine experience with ash-forming fuels, *J. Physique* 1993, IV, 3, 719–730.

Nagarajan, V., J. Stringer, and D. P. Whitle, The hot corrosion of cobalt-base alloys in a modified Deans Rig—I. Co–Cr, Co–Cr–Ta, and Co–Cr–Ti Alloys; II. Co–Cr–Al Alloys; III. Co–Cr–Mo Alloys, *Corros. Sci.*, 1982, 22, 407–427, 429–439, 441–453.

Natesan, K., High-temperature corrosion in coal gasification systems, *Corrosion*, 1995, 41, 646–655.

Park, C. O., and R. A. Rapp, Electrochemical reactions in molten Na_2SO_4 at 900°C, *J. Electrochem. Soc.*, 1986, 133, 1636.

Pettit, F. S., and C. S. Giggins, Hot corrosion, in *Superalloys II*, pp. 327–358, Eds. C. T. Sims, N. S. Stoloff, and W. C. Hagel, Wiley, New York, 1987.

Phase Diagram for Ceramists, Vol. II, American Ceramic Society, 1969.

Rapp, Robert A., Chemistry and electrochemistry of hot corrosion of metals, *Mater. Sci. Eng.*, 1987, 87, 319–327.

Rapp, R. A., Hot corrosion of materials: a fluxing mechanism? *Corros. Sci.*, 44, 2002, pp. 209–221.

Rapp, R. A., and K. S. Goto, Hot corrosion of metals by molten salts, in *Molten Salts I*, pp. 159–177, Eds. J. Braunstein and J. R. Selman, Electrochemical Society, Pennington, NJ, 1981.

Rapp, Robert A., and Y. S. Zhang, Hot corrosion of materials: Fundamental studies, *JOM*, December 1994, pp. 47–55.

Rhys-Jones, T. N., J. R. Nicholls, and P. Hancock, The prediction of contaminant effects on materials performance in residual oil-fired industrial gas turbine environment, in *Plant Corrosion: Prediction of Materials Performance*, p. 290, Eds. J. E. Strutt and J. R. Nicholls, Institution of Corrosion Science/Ellis Horwood, 1987.

Rosner, D. E., B. K. Chen, G. C. Fryburg, and F. J. Kohl, Chemically frozen multicomponent boundary layer theory of salt and/or ash deposition rates from combustion gases, *Combust. Sci. Technol.*, 1979, 20, 87–106.

Seybolt, A. U., Role of rare earth additions in the phenomenon of hot corrosion, *Corros. Sci.*, 1971, 11, 751–761.

Sidhu, T. S., S. Prakash, and R. D. Agrawal, Hot corrosion and performance of nickel-based coatings, *Curr. Sci.*, 2006, 90(1), 41–47.

Smialek, James L., Frances A. Archer, and Ralph G. Garlick, Turbine airfoil degradation in the Persian Gulf War, *JOM*, 1994, 12, 39–41.

Speranskaya, E. I., The system $MgO–V_2O_5$, *Bull. Head Sci. USSR, Inorg. Mater.*, 1971, 7, 605–608.

Stearns, C., F. Kohl, and G. C. Fryburg, *4th US/UK Navy Conference on Gas Turbine Materials in Marine Environments*, Vol. II, Naval Sea Command, Annapolis, MD, 1979.

Tschinkel, J. G., cited in J. A. Goebel and F. S. Pettit, Na_2SO_4-induced accelerated oxidation (hot corrosion) of nickel, *Metall. Trans.*, 1970, 1, 1943–1954.

Wortman, D. J., R. E. Fryxell, and I. I. Bessen, A theory of accelerated turbine corrosion at intermediate temperatures, *Proc. Third Conf. on Gas Turbine Materials in Marine Environment*, Session V, Paper No. 11, Bath, England, 1976.

Wortman, D. J., R. E. Fryxell, K. L. Luthra, and P. A. Bergman, Mechanism of low temperature hot corrosion: Burner rig studies, *Thin Solid Films*, 1979, 64, 281–288.

Chapter 6

OXIDATION- AND CORROSION-RESISTANT COATINGS

Superalloys used in high-temperature industrial processes are generally developed with optimized structural properties such as tensile, creep, and fatigue strength while maintaining microstructural stability over a wide temperature range. The attractive mechanical properties are achieved to some extent at the expense of environmental resistance. The environmental protection, therefore, has to be provided by compatible thin metallic coatings (Sivakumar and Mordike, 1989), which are not designed to carry load, but have constituents and microstructures to provide good oxidation and corrosion protection.

6.1 REQUIREMENTS FOR METALLIC COATINGS

At the heart of the successful environmental performance of metallic coatings lies the formation of thin oxide scales on the coating surface, that limit access of oxygen and corroding salts. For these scales to provide extended protection, a number of requirements have to be satisfied (Nicholls, 2000; Nicholls and Hancock, 1987). Many of the solutions to the requirements described below play multiple roles in protecting the coated article.

Oxidation/Corrosion Resistance

Thermodynamically stable, protective surface scale of uniform thickness

Slow growth rate of protective surface scale

Adherent surface scale

High concentration of scale former

Stability

No undesired phase changes within the coating

Low diffusion rate across interface at use temperature

Adequate compositional stability across interface

Minimized brittle phase formation

Adhesion

Good adherence of coating to substrate

Matched coating/substrate properties to reduce thermal stress

Minimized growth stresses (process parameters related)

Optimized surface condition (rough or smooth)

Structural Properties

Can withstand service-related creep, fatigue, and impact loading of surface without failure of function

Coating Constituents and Their Role

Coatings are tailored for specific applications by controlling their elemental composition, microstructure, and by selection of manufacturing process. Various elemental constituents and their influences are given in Table 6.1 (Nicholls, 2000).

Table 6.1 Elemental constituents of metallic coatings, their functions, and effects

Elemental Constituent	Beneficial Aspects	Detrimental Aspects
Ni	Major constituent of substrate alloy. Provides strength.	Prone to destructive interaction with sulfur.
Co	Major constituent of substrate alloy. Provides microstructural stability and strength.	Prone to destructive interaction with sulfur.
Al	Constituent of substrate alloy. Major contributor to providing strength. Contributes to oxidation resistance.	Large concentration lowers melting point.
Cr	Constituent of substrate alloy. Contributes to oxidation resistance to 1500°F (816°C). Reduces Al requirement for formation of alumina scale. Imparts resistance to hot corrosion.	Lowers creep strength.
Ta	Enhances hot corrosion and oxidation resistance. Improves strength.	
Si	Enhances oxidation and type II hot corrosion resistance.	Large concentration leads to formation of brittle phases.
Hf, Y, Y_2O_3, oxides of other reactive elements	Improves adherence of alumina and chromia scales.	Large amounts are detrimental.
Pt	Improves oxidation and hot corrosion resistance.	

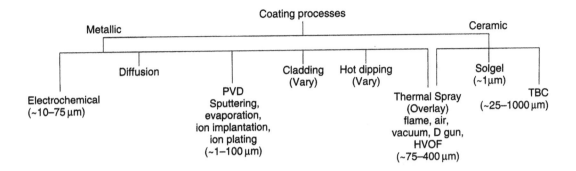

Figure 6.1   Some common coating processes.

6.2   COATING PROCESSES

A large number of coating processes are available to provide surface protection. The selection of the processes depends on the component design and the application. For example, coatings required for protection against hot corrosion may not be optimum for oxidation protection. Figure 6.1 lists some of the processes with typical coating thickness achieved.

6.3   DIFFUSION COATINGS

Diffusion coatings (Goward and Seigle, 1994; Goward, 1998) consist of a substrate alloy surface layer enriched with the oxide scale formers Al, Cr, Si, or their combination to a depth of 10 to 100 μm. These elements combine with the primary constituents of the substrate alloy to form intermetallics with significant levels of the oxide scale formers. For example, in Ni base superalloys, surface enrichment with aluminum forms nickel aluminide, NiAl (the β phase in the Ni–Al system), which is the predominant constituent of the coating. The substrate alloy participates in the formation of diffusion coatings. For oxidation protection, the diffusion coatings of choice are the aluminides, which form a protective alumina scale on high-temperature exposure in air. For protection against hot corrosion, incorporation of platinum in the aluminide, chromizing, and siliconizing are more beneficial. Because the β phase field in the Ni–Al system is quite broad in composition range, the Al content in diffusion aluminides can vary within a wide range with typical aluminides having Al $\leq$30 wt %. Higher Al content results in hyperstoichiometric composition. Such aluminides exhibit a bluish tint and are called "blue beta". Lower aluminum content, on the other hand, results in the hypo-stoichiometric composition. Compositions of diffusion coatings based on Cr and Si also may vary over a wide range. These coatings can be applied to components of complex shapes. The coating process involves exposure to high temperature. Additional postcoating heat treatment may be required to restore substrate properties such as creep and fatigue strength. As the heat-treatment temperature seldom exceeds 1100°C (2012°F), which can be done in traditional furnaces, the coating equipments and facilities do not require large capital investments. Diffusion coatings, which are essentially a type of surface enrichment in which vapors are deposited, can be produced by one of several methods described in Fig. 6.2. The basic process consists of the following steps:

- Generation of Al-, Cr-, or Si-containing vapors

- Transport of the vapors to the component surface

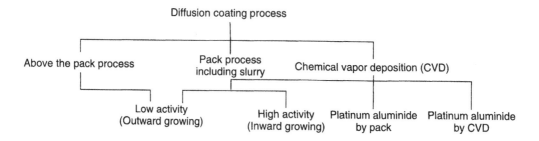

Figure 6.2 Various diffusion coating processes.

- Reaction of the vapors with the substrate alloy followed by associated diffusion processes within the alloy

- Additional heat treatments as necessary to achieve desired coating composition and coating as well as substrate properties

The microstructure, Al, Cr, or Si activity in the coating, and coating thickness depend on the substrate alloy, the process parameters including temperature and subsequent heat treatment.

Pack Coatings

Aluminiding of Ni base alloys

Aluminide coatings rely on the formation of βNiAl on the surface of the component alloys. There are three major processes (Fig. 6.2), by which the aluminide can be formed. These are the pack, above the pack, and chemical vapor deposition (CVD). Additionally, a variant of the pack process, called "slurry" coating, uses a slurry of the coating materials with a carrier fluid to inject into cavities to be coated or to paint accessible surfaces. One of the steps, which is common among the three processes, is the generation of vapors containing aluminum or the other metallic constituents of the coatings. The vapors are invariably halides because of their high vapor pressures. The vapors are transported to and react with the alloy constituents forming the aluminide coating. Additional heat treatment may be required to achieve desired composition, microstructure, and properties through diffusion processes. The difference among the processes depicted in Fig. 6.2 lies in the method by which the halide vapors are created and transported to the component. In the pack process, the component is imbedded in, and therefore in intimate contact with, a pack mix in a heated retort. The pack generates the halide vapors. In the above-the-pack process, the component to be coated is inside the retort but not in contact with the pack. The halide vapors are effectively plumbed on to the accessible internal and external surfaces of the component. In the CVD process, the halide vapor sources are external in individual generators and the vapors are plumbed on to the component held inside a reactor in a heated retort.

Pack Process

In this process, first the components to be coated are cleaned, generally by grit blasting with alumina grits to remove oxides and contaminants from the surface. The areas of the component to be protected from coating deposition are masked properly. The maskant should be capable

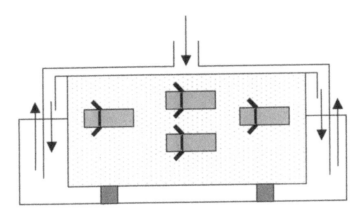

Figure 6.3 Schematic of the pack process.

of withstanding process temperatures without degrading. The masked components are buried in a "pack" mix (Fig. 6.3) contained in a sealed or semisealed retort. The retort is inserted into a furnace and heated in a protective atmosphere such as argon or hydrogen. The pack mix consists of the source of aluminum (or chromium or silicon for chromizing and siliciding, respectively); an "activator," which, on reacting with aluminum, generates the aluminum halide vapors; and an inert constituent such as alumina powder. The inert powder material provides interconnected porosity for vapor transport. It also inhibits sintering of the pack mix on continued exposure to high temperature. Good coating reproducibility and lower cost are the main advantages of the pack process. The disadvantages are that the flexibility of composition of the coating and the coating thickness are limited. Also, frequent pack-particle entrapment in the outer layer of the coating occurs in one variant of diffusion coatings. The entrapment of the pack particles is generally detrimental because they hinder the formation of a continuous protective scale. The limited coating thickness range achieved in this process arises from two factors. First, the process is diffusion limited. Therefore, progressively more time is required for the same thickness increase. Secondly, the pack continues to undergo depletion of the scale-forming constituents. Typical pack compositions (Nicholls, 2000; Angenete, 2002; Marijnissen, 1982) are given in Table 6.2.

Above-the-Pack Process

In this process (Fig. 6.4), the pack mix is held in trays. The masked components to be coated are positioned above the pack in a retort with the option of flowing inert gas. Plumbing is

Table 6.2 Typical Pack Compositions

Diffusion Coating	Pack Composition (wt %)	Coating Temperature Range, °C (°F)
Aluminide	1–15 Al, 1–3 NH$_4$Cl, balance Al$_2$O$_3$	750–1050 (1380–1920)
Chromize	48 Cr, 4 NH$_4$Cl, balance Al$_2$O$_3$	750–1050 (1380–1920)
Silicide	5 Si, 3 NH$_4$Cl, balance Al$_2$O$_3$	750–1050 (1380–1920)

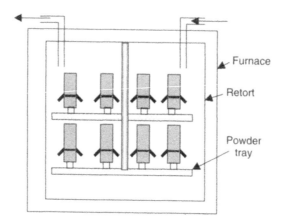

Figure 6.4 Schematic of the above-the-pack process.

so designed that the vapors, generated in the pack from the volatile halides, have access to both external and internal surfaces of the component to be coated. The retort is inserted into a furnace and held at the desired temperature for the selected duration. The process, which effectively involves chemical vapor deposition (CVD), typically yields low-activity (outward-growing) coating. We will discuss later what we mean by "low" and "high" activity coating.

The advantages of the above-the-pack process, also known as the gas-phase process, include the ability to uniformly coat internal passages and holes drilled to cool the components. An example of a component benefiting from such coatings is a turbine blade of a gas turbine engine. Turbine blades have internal passages as well as holes drilled through the blade walls for the flow and exit of cooling air. The above-the-pack process also provides a clean coating without pack particle entrapment. Diffusion coatings, which consist predominantly of β phase, are brittle below a transition temperature dependent on the coating composition and phase content. This temperature, considered in more detail in Section 6.6 in the subsection titled "Mechanical Properties of Coatings and Coated Materials," correlates well with the aluminum content of the β phase. The typical above-the-pack coating exhibits a lower transition temperature compared with the pack counterpart, due inherently to its lower aluminum content.

Pulse Aluminiding

A variant of the above-the-pack process, called pressure pulse aluminiding (Restall et al., 1980), uses subatmospheric but cyclic pressure. In a retort, the components are so arranged as to remain separated from the aluminum source. The chamber is filled with an inert gas such as argon. The pressure is pulsed by alternately evacuating and refilling to about 10 kPa, typically 8 to 10 times per minute. In addition to coating the external surface, the process allows finer internal passages to be coated. The ratio between thickness of external and internal coatings is in the range 2 to 1. The pulsing of the pressure improves the transport of the reactant vapors into and out of the internal passages.

Slurry Aluminiding

In another variant of the pack process, aluminum is deposited from a slurry or by thermal spray followed by subsequent diffusion. PWA 44 and SermeTel W are commercial slurry-based aluminides, the latter formed by diffusion heat treatment above 870°C (1598°F) after brushing with slurry. The slurry consists of aluminum in a water-based acidic solution of chromates and phosphates (Kircher et al., 1994).

Role of Activator

The activators are halide salts, generally chlorides and fluorides, that react with the metallic component of the pack to form metal halides having high vapor pressures. For aluminide coatings utilizing an ammonium chloride activator, the reactions involved in the formation of the coating composition are summarized in Fig. 6.5. The activator eventually creates AlCl vapor. Depending on the activity of Al and the temperature of the pack, the vapor reacts with Ni in the buried substrate, forming βNiAl or δNi$_2$Al$_3$. The latter subsequently decomposes into βNiAl by additional heat treatment. The gaseous products of the reaction include AlCl$_3$ vapor, which is recycled.

Microstructure and Mechanism of Coating Formation

The thickness increase and weight gains of diffusion coatings formed on the substrates with processing time have successfully been predicted by analyses of the underlying thermodynamics and kinetics of the pack process (Levine and Caves, 1974; Kandasamy et al., 1981). Important process parameters, that control these characteristics include the nature and concentration of the activators, the pack composition, the temperature, and the distribution of inert material, which aids in gas transport. A qualitative mechanism of the formation of diffusion aluminides on Ni base superalloys was first proposed by Goward and Boone (1971), followed later by Pichoir (1978). The model explained the microstructure of two major variants of diffusion aluminides on Ni base alloys. The important elements of the model are consistent with the diffusional analysis of intermetallics by Janssen and Rieck (1967) and by Shankar and Seigle (1978) (see Fig. 6.8c), who showed that the diffusivities of Ni (D_{Ni}) and Al (D_{Al}) are strong functions of the stoichiometry of NiAl. For example, $D_{Ni}/D_{Al} \approx 3.0$ for Ni-rich NiAl with Al content <50 at %, but for Al-rich NiAl at about 51.5 at % Al, $D_{Al}/D_{Ni} \geq 3$. The aluminum activity, therefore, plays a critical role in determining the predominant diffusing species. Depending on the content of aluminum in the pack and the processing temperature, the coating process is termed a "low-activity" or "high-activity" process. When the aluminum activity is low, and the temperature is in the high end of the temperature window (>1000°C, 1832°F), the predominant diffusing species is Ni, which diffuses out of the alloy, producing an "outward diffusion" coating. If the aluminum activity is high, as happens in the low end of the temperature range (<950°C, 1742°F), Al diffuses inward, resulting in an "inward diffusion" coating. Aluminum is also the predominant diffusing species in Ni$_2$Al$_3$. In order to understand the complexity of

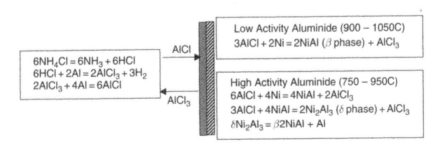

Figure 6.5 Reactions in which the pack activator participates.

the mechanism of coating formation, we need to look at differences in response among pure Ni, binary Ni–Cr alloy and more complex Ni base superalloys.

Low Activity, Outward Diffusion

Pure Ni: If the activity of Al in the pack is low and the coating temperature is in the 750–950°C (1380–1740°F) range, Ni preferentially diffuses out through the coating and combines with Al to form NiAl. The coating grows outward. Because in this case the outward diffusion of Ni is faster than the inward diffusion of Al, Kirkendall porosity is generated below the original interface, shown schematically in Fig. 6.6. The outward growth occasionally traps pack particles within the coating. These particles are observed only near contact surfaces, indicating a mechanical entrapment rather than formation by diffusion or vapor transport processes. The diffusing Ni flux appears to physically drag the particles outward, resulting in more inclusions closer to the surface

Ni–Cr alloy: The behavior of the alloy depends on its Cr content. For Ni10Cr (10 wt % Cr), Ni diffuses outward to form NiAl. The alloy near the interface loses Ni. The near interface region, therefore, becomes enriched with Cr. However, the Cr level is still within the solubility limit. As a result, no Cr precipitation occurs. On the other hand, for Ni20Cr, the situation is different. As before, Ni diffuses outward to form NiAl. The loss of Ni from the alloy at the interface again increases the Cr concentration. However, this alloy already has a high Cr content. As the increase in Cr concentration at the near-interface zone now exceeds the solubility limit, αCr precipitates in the alloy near the interface.

Ni base superalloy: As before, in the low-activity process Ni is the predominant diffusing species, which diffuses outward and combines with aluminum to form the external NiAl zone (Ni + Al = NiAl). Near the interface, the internal zone, which is also called the interdiffusion zone (IDZ), loses Ni. The loss of Ni from the alloy, which can be represented as a $\gamma + \gamma'$ system, results in the formation of NiAl ($\gamma + \gamma' - \mathrm{Ni} = \mathrm{Ni} + \mathrm{Ni_3Al} - \mathrm{Ni} = \mathrm{NiAl} + 3\mathrm{Ni}$). The NiAl so formed in the inner zone has very low solubility for many of the alloying constituents of the alloy. These constituents, therefore, precipitate out as shown later in Fig. 6.8a. The low-activity coating appears to have two zones, both of which are βNiAl phase. The total coating thickness includes both external and interdiffusion zones.

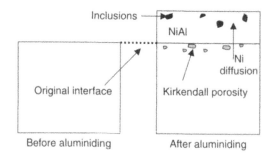

Figure 6.6 Diffusion aluminide formation on pure Ni.

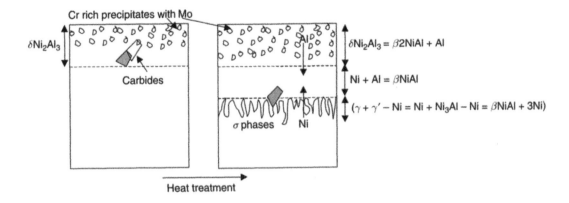

Figure 6.7 Microstructure formation in high-activity process.

High Activity, Inward Diffusion

Ni base superalloy: The high-activity coating process (Fig. 6.7) is characterized by Al as the predominant diffusing species. The consequences of higher inward diffusion of Al relative to outward Ni diffusion are the absence of Kirkendall porosity, elimination of embedded pack particles, and the original surface becoming the external surface of the coating. The microstructural detail and the coating composition depend on the Al activity. Qualitatively, three possibilities exist:

(a) Very high Al activity results in the formation of a δNi$_2$Al$_3$ phase (see Fig. 2.8).

(b) High Al activity results in a δNi$_2$Al$_3$ outer layer with adjacent βNiAl of high Al content.

(c) Moderate Al activity results in βNiAl of high Al content.

To form acceptable coatings, cases (a) and (b) require additional heat treatment to convert the brittle δNi$_2$Al$_3$ phase into less brittle, Al-rich "hyper-stoichiometric" ($>$50 at %) βNiAl. Whereas Al is the predominant diffusing species in hyperstoichiometric βNiAl and δNi$_2$Al$_3$, Ni diffusion dominates in "hypostoichiometric" ($<$50 at % Al) composition. Therefore, Al diffuses inward in the top third of the coating while Ni diffuses outward in the bottom third, the interdiffusion zone. The NiAl in the middle zone forms because of the combination of the Ni moving out of the inner zone and the Al moving in from the outer zone. The growth of the coating is diffusion controlled. The thickness, therefore, increases roughly as the square root of exposure time at the processing temperature. The coating thickness can be increased to some extent by raising the process temperature. However, the possibility of interface melting and depletion of the strength of the pack material become matters of concern.

The growth mechanism and the formation of coating microstructure of the low- and high-activity aluminides are compared in Fig. 6.8 (Goward, 1970; Shanker and Seigle, 1978).

Aluminiding of Co Base Alloys

In Ni base alloys, coherent γ' (Ni$_3$Al) precipitates provide the main strengthening mechanism by impeding dislocation motion. However, because of their structure, Co base alloys lack such coherency for γ' precipitates. This is the primary reason why Co alloys are not strengthened

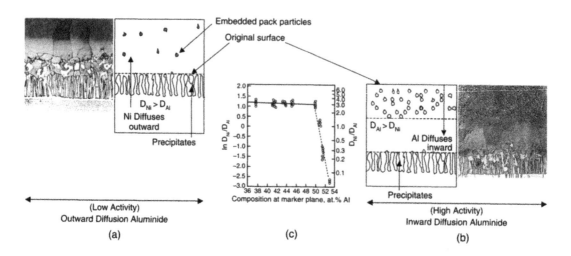

Figure 6.8 Microstructure of (a) low- and (b) high-activity aluminide coatings (G. W. Goward, Current research on the surface protection of superalloys for gas turbine engines, *JOM,* Oct. 1970, pp. 31–39). Reprinted with permission from The Minerals, Metals & Materials Society, (c) Dependence of Ni and Al diffusivities as a function of NiAl composition (S. Shanker and L. L. Seigle, Interdiffusion and intrinsic diffusion in the NiAl (β) phase of the Al–Ni system, *Metall. Trans.,* 1978, 9A, 1468–1476). Reprinted with permission from ASM International.

this way and do not contain any significant Al. The strengthening is achieved through solid solution and carbide precipitate formation.

Typical Co base alloys contain 20–30 Cr, 5–10 W, and 0.5–0.6 C. The alloys are protected against oxidation by forming low-activity aluminides on the surface. During the development of coating microstructure, Co diffuses out and forms CoAl. As a result Co depletes in the alloy near the interface. Many of the alloying elements with low solubility in the Co-depleted zone, therefore, precipitate out in the alloy at the interface. These precipitates are typically carbides of tungsten and chromium, which often form a continuous layer. Unlike aluminides on Ni base alloy, there is no internal zone (Fig. 6.9), because of the absence of aluminum and the lack of the associated diffusion process. Because diffusion in Co is slower than in Ni, coating thicknesses are smaller, of the order of 25 to 40 µm. Typical process temperatures are about 30°C (54°F) higher than for Ni alloys. This increase in temperature is needed to compensate for slower diffusion in Co alloys.

Chromium-Modified Aluminide Coating for Ni Base Alloys

Chromizing or chromizing followed by aluminiding of high-temperature alloys is used to improve resistance to hot corrosion and high-temperature oxidation (Sivakumar, 1982; Streiff and Boone, 1984). Chromium is added either through codeposition with aluminum (Marijnissen, 1982) or more effectively in a two-step pack process (Godlewska and Godlewski, 1984) in which pack or gas phase chromizing is followed by pack aluminiding. The chromizing pack consists of Cr, NH_4Cl, and the inert constituent Al_2O_3. Pack chromizing results in either a graded Cr-enriched layer or the formation of a brittle α Cr layer. Additional overaluminiding by low-activity process provides an outer βNiAl layer grown outwardly with Cr present in the β layer to the extent of its solubility limit. If overaluminiding is done by a high-activity process, the

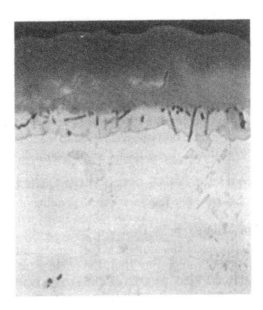

Figure 6.9 Microstructure of aluminided Co base alloy (C. Duret and R. Pichoir, Protective coatings for high temperature materials: Chemical vapor deposition and pack cementation processes, in *Coatings for High Temperature Applications*, Ed. E. Lang, Applied Science Publishers, 1983, pp. 33–78). Reprinted with permission from Elsevier.

coating grows inward with Cr present as inclusions of Cr-rich phases (Mevrel et al., 1986). The latter process generally provides better performance. The growth rate of a Cr pack coating is about an order of magnitude lower than that of an Al pack coating.

Siliconized Coating for Ni Base Alloys

Silicon is beneficial in improving both resistances to oxidation as well as to type II hot corrosion. It can be incorporated as a silicide coating by a pack process with metallic Si as the source, NH_4Cl as activator, and Al_2O_3 as filler (Fitzer et al., 1979). It can also be introduced during or prior to the aluminiding process. Introduction during the aluminiding process is difficult because of the large difference in the vapor pressures of the halides of Al and Si. In one of the aluminiding processes (Young et al., 1980), the Ni base substrate is coated with metallic silicon using a slurry or by spraying, followed by aluminiding at 1100°C (2012°F) for 16 hours. In commercial slurry-based coatings, a slurry containing 12% Si, the remainder being Al, is brushed on the component, followed by diffusion heat treatment. It forms an aluminide containing about 10% Si (Kircher et al., 1994). One important issue with silicide coatings is the formation of low-melting silicon-containing eutectics as well as brittle, complex intermetallic compounds formed with the alloying constituents that render the coating prone to cracking during thermal cycling (Nicholl et al., 1979).

Platinum-Modified Aluminide Coating for Ni Base Alloys

Incorporation of noble metals, particularly platinum, in aluminide coating improves resistance to oxidation and hot corrosion. The first platinum-containing commercial coating was designated LDC-2. It improved oxidation capability by a factor of 4 and resistance to type I hot corrosion by a factor of 2 (Lehnert and Meinhardt, 1972/73; Deb et al., 1986) relative to plain aluminide. The early patents describing the details of Pt aluminides are by Cape (1963) and Todd (1970). The commercial process for LDC-2 has been described by Lehnert and Meinhardt (1972/73). It consists of deposition of a thin layer of platinum (2 to 10 μm) on the component by either electroplating or physical vapor deposition, followed by diffusion heat treatment (vacuum or inert atmosphere, 850–1050°C, 1560–1920°F, for a few hours) to improve adherence of platinum with the substrate alloy. The component is subsequently aluminided by pack, above-the-pack (950–1000°C, 1742–1832°F, ~20 hours), or chemical vapor deposition (CVD) processes. After the completion of the coating step, additional heat treatments are generally done to achieve the required microstructure and phase distribution in the coating. The CVD process will be considered in a later section. Depending on the processing conditions, Pt enters the coating as single-phase $PtAl_2$, as two-phase $PtAl_2$ and solid solution (Pt, Ni)Al, or only as single-phase solid solution (Pt, Ni)Al.

The incorporation of platinum in the aluminide coating was initially intended to create a diffusion barrier against aluminum migrating from the coating into the substrate. However, subsequent analyses showed that Pt did not act as a diffusion barrier. A number of other mechanisms have been offered to explain the role of Pt in improving the environmental resistance of Pt-containing aluminide coating, although Pt itself has not been found directly to participate in the formation of the protective oxide scale:

- It eliminates precipitation of the Cr-rich σ phase within the outer layer of the coating. As opposed to the beneficial effects when present in solid solution, Cr dispersed as precipitates reduces oxidation resistance. Platinum also suppresses the diffusion of refractory elements such as Mo, V, and W from the alloy to the coating outer layer and improves the stability of the coatings against interdiffusion between the substrate and the coating (Tawancy et al., 1991).

- It encourages formation of pure alumina scale and reduces scale growth rate (Tawancy et al., 1995).

- It acts as a β phase stabilizer and prevents the detrimental conversion of β to γ' on thermal cycling.

- It improves alumina scale adherence, including mechanical keying, as a possible mechanism due to the formation of an irregular oxide–alloy interface (Felten and Pettit, 1976).

- It improves self-healing of the alumina scale after modest spalling (Felten and Pettit, 1976).

- It suppresses interfacial void formation.

- It ties up Al as Pt–Al compounds, thus lowering Al activity, which in turn reduces the driving force for the diffusion of Ni from the substrate into the coating.

Degradation of coating occurs eventually because of depletion of Al as a result of consumption by the oxide scale growth and because of some diffusion into the substrate alloy.

Ruthenium-Modified Aluminide Coating for Ni Base Alloys

The melting point of RuAl is 400°C (720°F) higher than that of NiAl, although both have the same β crystal structure. It is therefore not surprising that RuAl has higher creep strength compared with NiAl. Tryon et al. (2006) have utilized, on an experimental basis, the increased creep strength and consequent resistance to rumpling of Ru-containing NiAl to form creep-resistant bond coats for thermal barrier coatings (TBCs), a subject discussed more fully in Chapter 7. The bond coat produced for experimental studies appeared promising. The concept has been further extended to include Pt.

Chemical Vapor Deposition

In the CVD process for diffusion coatings, halide vapor such as $AlCl_3$ is created in external gas generators by passing HCl or Cl_2 gas over pellets of an Al alloy at temperature around 300°C (570°F) (Lee and Kim, 1999). The dedicated generators for the halide vapors are geographically separated from the chamber housing the components. The parts to be coated are fixtured inside a retort in a CVD furnace, typically held at a temperature range of 1000–1100°C (1830–2010°F). Figure 6.10a (Warnes, 2001) shows a schematic of a CVD setup. As shown, Al halide vapor is created in the external reactor by a chemical reaction of flowing HCl gas over Al alloy chips. For high-activity aluminide, additional Al chips are placed inside the retort in the path of flowing gases (Fig. 6.10b) (Warnes and Punola, 1997). If reactive elements such as Y, Hf, and Zr are to be incorporated into the aluminide coating to improve performance, additional reactors are connected to the retort. We know that $AlCl_3$ has a high vapor pressure (it sublimes at 178°C, 350°F). Through appropriate plumbing of the components held in the reactor, the halide vapor is transported to the external surface, and, if required, to the internal passages of the components by a carrier gas consisting of a mixture of hydrogen and argon. In the presence of H_2 gas, $AlCl_3$ vapor reacts with more Al to form AlCl, which in turn reacts with Ni on the surface of the component to form a βNiAl coating according to the overall reactions:

$$6HCl + 2Al = 2AlCl_3 + 3H_2,$$

$$2AlCl_3 + 4Al = 6AlCl,$$

$$6AlCl + 6Ni + 3H_2 = 6NiAl + 6HCl.$$

The H_2 concentration in the carrier gas is an important parameter because it controls whether the reaction just shown goes forward or backward.

The CVD process has several advantages over its pack and above-the-pack counterparts. The coating composition can be varied by the introduction of multiple elements with their independent reactors. Also, the coating deposition rate can be controlled independently. The postcoating heat treatment can be done in the reactor in a single step without removing components, thus reducing processing cost. The coating produced by the CVD process is generally an outward-growing "low-activity" type in which the outward diffusion of Ni through the diffusion zone is the rate-limiting step.

The coating growth rate follows the diffusion-based relation $x^2 \approx Dt$, where x is the coating thickness, D is the relevant diffusion coefficient, and t is the duration of aluminiding. The diffusion coefficient is strongly dependent on temperature and the substrate alloy composition.

Issues encountered in production for large batches of hardware typically involve nonuniform temperature distribution in the retort and reactant starvation. It is, therefore, important to eliminate large temperature gradients and have uniform gas flow in the vicinity of the components.

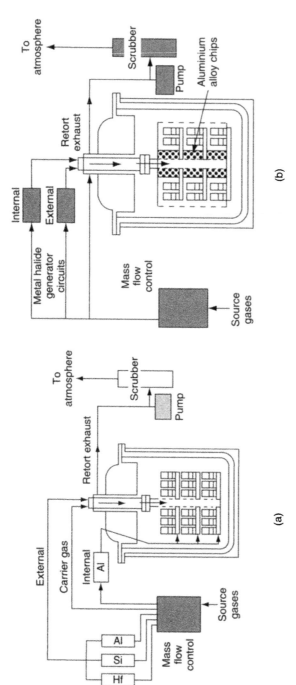

Figure 6.10 CVD process for diffusion coatings: (a) low-activity codeposition of Si, Hf, and Al (Bruce Michael Warnes, Reactive element modified chemical vapor deposition low activity platinum aluminide coatings, *Surf. Coat. Technol.*, 2001, 146–147, 7–12), (b) high-activity plain aluminide (Bruce M. Warnes, and David C. Punola, Clean diffusion coatings by chemical vapor deposition, *Surf. Coat. Technol.*, 1997, 94–95, 1–6). Reprinted with permission from Elsevier.

Platinum Aluminide by Chemical Vapor Deposition

As described in the earlier subsection titled "Platinum-Modified Aluminide Coating for Ni Base Alloys," the first step in the production of platinum aluminide coating is the deposition of a thin layer of platinum on the external surface of the component, which will be exposed to high temperature. Whereas coatings do improve the environmental life of components, they have a negative impact on fatigue life, as will be discussed in Section 6.6. Areas of the components where a debit in fatigue life is unacceptable need to be masked to avoid coating deposition. For example, reduction in attachment fatigue life of the root sections of turbine blades may not be acceptable. The roots, therefore, have to be masked against platinum deposition. The masking is usually achieved by applying coats of resins to the roots. The Pt deposition is done electrolytically at temperatures between 80 and 90°C (175–195°F), with a platinum anode, which goes into solution, and the part to be coated as the cathode. To achieve uniform Pt distribution, the anode is appropriately shaped to conform to the part surface profile and remains approximately equidistant from all points of the cathode surface to be coated. Suitable electrolytes (Albon et al., 1992) are chloroplatinic acid, dinitroplatinic sulfate, hydrogen hexachloroplatinate, and an amine complex of platinum hydrogen phosphate known as the Q salt in a hydrogen phosphate buffer solution. Electrolytes are prepared in deionized water. In order to improve ionic conductivity, phosphate salts are added to the electrolyte. Phosphorus is, therefore, a common impurity of the electrolytic process. Typical platinum layer thickness lies between 2 and 10 microns. After the plating step, the components are heat treated to diffuse Pt into the substrate to improve adherence. The diffusion step is generally combined with the subsequent aluminiding process, which is done in a retort. The components are assembled in the retort as shown in Fig. 6.10, and the aluminiding process is initiated. If the internal passages of components such as turbine blades need coating for protection against oxidation and corrosion, appropriate plumbing should be established to carry the halide vapors and the carrier gas. Gas flow, temperature, and time are critical parameters for successful aluminiding. Once aluminiding is completed, postcoating heat treatment may be required to achieve appropriate microstructure and properties. This heat treatment is generally completed in the retort prior to disassembly. The opportunity to combine heat treatment with the aluminiding steps is an advantage of the CVD process in terms of reducing cost.

Role of Reactive Elements in Diffusion Coatings

Reactive elements (REs) such as Y, Hf, Zr, Ce, La, and Si, when present in small concentrations as alloying constituents in the coating, improve the adherence of Al_2O_3 and Cr_2O_3 scales and significantly reduce susceptibility to spallation during thermal cycling. One of the principal sources of the RE for diffusion aluminides is the substrate alloy (Aldred, 1975; Tawancy et al., 1992). During high-temperature exposure, the RE from the substrates migrates through the coating to the oxide scale–coating interface. The incorporation of the RE in aluminides from external sources, however, has generally been difficult because of its low solubility in βNiAl. Several approaches have been pursued (Lee and Kim, 1999) to overcome this drawback. These include pack aluminiding followed by additional processing in a pack containing Y using Y_2O_3 as a filler, codeposition of Al and Hf by pack process, and incorporation of Hf through a CVD process.

In the CVD process the codeposition is achieved by attaching additional independent reactors (Warnes, 2001) as shown in Fig. 6.10. The reactors produce $SiCl_4$, $HfCl_4$, and $ZrCl_4$ by flowing HCl gas over the respective metals. Warnes gives the relative concentrations used for experimental codeposition on IN 738 alloy to be 3.5% $AlCl_3$, 0.5% $SiCl_4$, 1.5% $HfCl_4$, plus $ZrCl_4$, 15% argon, and the balance H_2.

Microstructure of Platinum Aluminides

The microstructure and phase distribution in platinum aluminide coatings depend on the relative activity of aluminum which is controlled by the coating process, the process parameters, and subsequent heat treatment. Generally three variants of phase distribution have been observed (Fig. 6.11) (Smith and Boone, 1990): (a) external single-phase $PtAl_2$, (b) external two-phase $PtAl_2 + NiAl$, and (c) external single-phase solid solution (Pt,Ni) Al. As can be seen from the microstructures, the coating may consist of two to four zones including the interdiffusion zone (IDZ). The IDZ consists primarily of NiAl matrix with elongated precipitates of Cr-rich σ phases and refractory elements, which have low solubility in the β NiAl phase. The composition profile of the single-phase coating (c) is also shown in Fig. 6.11. Typical coating thickness including the diffusion zone is between 50 and $100\,\mu m$. It is evident that Pt is predominantly concentrated in the outer part of the coating. The concentration profile also shows that Pt does not provide a physical diffusion barrier to Al diffusion.

The presence of reactive elements modifies the microstructure as shown in Fig. 6.12. The outer zone is the solid solution (Pt,Ni) Al in which Hf, Zr, and Si are in solution. In addition, the solid solution is dispersed with a second phase precipitate consisting of silicides of Hf and Zr. The inner IDZ structure and its composition are very similar to those observed with plain Pt aluminides.

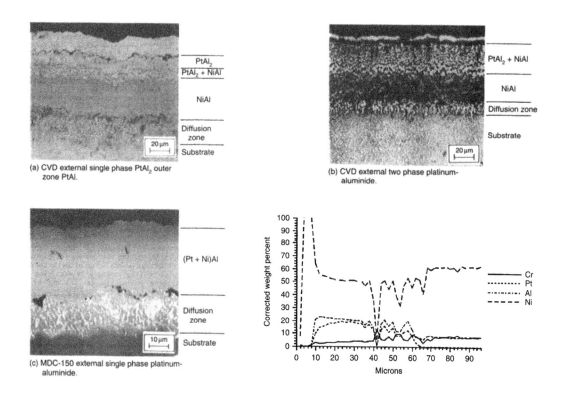

Figure 6.11 Microstructure of Pt aluminide coatings made by the CVD process (J. S. Smith, and D. H. Boone, Platinum modified aluminides—present status, *ASME Paper No.* 90-GT-319, 1990, p. 6). Reprinted with permission from American Society of Mechanical Engineers International.

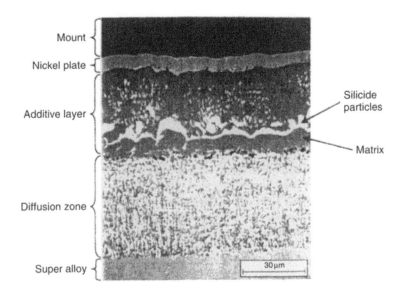

Figure 6.12 Back scatter image of Pt Aluminide made by CVD codeposition containing Si, Hf, and Zr (Bruce Michael Warnes, Reactive element modified chemical vapor deposition low activity platinum aluminide coatings, *Surf. Coat. Technol*, 2001, 146–147, 7–12). Reprinted with permission from Elsevier.

Manufacturing Aspects of the Coating Process

Typical manufacturing steps in a pack process are shown in Fig. 6.13. Because of their brittle nature, the application of coatings results in debits in some of the mechanical properties of the coated article. Critical areas of hardware exposed to high stresses but not to very high temperature or hot corrosion during operation and, therefore, not requiring coating need to be masked during the coating process.

```
              Surface cleaning
              (Masking roots)
                (Cleaning)
               (Plating of Pt)
             (Removal of mask)
                (Cleaning)
                (Heat treat)
                  Masking
        Fixturing of parts for loading
              Furnace loading
                  Coating
                 Heat treat
                 Cooldown
                 Unloading
              Removal of mask
                 Cleaning
             Quality/inspection
```

Figure 6.13 Manufacturing process steps for coating articles.

Commercial Diffusion Coatings

Diffusion coatings are commercially produced by a number of suppliers and manufacturers, particularly gas turbine engine makers. A number of such coatings are listed in Table 6.3.

Coating–Substrate Interdiffusion Effects

In general coatings are not in thermodynamic equilibrium with the substrate alloys. The compositional difference between the coating and the substrate provides the driving force for interdiffusion of various elements when the coated alloy is exposed to high temperatures. Migration of many of the elements across the coating–substrate interface results in the formation of detrimental phases, modifications of the microstructure and coating properties. Some of the changes include:

Table 6.3 Commercial Diffusion Coatings

Coating Designation	Producer	Composition	Comments
PWA 73	Pratt & Whitney	Aluminide	High-activity pack
PWA 273	Pratt & Whitney	Aluminide	Gas-phase Cr, low-activity pack
PWA 275	Pratt & Whitney	Aluminide	Low-activity above-the-pack
PWA 70 - 73	Pratt & Whitney	Chromized + aluminized	High-activity pack
MDC 200	Howmet	Aluminide	High activity
MDC 210	Howmet	Aluminide	Low activity
RT 21	Chromalloy	Aluminide	High activity
RT 22	Chromalloy	Pt aluminide	High-activity pack
RT 44	Chromalloy	Pt–Rh aluminide	High-activity pack (Co base)
JML 1	Johnson Matthey	Pt aluminide (fused salt deposition + aluminized)	High-activity pack
SS 82A	Turbine Component Corp.	Pt aluminide	High-activity above-the-pack
LDC 2E	Howmet	Pt aluminide	High-activity pack
MDC 3V	Howmet	Chromized	High-activity pack
MDC 150	Howmet	Pt aluminide	High-activity CVD
MDC 150L	Howmet	Pt aluminide	Low-activity CVD
MDC 151L	Howmet	Pt aluminide + Hf, Si, Zr	Low-activity CVD
SermaTel W	Sermatech	Slurry aluminide	High-activity pack
SermaLoy J	Sermatech	Slurry aluminide with Si	High-activity pack
Elcoat 360	Elbar	Slurry Ti with Si	High-activity pack
Jo Coat	Pratt & Whitney	Slurry aluminide with Si	High-activity pack
CODEP	General Electric	Aluminide	High-activity pack
ALPAK	Rolls Royce	Aluminide	High-activity pack
SermAlcote	Sermatech	Pt aluminide	Plating + slurry + diffuse
Snecma C1A	SNECMA	Chromized + aluminized	Low-activity pack

- Diffusional loss of Al from the coating to the outer surface during oxidation is well understood. However, the loss of Al due to inward diffusion into the alloy has been the subject of serious debate.

- Oxidation resistance of the scale is reduced due to doping of the oxide with species from the coating, which increase oxide growth rates of the scale. For example, Ti from Ti-containing alloy migrates to the coating–aluminum oxide scale interface and incorporates as Ti^{4+} ion into the aluminum oxide crystal structure to replace Al^{3+}. To maintain charge balance, vacancy sites are created, aiding in increased diffusivity and enhanced Al_2O_3 scale growth rate.

- Hot corrosion resistance is reduced. Particularly, Mo and V from alloy substrate migrate to the coating and increase susceptibility to hot corrosion.

- Precipitation of deleterious phases such as αCr within the coating reduces ductility and depletes the coating of the oxide scale formers.

- The interdiffusion zone results from conversion of the original γ–γ' microstructure of the alloy to the β NiAl phase. The latter has low solubility of many of the elements dissolved in the superalloy.

- Grain growth occurs within the coating, the IDZ, and the substrate at the interface.

- Kirkendall porosity forms in some systems because of unbalanced diffusion of species into and out of the alloy.

- For higher generation single-crystal superalloys strengthened by refractory metals, secondary reaction zones (SRZs) with large-angle grain boundaries form (Walston et al., 1996). These SRZs reduce the load-bearing cross-section of the component and may act as a source of fatigue crack generation.

Aluminum is one of the major constituents of the substrate alloy and the coating. The transport of Al affects both the oxidation resistance and the overall phase stability of the coating. During high-temperature exposure of the coated article in an oxidizing environment, Al diffuses from within the coating toward the coating surface to form or replenish the protective alumina scale. The possibility of Al diffusion inward from the coating into the alloy has also been investigated. However, this diffusion is under debate, particularly in light of the fact that in the hypostoichiometric (Ni-rich) inner region of the aluminide coatings $D_{Al} << D_{Ni}$. This leads to the belief that the primary diffusing species in the inner zone is Ni migrating outward from the alloy to the coating (Goward and Boone, 1971). Smialek and Lowell (1974), however, have shown that inward diffusion of Al from the coating into the substrate alloy on high-temperature exposure does indeed occur, and is a significant coating degradation mode. They compared bulk NiAl and diffusion-annealed pack-coated samples of IN 100 and columnar grain Mar-M 200 in oxidation tests. Oxidation of diffusion-annealed samples was significantly poorer than that of the as-coated samples as well as bulk NiAl. The authors, therefore, concluded not only that Al diffused into the substrate, but also that the loss was one of the major factors for degradation in oxidation. The other alloying constituents such as Cr also diffuse following their concentration or activity gradients. An example of microstructural evolution of a high-activity pack aluminide-coated Ni–10 at % Cr–17 at % Al (composition in atomic percent) alloy substrate on isothermal exposure at 1150°C (2100°F) for various lengths of time is shown in Fig. 6.14 (Basuki et al., 2000). Table 6.4 lists the growth or changes of various phases formed as a function of exposure time.

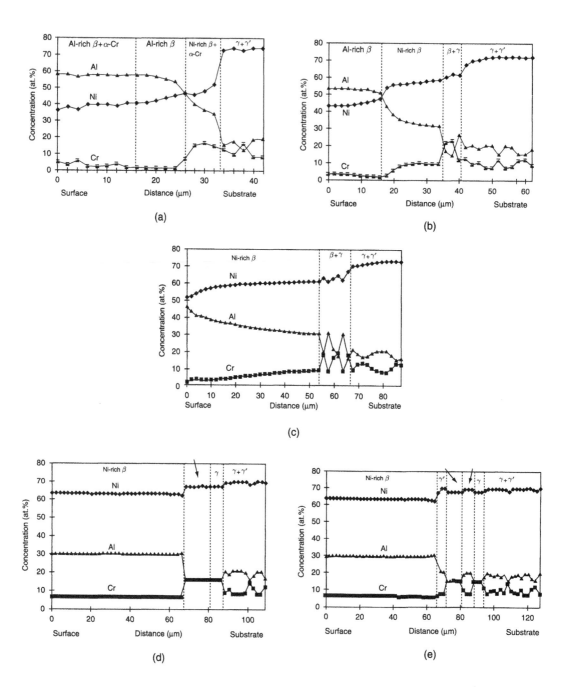

Figure 6.14 Concentration profile of high-activity aluminide coated $\gamma + \gamma'$ alloy after exposure at 1150°F for (a) 0 min, (b) 5 min, (c) 50 min, (d) 500 min, (e) 1000 min (E. Basuki, A. Crosky, and B. Gleeson, Stages of interdiffusion behavior in an aluminide coated $\gamma - Ni + \gamma'$-Ni3Al Alloy at 1150°C, in *Proc. International Symposium on High-Temperature Corrosion and Protection 2000*, Eds. Toshio Narita, Toshio Maruyama, and Shigeji Taniguchi, Sept. 17–22, 2000, Hokkaido, Japan, pp. 315–322). Reprinted with permission from *Science and Technology Letters*.

Table 6.4 Evolution of microstructure of aluminide coated γ-Ni + γ'-Ni$_3$Al alloy on thermal exposure at 1150°C (2100°F)

Exposure Time (min)	Phases	Thickness of Layers[a] (μm)
0	Al-rich β + αCr	O 14
(as coated)	Al-rich β	M 16
	Ni-rich β + αCr	I 7
5	Al-rich β	O 18
	Ni-rich β + αCr	M 18
	β + γ	I 3
50	Ni-rich β	O 54
	β + γ	I 14
500	Ni-rich β + αCr	O 68
	β + γ	M 12
	γ	I 8
1000	Ni-rich β	O 65
	γ'	5
	γ + γ'	8
	γ	I 5

[a] Thickness indicates outer, intermediate, and two phase inner zones, respectively. O, outer; M, middle; I, inner.

Depletion of Al due to migration to the coating surface and interdiffusion of Al and Ni between the coating and the alloy substrate follow the general sequence of reactions

Al-rich β converts to Ni-rich β,

Ni-rich β converts to β + γ',

β + γ' converts to γ' + γ,

γ' + γ converts to γ.

For Pt aluminide coating, the microstructural evolution due to diffusion into and out of the substrate as well as to the coating–oxide scale interface follows similar trends. For example, high-temperature (950 and 1100°C; 1740 and 2010°F) thermal exposure of a two-phase PtAl$_2$ and (Pt, Ni)Al coated third-generation nickel base single-crystal superalloy CMSX-10 (Reid et al., 2003) exhibited the following features:

- Initially ($t < 200$ hrs) the Ni content of the outer region increased due to diffusion from the inner region of the coating. The Pt content decreased. The coating changed from a two-phase to a single-phase microstructure due possibly to the reaction

PtAl$_2$ and (Pt, Ni)Al + Ni = Ni-rich β(Pt, Ni)Al.

- At intermediate times ($t \sim 200$ to 750 hrs), Ni from the interface region and the substrate diffused into the coating and Al diffused from the coating into the substrate. Göbel et al. (1994) studied interdiffusion between Pt aluminide RT 22 and advanced single-crystal Ni base superalloys CMSX 6 and SRR 99. They found inward diffusion of Al into the alloys to be an important degradation mode. Because of the interdiffusion of Ni and Al, composition shifts to a $\beta + \gamma'$ phase field according to the reaction

$$\beta(\text{Pt}, \text{Ni})\text{Al} + \text{Ni} = \beta(\text{Pt}, \text{Ni})\text{Al} + \gamma'.$$

The γ' phase generally appears at the grain boundaries.

- Several studies conducted at higher temperatures or for longer times show further shift in composition in which Pt diffuses into the substrate, initially leaving a γ' phase that eventually converts to γ with the possible reaction sequences

$$\beta(\text{Pt}, \text{Ni})\text{Al} + \gamma' = \text{Pt (in alloy)} + \gamma',$$

$$\gamma' = \gamma + \text{Al (in alloy)}.$$

- For some alloys (Chen and Little, 1997) at longer times, Pt diffusion into the substrate has been found to cause precipitation of topologically close packed (TCP) phases containing high fractions of the alloy refractory elements such as Re, W, Cr, and Mo. The precipitation of TCP phases is detrimental to the mechanical properties of the alloy.

For advanced superalloy substrates containing high levels of refractory metals such as Mo, W, Re, and Ru, interdiffusion between the coating and the substrate results in the formation of SRZs (Walston et al., 1996). These SRZs form colonies below the interdiffusion zone and have a distinctive microstructure consisting of three phases: a TCP phase of composition such as $(\text{Cr}, \text{Mo})_x(\text{Ni}, \text{Co})_y$, where x and y vary between 1 and 7 (Ross and Sims, 1987), strings of γ in a γ' matrix. The SRZ colony is separated from the superalloy structure by large-angle grain boundaries. The TCP phase formation depletes the alloy of refractory elements. The loss of strengthening elements and the precipitation of brittle phases with platy or needle morphology reduces substrate creep and fatigue strength.

Coating Phase Stability

For Ni-rich βNiAl including Pt-containing NiAl, cooling from temperatures in excess of 1100°C (2010°F) results in a reversible martensitic transformation. This type of transformation does not require diffusion but takes place by the coordinated displacement of atoms. The high-temperature B2 (body-centered cube) phase changes to an $\text{L}1_0$ (face-centered tetragonal) lower temperature phase. The morphology of the low-temperature martensitic phase is plate-like, as shown in Fig. 6.15 (Zhang et al., 2003). Several conditions must be satisfied for the martensitic transformation to occur. The Al content needs to be less than about 37 at %, the β phase should not be stabilized by the presence of other elements such as Cr, Mo, W, Ta, Ti, and Si, and the coating needs to be exposed to high temperatures above 1100°C (2010°F). The temperature at which this transformation occurs depends on the Ni content and ranges between −200°C (−330°F) and 900°C (1650°F). The martensitic transformation results in a volume change $\Delta V / V$ of approximately 2%, where ΔV is the change in volume V. The associated linear strain $\varepsilon = 1/3 \, (\Delta V / V)$ is about 0.7%, which is comparable to the thermal expansion mismatch strain between the coating and the substrate. The transformation-induced strain provides driving force for creep of the coating, known as rumpling, and possibly spallation of thermal barrier coatings (TBCs) if the transformable metallic coating is used as the

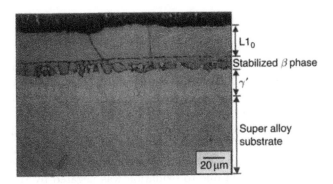

Figure 6.15 Etched cross-sectional microstructure of NiAl coating formed by CVD process after 100 hours of exposure at 1150°C (2012°F). Note the plate-like martensite in the coating layer (Y. Zhang, J. A. Haynes, B. A. Pint, I. G. Wright, and W. Y. Lee, Martensitic transformation in CVD NiAl and (Ni,Pt)Al bond coatings, *Surf. Coat. Technol.*, 2003, 163–164, 19–24). Reproduced with permission from Elsevier.

bond coat on which the TBC is deposited. It should be noted that the martensitic phase transformation of the metallic coating is not the only mechanism for rumpling of coating or TBC spall.

Platinum Modified Gamma + Gamma Prime Coating

As demonstrated by Ni base superalloys (see Section 3.6), $\gamma + \gamma'$ microstructure exhibits high creep strength. A metallic coating based on this microstructure is expected, therefore, to resist creep-induced "rumpling." A Pt-containing coating with a $\gamma + \gamma'$ phase structure has recently been developed (Rickerby and Wing, 1999; Rickerby et al., 1999; Adesanya et al., 2001; Nicholls, 2003; Gleeson et al., 2004). The attractive characteristics of this coating include improved alumina scale adherence resulting in good oxidation resistance, in spite of the fact that the Al content is significantly lower than that in the traditional β(Pt,Ni)Al, reduced creep rumpling, and improved stability against formation of SRZs when applied on refractory metal–containing superalloys. The coating is produced by electroplating of Ni base substrate alloy with an 8- to 10-μm layer of Pt, followed by diffusion heat treatment of a few hours in a vacuum at 1150–1190°C (2100–2140°F). Interdiffusion between the substrate alloy and the coating results in elements such as Ta and Ti migrating into the γ' phase while W and Re partition into the γ phase. Comparative oxidation testing (Gleeson et al., 2004) has shown that the coating of composition Ni 22Al 30Pt + RE, in which RE is reactive elements such as Hf, has exceptional oxidation resistance. Interdiffusion studies conducted on superalloys coated with Pt-containing $\gamma + \gamma'$ coating reveal that the system exhibits uphill diffusion of Al, which migrates from the substrate (with 13 to 19 at % Al) to the coating (with 22 at % Al), as long as the Pt content of the coating was greater than 15 at %. Although the Al content of the coating is higher than that in the alloy, the presence of Pt in the coating effectively reduces the Al activity. Diffusion, therefore, occurs along an activity gradient. When used as part of a thermal barrier coating system (see Chapter 7), $\gamma + \gamma'$ metallic coatings are expected to enhance coatings life significantly compared with traditional Pt aluminides.

Oxidation Resistance of Diffusion Coatings

Although plain aluminides have a large reservoir of aluminum to form and regenerate the protective alumina scale, the adherence of such scale is not as good as in Pt-containing aluminides. Relative cyclic oxidation life of Pt aluminides made by a high-activity pack process (LDC 2E), a high-activity gas-phase process (MDC 150), and a low-activity CVD process (MDC 150L) are shown in Fig. 6.16 (Warnes and Punola, 1997). The Pt aluminides on a IN 738 Ni base alloy were exposed in cyclic oxidation, 50 min at 1100°C (2010°F) and cooled to room temperature for 10 min. The test criterion involved specific weight gain or loss. Life was defined as number of cycles after which total weight of the sample was the same as the starting weight, that is, weight gain due to oxidation was negated by the weight loss due to spalling of the oxide scale. The longest life is exhibited by the low-activity outward-grown CVD process. The relative lives are found to scale with the purity of the coatings and how "clean" the microstructures are.

Because of the low solubility of reactive elements in the β NiAl phase, it is difficult to incorporate them into aluminides. However, on an experimental basis, Warnes (2001) has been successful in introducing RE into Pt aluminides. Warnes' study showed (Table 6.5) that the presence of reactive elements in MDC 151L improves oxidation life significantly relative to MDC 150 and 150L, which do not contain reactive elements.

The effect of platinum concentration on the oxidation resistance of single-phase platinum aluminide (Pt, Ni) Al formed on nickel base polycrystalline (IN 100, Mar-M247) and single-crystal (PWA 1480) superalloys by a low-activity CVD process has been reported by Purvis and Warnes (2001). The single-phase coating with concentration of Al around 15–20 wt % and Pt around 20–25 wt % is less brittle than the two-phase coating containing the brittle phase $PtAl_2$. Also, unlike the two-phase alloy, which undergoes a detrimental volume change on dissolution of $PtAl_2$, the single-phase aluminide is phase stable. The concentration of Pt in the single-phase coating deposited on the three alloys varied between 8.0 and 37.7 wt %, with the thickness

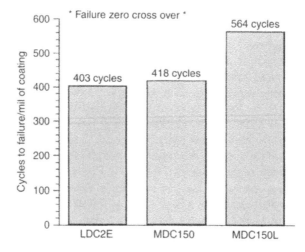

Figure 6.16 Oxidation life of Pt aluminide coatings formed on nickel base superalloy IN 738 substrate in cyclic furnace testing between 1100°C and room temperature (Bruce M. Warnes and David C. Punola, Clean diffusion coatings by chemical vapor diffusion, *Surf. Coat. Technol.*, 1997, 94–95, 1–6). Reprinted with permission from Elsevier.

Table 6.5 Relative Cyclic Oxidation Life of Pt
Aluminides in Cyclic Furnace Testing between
1100°C (2012°F) and Room Temperature

Coating	Relative Life
LDC 2E	1.0
MDC 150	1.06
MDC 150L	1.35
MDC 151L[a]	2.23

[a] Contains 0.41% Hf, 0.20% Zr, and 0.60% Si.

between 60.0 and 92.1 μm. A cyclic oxidation test was conducted at 1175°C (2150°F), each cycle with 50 minutes exposure at temperature followed by 10 minutes at room temperature. Oxidation life was defined as the number of cycles at the end of which the weight gain due to oxidation was exactly compensated by the weight loss due to oxide scale spall. The test data showed that the oxidation life improved with increase in Pt concentration. Additionally, it was found that the presence of Ti (PWA 1480) and, to a lesser extent, Ta in the substrate (IN 100) had negative effect on oxidation life for the coatings with same level of platinum. The negative effect of Ta in the substrate on the oxidation resistance of CoNiCrAlY-coated alloys has also been observed by Fox and Tatlock (1989). There are a few other alloys such as B1900 and coatings such as NiCoCrAlY in which Ta exerts beneficial effects toward oxidation. In the case of Ti-containing alloys, as discussed before, Ti migrates to the Al_2O_3 oxide scale and replaces some of the trivalent Al^{3+} with tetravalent Ti^{4+}, creating aluminum vacancies to maintain charge balance. The additional vacancies contribute to increased diffusivity and oxide scale growth rate.

Corrosion Resistance of Diffusion Coatings

Plain aluminides, which are alumina formers, do not offer any significant resistance to hot corrosion (Streiff and Boone, 1988). Four avenues have been explored to improve the hot corrosion performance of aluminides: incorporation of platinum, chromizing of the aluminide, chromized diffusion coating, and siliconized diffusion coating:

Platinum aluminide: Generally platinum aluminides provide resistance to type I hot corrosion characterized by basic or neutral fluxing of oxide scales. They offer little protection against the lower temperature type II hot corrosion, which involves acidic fluxing. However, some studies (Barkalow and Pettit, 1978) have shown that if the Pt-containing phase, possibly $PtAl_2$, is continuous at the salt–coating interface, the coating provides excellent protection against type II hot corrosion. Aluminides containing other precious metals such as rhodium with or without platinum generally have not fared as well as platinum aluminide.

Chromized diffusion coating: Chromized diffusion coatings are effective against type II hot corrosion. As discussed in Section 6.3, chromized coatings are generally thin. However, at the low temperature characteristic of type II corrosion, thickness is seldom an issue. In a comparative rainbow test of various coatings on Ni base alloy Udimet 520, done in early 1980 by Kraftwerk Union AG in Germany (Czech et al., 1986), chromized diffusion coatings performed exceptionally well on blades in an industrial gas turbine engine in a heavily polluted industrial environment. The coated blades were run within the type II temperature regime. The effect of chromium lies in its multiple valence state, which, as discussed in Section 5.2, results in a positive solubility gradient.

Two-step, chromizing followed by aluminiding: The two-step process optimizes resistance to high-temperature oxidation because of the presence of aluminum and to low-temperature corrosion by the presence of chromium. This coating performs better than plain aluminides against type II corrosion. The first step is to form the chromized coating followed by the second step of aluminiding.

Siliconized diffusion coatings: Because silica has low solubility in molten sulfates and the solubility is somewhat independent of acidity of the melt as shown in section 5.0, it is expected that siliconized diffusion coatings would do well against type II hot corrosion. Silica scales, however, are not resistant to basic fluxing leading to type I hot corrosion. Jo Coat (Pratt & Whitney) and Sermaloy J (Sermatech) are a couple of the commercially used Si-based coatings finding application on gas turbines for marine application. However, such coatings are rarely used on Ni base alloys because of the formation of brittle or low-melting phases. A more attractive way to utilize the benefits of Si is to tie it to Ti in a diffusion coating as in Elcoat 360 (Elbar). This coating has performed well against type II hot corrosion in the laboratory as well as in engine tests.

Mechanical Properties of Platinum Aluminides

Metallic coatings based primarily on the β phase in the Ni–Al system are brittle at lower temperature and become increasingly ductile when exposed above the ductile-to-brittle transition temperature (DBTT), as depicted in Fig. 6.17. The DBTT is the temperature at which the slope of the strain-temperature curve changes significantly. For the coated systems in Fig. 6.17a, the strain resulting from the thermal expansion mismatch between the coating and the substrate imposes a small tensile component, which needs to be accounted for in determining the strain at fracture.

For a single-phase (Pt, Ni)Al coating produced by a CVD process, Pan et al. (2003) measured the DBTT to be around 600°C (1110°F), which did not change significantly on furnace cycle exposure equivalent to 28% of cyclic life.

6.4 OVERLAY COATINGS

The behavior of diffusion coatings is strongly dependent on the composition of the substrate alloy because the alloy participates in the formation of the coating. These coatings are based on the β NiAl phase of the Ni–Al alloy system. NiAl has poor solubility of other elements. As a result, these coatings do not offer wide flexibility for the incorporation of minor elements. In order to address this limitation, a new class called "overlay" coatings has been developed with minimal direct contribution of the substrate alloy. The overlay coatings have a typical composition represented by MCrAlX, where M stands for Ni, Co, and occasionally Fe, and X represents oxygen-reactive elements such as Zr, Hf, Si, and Y. The composition is so selected that the microstructure consists of varying amounts of β phase in a γ matrix. Phase stability, resistance to oxidation and corrosion, and susceptibility to cracking are controlled by adjusting the composition and the β/γ ratio. One of the requirements in selecting the composition of

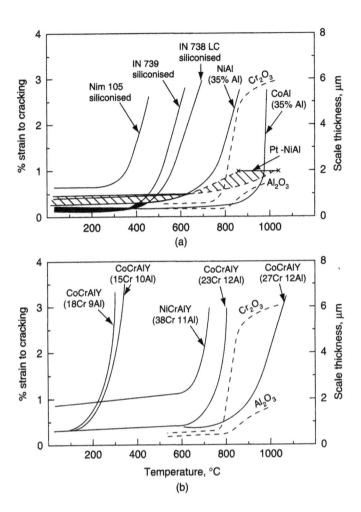

Figure 6.17 Strain to crack plots of several coatings showing the ductile-to-brittle transition temperature. (a) Aluminides, (b) overlays (S. R. J. Saunders and J. R. Nicholls, Coatings and surface treatments for high temperature oxidation resistance, *Mater. Sci. Technol.*, 1989, 5(8), 780–798, based on P. Hancock and J. R. Nicholls in *Coatings for Heat Engines* (Workshop Proc.), Eds. R. L. Clark et al., pp. 31–58, U.S. Dept. of Energy, Washington, DC; 1984). Reproduced with permission from Maney Publishing.

the overlay coatings is to avoid the detrimental phase transformation $\gamma + \beta = \alpha Cr + \gamma'$, as explained in the next section.

6.5 OVERLAY COATINGS BY SPRAY AND ARC PROCESSES

The majority of overlay coatings deposited by a spray process use raw materials in the form of alloy powders of appropriate particle morphology, particle sizes, and size distribution. Arc processes, on the other hand, use materials in the form of ingots. The composition of the raw material is critical in achieving the desired composition of the final coating, and in maintaining microstructural stability and adequate oxidation and corrosion resistance.

Beta–Gamma System Phase Stability

The microstructure of NiCoCrAlY coatings consists of both β and ν phases, the relative volume fractions being dependent on the actual composition. In the absence of Co, and minor elements such as Y, the composition of a typical coating can be represented on a Ni–Cr–Al ternary phase diagram. The desired composition at high temperature is located within the $\beta + \gamma$ field (Fig. 6.18a, indicated by the inset square). The coating, when cooled from 1150°C to 850°C (2102°F to 1562°F), undergoes transformation from $\beta + \gamma$ to $\alpha + \gamma'$. The composition at low temperature is located in the phase field $\alpha + \gamma'$ of Fig. 6.18b. It is indicated by the inset square. The transformation $\beta + \gamma$ to $\alpha + \gamma'$ is accompanied by a significant change in volume as shown by the kink in the thermal expansion profile (Fig. 6.19) (Mevrel, 1989;

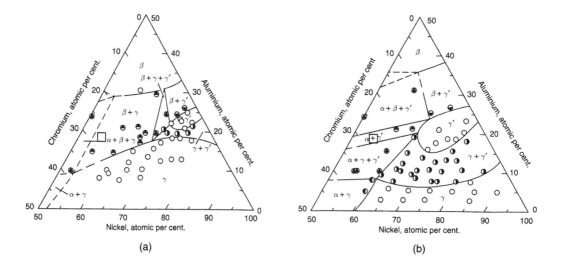

Figure 6.18 Isothermal Ni–Cr–Al ternary phase diagram showing the various phases as a function of composition, (a) at 1150°C (2102°F), (b) at 850°C (1562°F) (A. Taylor and R. W. Floyd, The constitution of nickel rich alloys of the nickel–chromium–aluminum system, *J. Inst. Metals*, 1952–53, 81, 451–464). Reprinted with permission from Maney Publishing.

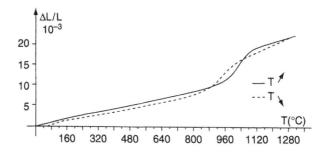

Figure 6.19 Thermal expansion of NiCrAl as a function of temperature. The large increase around 1000°C (1832°F) corresponds with the $\beta + \gamma = \alpha + \gamma'$ transformation (R. Mevrel, State of the art on high-temperature corrosion-resistant coatings, *Mater. Sci. Eng.*, 1989, A120, 13–24; P. Choquet, Doctoral Thesis, Orsay, 1987). Reproduced with permission from Elsevier.

Choquet, 1987). The phase change is, therefore, detrimental to the mechanical integrity of the coating. On thermal cycling during use, the volume change associated with the phase transformation appears as a large repeated inelastic strain that eventually induces fatigue cracking. In order to impart cracking resistance, the coating has to be stabilized against the phase transformation. Stability is achieved by adding an appropriate level of Co, the presence of which eliminates the phase transformation. Additionally, Co present in the 20 to 26 wt % range improves ductility of the coatings.

The overlay coatings are usually deposited either by electron beam physical vapor deposition (EB- PVD) or by spray processes.

A "cold spray" process, which uses only kinetic energy without the input of thermal energy, is currently being developed to minimize thermal interaction between the coating and the substrate. The cold spray process, discussed next, has been used on an experimental basis to deposit overlay coatings.

A cathodic vacuum arc process has also been used to deposit overlay coatings (Vetter et al., 1994). This process uses electrical arcing to eject coating material from ingots held as a cathode onto substrates kept as anodes.

The specific processes used to deposit overlay coatings have a significant effect on the details of the microstructure and mechanical as well as environmental properties. For example, NiCoCrAlY deposited by electron beam vapor deposition exhibits a columnar texture with "leaders" running perpendicularly from the surface of the coating to the coating–substrate interface. These leaders are grain boundaries oriented normal to the deposition surface. Because of their particular orientation, the leaders may affect fatigue behavior of coated components unless the leaders are eliminated through postcoating processing, including diffusion annealing.

Spray Coatings

The basic principle of the spray process to produce overlay coatings consists of three primary steps: (1) creating coating material of appropriate composition in powder, wire, or rod form, (2) imparting sufficient kinetic and thermal energy to create a confined high-energy particle stream, and (3) propelling the energetic particles toward the substrate using high-pressure carrier gas. The particles plastically deform on impact with the substrate or with each other. Partly because of the energy of impact, the depositing particles form cohesive bonds with each other and adhesive bonds with the substrate. If all the energy is kinetic in nature with absence of any significant thermal contribution, the process falls within "cold spray" technology. If a combination of kinetic and thermal energy is involved, the process is in the category of "thermal spray" technology. Because of the nature of the formation and deposition of the coating material, spray coatings constitute a line-of-sight process. The relative contribution of kinetic and thermal energy in various spray processes is shown in the temperature-velocity diagram of Fig. 6.20 (Papyrin, 2001).

Cold Spray

As the name implies, the cold spray process produces coatings at relatively low temperatures compared with other spray processes. Kinetic rather than thermal energy imparted to the coating feed stock drives the formation of the coating. The method was originally developed in Russia in the mid 1980s (Papyrin et al., 1990; Blose et al., 2003). The equipment for cold spray (Fig. 6.21) consists of a convergent–divergent (CD) nozzle. The combination of convergence

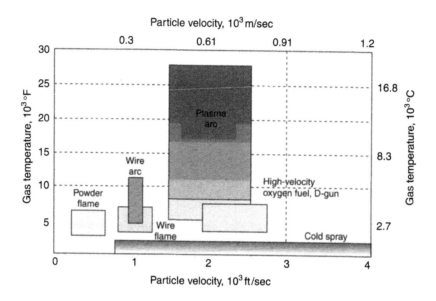

Figure 6.20 Gas temperature–particle velocity regimes for various thermal spray processes and the cold spray process (Anatolii Papyrin, Cold spray technology, *Adv. Mater. Proc.*, 2001, 159(9), 49–51). Reprinted with permission from ASM International.

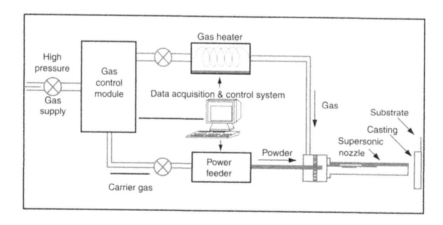

Figure 6.21 A schematic of a typical cold spray process (R. E. Blose, T. J. Roemer, A. J. Meyer, D. E. Beatty, and A. N. Papyrin, in *Thermal Spray 2003*, Eds. C. Moreau and B. Marple, ASM International, Materials Park, OH, Vol. 1, pp. 103–111; also Fig. 1 in Anatolii Papyrin, Cold spray technology, *Adv. Mater. Proc.*, 2001, 159(9), 49–51). Reprinted with permission from ASM International.

and divergence of the nozzle design allows the flowing gas to be first compressed, followed by fast expansion. This results in gas velocities achieving or exceeding supersonic levels. Coating material, in the form of powders with particle size ranging between 1 and 50 μm, is injected into the nozzle by a carrier gas under pressure of 1 to 3 MPa (150 to 450 psi). The gas, typically air, nitrogen, helium, or their mixture, may be resistively heated to temperatures up to 700°C (1292°F) to increase the propulsion velocity. However, the temperature is always kept well below the melting point of the coating material. The supersonic jet of exiting gas, at Mach numbers between 2 and 4, propels the powder particles to velocities between 500 and 1000 m/s. On exiting the nozzle, the particles ballistically impact on the substrate to be coated. Upon impact, the particles undergo extensive plastic deformation. The deformed particles form bonds with the substrate by a process akin to explosive welding but on a much finer scale. For a given material and particle size, the impact velocity of the particles needs to exceed a critical value to form a well-bonded coating. A large number of materials and coatings have successfully been formed by the cold spray process. These include metal matrix composites, wear coatings, and NiCoCrAlY coatings (Voyer et al., 2003). Typical bond strengths of cold spray coatings, without subsequent heat treatment, are in the range of 34.5–48.9 MPa (5 to 10 ksi), comparable to the values obtained for thermally sprayed coatings. Some of the advantages of the cold spray process lie in its high deposition rate, lower oxide content for metallic coatings, and the absences of oxidation, phase change, decomposition, grain growth, and postcoating component distortion.

Thermal Spray

In the thermal spray process, thermal energy is used in the deposition of coatings. As indicated in the section on spray coatings, the process consists of three steps. First, coating material of the right composition is produced in the appropriate form, generally as powders, but also in wire or rod form. Second, the coating material is melted using thermal energy. Third, the molten material is propelled to the substrate to form the coating deposit. The thermal energy to melt the coating material is derived from one of several sources:

- Detonation of combustion gases (Detonation Gun)

- Flame created by combustion of gases (flame spray, high-velocity oxygen fuel (HVOF))

- Sustained plasma created by electrical discharge (plasma spray)

- Electric arc (arc spray)

Detonation Gun Process

The "Detonation Gun," also known as the D-Gun (Registered Trademark of Praxair Technologies), was developed by Union Carbide in 1955. In principle, the D-Gun process consists of establishing high-pressure supersonic shock waves in a flammable gaseous mixture of fuel such as acetylene, butane, propane, and oxygen (Kadyrov and Kadyrov, 1995) in a chamber. The resulting high temperature melts the coating material contained in the same chamber. The expanding gas from the shock wave propels the molten particles toward the substrate to form the coating. The coating equipment (Cashon, 1975) consists of a water-cooled tubular gun barrel, typically 25 mm in internal diameter and 1 meter in length. It is fitted with poppet valves to permit entry of oxygen, nitrogen, and fuel gas (Fig. 6.22). Through a dedicated port, a carrier gas injects into the barrel the coating material in the form of powders. In some designs, the combustible mixture of fuel and oxygen is created in a mixer and metered into the barrel through the poppet valve. In other designs, the gases are metered separately. Nitrogen is introduced into the barrel to surround the poppet valves in order to protect them against hot gases. Once the gases are introduced into the barrel, the fuel gas mixture is detonated by ignition from a spark plug. The detonation creates temperatures as high as 4500°C (8132°F), high enough to melt the coating materials. It also creates high-pressure shock waves with velocities between 3000 and 4000 m/s. Following the detonation, the expanding product gases

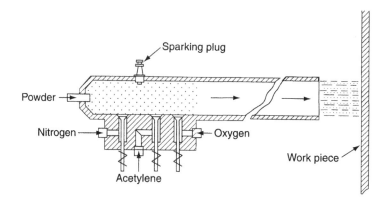

Figure 6.22 Detonation Gun (Trademark Praxair Technologies) operation schematic (E. P. Cashon, Wear resistant coatings applied by the detonation gun, *Tribol. Int.*, June 1975, pp. 111–115). Reprinted with permission from Elsevier.

propel the molten materials at velocities around 800 m/s toward the substrate. After each detonation and release of coating material, the gun is purged with nitrogen, completing one cycle (Fig. 6.23) (Kadyrov, 1996). Each cycle deposits coatings in the form of a circular disk a few millimeters in thickness. Overlap of circular disks is achieved through automation to form the final coating. Typically, detonation cycles are repeated between 1 and 15 times per second. The temperature and the particle velocities depend on the gas mix, the ratio of fuel to oxygen

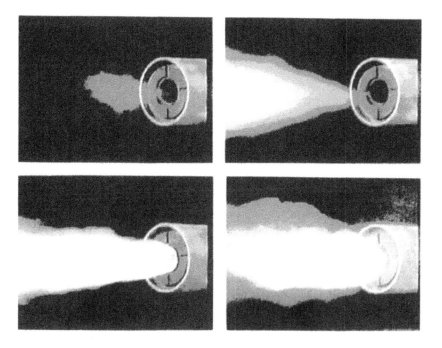

Figure 6.23 Detonation product exiting D Gun, photo at intervals of 50 µs (E. Kadyrov, Gas–particle interaction in detonation spraying system, *J. Therm. Spray. Technol*, 1996, 5(2), 185–195). Reprinted with permission from ASM International.

Table 6.6 Effect of Gas Additions on Detonation Wave Properties (Kadyrov and Kadyrov, 1995)

Gas Addition to $2H_2 + O_2$ Mixture	P_p/P_o*	T_p (K)	V_d (m/s)
No addition	18.0	3583	2806
Oxygen	17.4	3390	2302
Hydrogen	16.0	2976	3627
Nitrogen	17.4	3367	2378
Argon	17.6	3412	2117

P, Pressure; T, temperature; V, velocity; subscripts: p, detonation product; o = front of reaction zone; d = detonation wave.

and the gun dimensions (Table 6.6) (Kadyrov and Kadyrov, 1995). The D-Gun process has been used mostly to deposit wear coatings such as WC in a cobalt or NiCr matrix. Typical coating thickness ranges between 0.05 and 0.4 mm. By the nature of the process, the D-Gun creates high noise level, around 150 dB. On a comparative basis this is equivalent to the noise of jet engines taking off. On the manufacturing floor, the equipment is, therefore, enclosed in a soundproof chamber and is operated from outside.

Flame Spray Process

Flame spray, a century-old process, consists of a gun nozzle (Fig. 6.24) in which a mixture of oxygen and fuel such as acetylene, propane, or hydrogen is injected. The gas mixture is combusted in the front of the nozzle to create a flame external to the nozzle. Depending on the ratio of oxygen to fuel, the temperature of the flame ranges between 3000 and 3300°C (5430 to 5970°F).

The selection of the nozzle design depends on the coating material, which can be in the form of either powder or wire. A cold air sheath is used to collimate the flame to control the spray pattern. In the powder feed system, the coating material is gravity fed to the gun from a hopper. Aspirating gas carries the powder through the flame. The powder particles melt during transit through the flame and deposit on the substrate. In the wire feed system, the wire tip is continually fed into the flame by a roller drive mechanism. The molten material at the tip of the wire is fragmented and propelled to the substrate by compressed air injection. The wire diameter and feed rate, air and gas flow rates, flame temperature, and gas pressure determine the coating properties. The deposition rates of the coating depend on the material and vary between 15 and 500 g/m, with coating thickness ranging from 50 to 1000 μm. Any material that can be melted at the combustion temperature can be sprayed by the flame spray process. Materials deposited by this method typically include Ni- and Co-based alloys, some refractory metals, Cr carbide in a NiCr matrix, alumina, zirconia, and titania. The advantage of the spray process lies in its versatility, portability, and low capital cost.

High-Velocity Oxygen Fuel (HVOF)

The HVOF technology has evolved from the D-Gun process. In one of the gun designs (Fig. 6.25), the HVOF nozzle system consists of an internal combustion chamber attached to a two-dimensional converging diverging (2DCD) Laval nozzle (Sturgeon et al., 1995). A combustible mixture of fuel and oxygen under high pressure is ignited in the combustion chamber

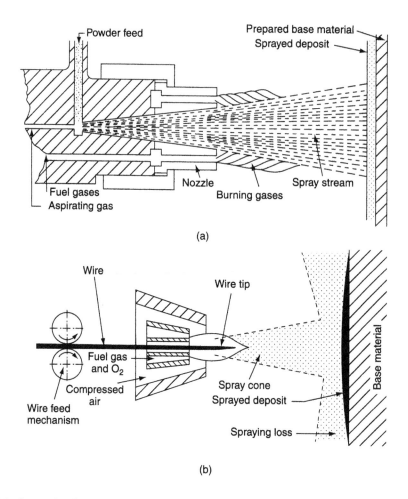

Figure 6.24 Principle of the flame spray process. (a) Powder feed, (b) wire feed (H.-D. Steffens, Spray and detonation gun technologies, in *Coatings for High Temperature Applications*, Ed. E. Lang, Applied Science Publishers, London, 1983; W. A. Saywell, Flame sprayed coatings, *Electroplating Metal Finishing*, 1973, 26, 12–16). Reprinted with permission from Elsevier.

to create a continuous flame. The combustion products exit the nozzle at supersonic velocity with associated "shock diamonds." Coating materials in the form of powder are injected into the flame axially or radially. The expanding jet of gas carries the heated powder particles, which on impact with the substrate plastically deform, cool, and solidify.

The design and the dimensions of the HVOF gun determine the types of fuel to be used, the gas temperature, and the resulting particle velocities. Typical gaseous fuels include propylene, acetylene, propane, methyl acetylene–propadiene mixture (MAPP), and hydrogen. K-1 kerosene is a common liquid fuel. Oxygen is the most widely used oxidizer to initiate and maintain combustion. The high-velocity air fuel (HVAF) system is a variant of the HVOF process and uses a modified design of the gun, which accepts air as the oxidizer. The fuel–oxidizer mixture is injected at high pressure (0.5–3.5 MPa, 80–500 psi) at flow rates of up to $0.16\,m^3/s$ (Irving et al., 1993). Some of the specific characteristics of the HVOF process are described here. A 2DCD nozzle of length between 75 and 305 mm (between 3 and 12 inches) generates gas velocities in the range 1370–2930 m/s (4500–9600 ft/s). Powders from

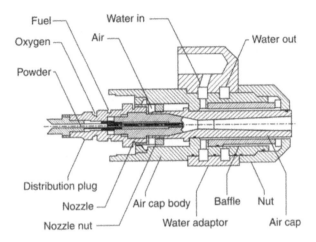

Figure 6.25 Nozzle of an HVOF gun (Hybrid Nozzle Diamond Jet 2600), Courtesy of Sulzer Metco.

pressurized feeders, injected downstream of combustion chamber, achieve supersonic velocities in the range 480–1020 m/s (1570–3350 ft/s) (velocity of sound $v \sim (C_p R T / C_v)^{1/2} = (1.4RT)^{1/2}$ for air = 340 m/s at 15°C, 60°F). Depending on the type of fuel used, the gas temperature ranges between 1650 and 2760°C (3000 and 5000°F). Here C_p and C_v are the heat capacity at constant pressure and volume, R is the gas constant, and T is the absolute temperature. The high particle velocity reduces the residence time of the particles in the flame before deposition, which in turn reduces the oxide content of the coating. Further reduction of oxide content, particularly for oxygen-sensitive materials, is possible with an inert gas shroud. The high particle velocity and momentum combined with high gas temperature increase the coating density. Best results in terms of coating properties are achieved with spherical powder particles in the size range −45 and +15 μm. Finer size particles tend to melt and deposit on the inner walls of the nozzle. Typical spray rates range between 2.25 and 12 kg/hr (5 and 25 lb/hr).

HVOF has been used successfully to deposit wear-resistant coatings WC/Co and Cr_3C_2/NiCr. Because of the lower flame temperature in HVOF than in plasma spray jets, decarburization of carbide-containing wear coatings is reduced significantly. Other materials deposited by this process include nickel base superalloy Inco 718, cobalt-based tribology coating Triboloy 800, refractory metals such as Mo, and high-temperature oxidation-resistant MCrAlY coating where M stands for Ni, Co, and Fe.

The microstructure of HVOF-sprayed coatings is characterized by low porosity (<2%), low oxide content (<2%), and a clean coating substrate interface (Irons and Zanchuk, 1993). MCrAlY coatings deposited by HVOF exhibit microstructure similar to that obtained in the low-pressure plasma spray (LPPS) process discussed in the next section. With postdiffusion heat treatment, bond strengths in excess of 65.5 MPa (9.5 ksi) have been achieved.

Plasma Spray Process

Plasma, known as the fourth state of matter, is a collection of charged and neutral particles (Braithwaite and Graham, 1993). It is created by the ionization of gases. The ionization results in the separation of some of the outer shell electrons of the gas atom, making it a positively charged ion. As a result, the gas becomes a collection of electrons, positive ions, and some neutral atoms. Because of the large attractive forces between the electrons and the positive ions, they eventually recombine by a process akin to a chemical reaction. The recombination releases large amount of enthalpy, providing kinetic energy to the species in the plasma. The kinetic energy involved in the process is large enough to generate high-temperature plasma. In a thermal plasma created with a gun power of 40 kW, the temperature could be as high as 15,000 K (26,500°F). One attribute of the plasma is that the temperature drops off very rapidly from the center to the outer periphery of the jet, thus requiring no containment.

The Spray Process   In this process, a plasma jet melts the coating feed stock, which is in the form of powder. The plasma is created in a plasma gun. The internal details of a typical gun are shown in Fig. 6.26 (English, 1985). The details of the process (Herman, 1988) are as follows:

• A direct current between a stick-type copper cathode (negative polarity) with thoriated tungsten tip located along the gun axis and a water-cooled copper anode (positive polarity) creates electrical arcs (Figs. 6.27 and 6.28). Being hard, the thoriated tungsten tip reduces wear of the cathode.

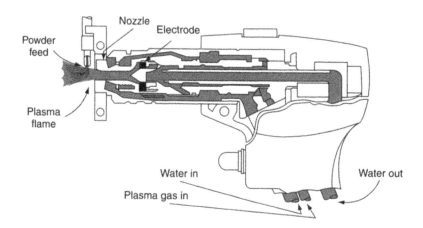

Figure 6.26   Internal features of a Metco plasma gun (Lawrence K. English, Coatings technology for wear protection, *Mater. Eng.,* Sept. 1985, pp. 53–56). Courtesy of Sulzer Metco and ASM International.

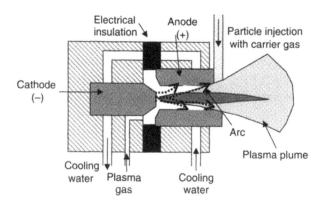

Figure 6.27   Schematic of a plasma spray nozzle and the internal process.

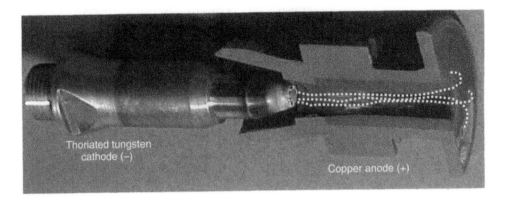

Figure 6.28 The nozzle of a Sulzer Metco plasma gun, schematically showing path of the electric arcs. Courtesy of Sulzer Metco.

- The arcs ionize a high-pressure plasma gas flowing through the center of the gun. The plasma is typically created in an inert gas such as argon or helium. Hydrogen is added to the gas to provide increased enthalpy and higher plasma temperature. The additional enthalpy is derived from the dissociated hydrogen atoms recombining into a molecule. The energy release associated with hydrogen recombination, $2H^+ + e = 2H = H_2$, is five times that resulting from argon recombination, $A^+ + e = A$; here e is an electron. The flowing ionized gas generates a plasma wherein ions and electrons recombine to create an electrically neutral hot plume, which is unlike a fuel-generated flame. As indicated before, for a plasma gun operating at a power rating of 40 kW, the plasma temperature just outside the nozzle is very high. Depending on the electrode shapes, power input, the gas pressure and distribution, the plasma plume length varies between 3 and 15 cm. The gas velocity within the plasma ranges between several hundred and 1000 m/s.

- The coating material, in the form of a powder, is injected into the hottest zone of the plasma by a high-pressure carrier gas, usually argon. A typical trajectory of the powder particles is shown in Fig. 6.29 (Montavon et al., 1997).

During transit through the hot plasma, the coating material melts. Propelled by the high-pressure gas, the molten particles continue in their flight until they impact the substrate to form the coating. The details of the microstructure formation of plasma spray coating are described in Chapter 7 on thermal barrier coatings. If this process is conducted in air, it is called air plasma spray (APS). The inevitable exposure to air during the APS process, particularly during flight of the molten particles, results in significant oxidation of the coating material. Consequently, APS-processed metallic coatings have a high oxide content, resulting in reduced capability to protect substrate alloys against oxidation and corrosion. Process-related oxidation of the coating is eliminated in "inert gas shrouded" plasma spray, and more effectively in low-pressure plasma spray (LPPS). In the shrouded plasma spray process, inert gas is injected surrounding the plasma plume to create a sheath. This sheath limits access of air to the molten particles. A more robust process to eliminate oxidation is to conduct the process in the absence of air, as is done in low-pressure plasma spray (LPPS). The oxidation resistance of metallic NiCoCrAlY coatings deposited by the LPPS process is about twice that of an equivalent coating composition deposited by APS.

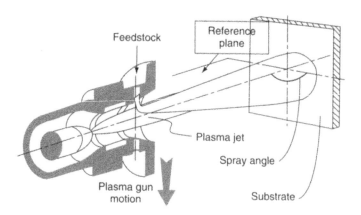

Figure 6.29 Details of plasma jet and powder trajectory (G. Montavon, C. C. Berndt, C. Coddet, S. Sampath, and H. Herman, Quality control of intrinsic deposition efficiency from controls of the splat morphologies and deposit microstructure, *J. Therm. Spray Technol.*, 1997, 6(2), 153–166). Reprinted with permission from ASM International.

Low-Pressure Plasma Spray (LPPS)   As described before, one of the major shortcomings of the atmospheric air plasma spray process for deposition of metallic coatings is the exposure of the molten metal droplets to oxygen during their transit in air. Consequently, there is significant internal oxidation of the deposited coating tying up aluminum, for example, as oxide. A significant fraction of the aluminum content is no longer available to form oxide scale and provide adequate oxidation protection.

The low-pressure plasma spray process is designed (Muehlberger, 1998) to eliminate exposure to oxygen in air during the deposition of the coating. This is accomplished by confining the plasma spray gun as well as the work piece to be coated within a chamber pumped down to a partial vacuum and back-filled with inert gas. The LPPS design offers further capability of preheating and cleaning of the parts by the process of reverse transferred arc (RTA) prior to the initiation of coating deposition. A schematic of the LPPS system is shown in Fig. 6.30 (Steffens, 1983). It consists of the following components:

- A water-cooled vacuum spray chamber, which houses the plasma gun.

- A plasma gun modified to operate under low ambient pressure. The system also has the associated power supply and control units.

- Vacuum pumps.

- Power supply for the transferred arc.

- Single or multiple shafts called "stings" to hold parts, and associated hardware to drive the stings.

- A load lock chamber, isolated from the coating chamber, in which parts are held in vacuum prior to insertion into the main chamber for the coating process. This step eliminates the need for frequent breaking of vacuum of the coating chamber.

- Powder hoppers containing the raw material powder and feeder systems to transport them.

- A multiaxis spray gun mount, and in some design, the part manipulation system.

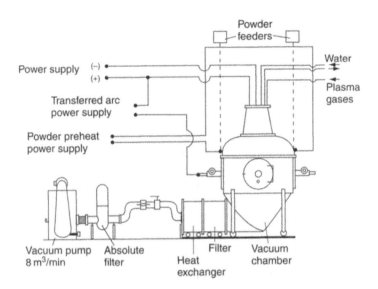

Figure 6.30 Sketch of an LPPS Coating system (H.-D. Steffens, Spray and detonation gun technologies, in *Coatings for High Temperature Applications*, Ed. E. Lang, Applied Science Publishers, 1983, pp. 121–138). Reprinted with permission from Elsevier.

The coating chamber is evacuated to a pressure of 0.2 to 0.5 mbar (0.15 to 0.38 mm Hg) followed by back filling with an inert gas such as argon to a pressure between 50 and 100 mbar (38 to 76 mm Hg) depending on the details of the design of the LPPS unit. Once the plasma operation is initiated, the chamber pressure is adjusted to a level commensurate with stable gun operations and optimum coating properties, microstructure, and yield. The plasma gun consists of a two-dimensional convergent–divergent nozzle (2DCD) to achieve exiting gas velocities in excess of 3500 m/s. A typical plasma gas consists of a mixture of 80% argon and 20% helium. Coating material in the form of powders is injected into the plasma with a carrier gas of argon. The powder particles leave the nozzle with velocity around Mach 3 and melt during flight through the plasma jet. The gas and particle velocities are strong functions of the chamber pressure, initially increasing as the pressure decreases. However, the density of the plasma gas also decreases as the pressure is reduced, affecting particle velocity. The two phenomena eventually balance each other, resulting in a maximum velocity at an intermediate pressure (Smith and Dykhuizen, 1988). Gas temperatures in the low-pressure plasma, measured with enthalpy probes, have been found to exceed 4500 K (7640°F). The high gas and particle velocities in the LPPS process increase deposition rate. Particles with increased kinetic energy form better adhesive bonds on impact with the substrate. This also results in lower porosity of the spray material. Because of low pressure in the coating chamber, the plasma jet leaving the gun expands considerably, attaining lengths of 40 to 50 cm. For comparison, plasma jets in atmospheric air plasma spray have lengths between 4 and 5 cm (Fig. 6.31) (Sodeoka et al., 1996). The reduction in pressure also results in a larger, yet more uniform, plasma jet cross section. The increased length and enlarged cross section of the plasma jet allows uniform heating and spraying of larger areas. Also, a minor variation in spray distance has minimal influence on coating quality. However, increases in jet length and cross section result in lower power density and increased loss of coating powder.

The powder materials for spraying coating are generally obtained by atomization in an inert gas environment. The composition of the powders can be varied and controlled relatively

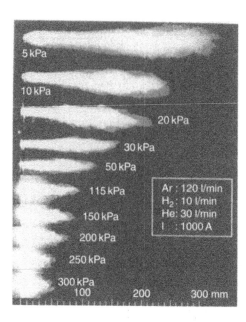

Figure 6.31 Length of plasma plume as a function of chamber pressure (S. Sodeoka, M. Suzuki, and K. Ueno, Effects of high-pressure plasma spraying for yttria-stabilized zirconia coatings, *J. Therm. Spray Technol.*, 1996, 5(3), 277–282). Reprinted with permission from ASM International.

easily. The particle size distribution is one of the key parameters to achieve good coating microstructure. Because the whole coating process is conducted in a closed low-pressure chamber, all process-related noise and particulate emissions are reduced significantly or eliminated altogether.

One of the important features of the LPPS system is the capability to clean and preheat parts by the process of reverse transfer arc prior to the coating step (Fig. 6.32) (Steffens et al., 1980; Takeda et al., 1990). In the RTA process, the part is temporarily biased negatively to a cathode polarity relative to the nozzle being the anode. Thus, some of the arcing between the cathode and anode is transferred to the part in the form of a diffused arc. The arcing ejects contaminant atoms from the cathodic part surface to clean it. RTA also provides additional energy for heating of the part and for melting of the coating powder particle.

In a manufacturing setup, the sequence of the coating process by LPPS consists of the following steps:

- Parts to be coated are degreased and cleaned by wet or dry grit blasting.

- Parts are appropriately masked to protect areas that do not require coating. Masked parts are mounted on "sting" manipulators and moved to the load lock chamber.

- The load lock chamber is evacuated.

- Parts are transferred to the coating chamber, which has already been evacuated and back filled with inert gas to the desired pressure level.

- Prior to coating, parts are preheated to 900–1000°C (1650 to 1830°F) depending on the material. Preheat improves coating adherence. The surfaces to be coated are cleaned by RTA.

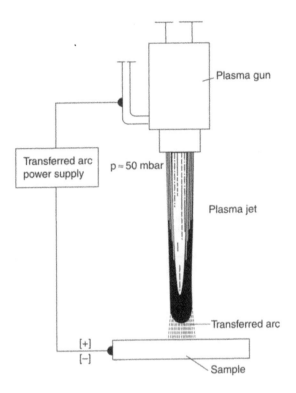

Figure 6.32 The principle of reverse transfer arc (H. D. Steffens, H. M. Hohle, and E. Erturk, Low pressure plasma spraying of reactive materials, *Thin Solid Films*, 1980, 73, 19–29). Reprinted with permission from Elsevier.

- The flow of the coating powder to the spray gun is initiated.

- The part is manipulated in the plasma jet and is coated appropriately to meet the required thickness distribution. If additional heating is required to achieve better coating properties, the part is biased positively relative to the nozzle. A representative heating profile to deposit coating is shown in Fig. 6.33 (Shankar et al., 1981).

- Optimization of the coating substrate bonding as well as coating and substrate properties may require additional heat treatment.

Representative values of the process parameters for LPPS coating and times are given in Table 6.7.

Because of the high coating deposition rate, the process and cycle time for LPPS is low for typical manufacturing processes.

LPPS Coating Deposition Profile and Microstructure

Because the plasma plume in the LPPS process is significantly larger than in APS, the particle residence time in the plasma is longer. Therefore, the depositing particles stay longer at higher temperature. Additionally, unlike in the APS, the substrate in LPPS does not experience convective cooling. Consequently the substrate reaches a considerably higher temperature. The

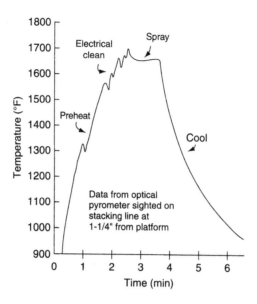

Figure 6.33 Temperature profile of a typical LPPS Process (S. Shankar, D. E. Koenig, and L. E. Dardi, Vacuum plasma sprayed metallic coatings, *JOM*, Oct. 1981, pp. 13–20). Reprinted with permission from The Minerals, Metals & Materials Society.

Table 6.7   Representative Values of LPPS Parameters for Coating Deposition

Arc chamber pressure	~10 mbar
Coating chamber pressure	~30–50 mbar
Arc gas	Argon/helium, flow rate: 50 to 200 L/min
Power	90 kW, current: 1000 A
Gas velocity at exit	3650 m/s, powder velocity: 1000 m/s
Gas temperature	~4225°C (7640°F)
Feed rate of powder	15 kg/hr,
Plasma jet size	20–50 cm
Pump down to 0.5 mbar	~1 min
Preheat	~3 min
RTA	~0.25 min
Spray Process	~0.75 min
Total cycle time	5 min

molten coating particles passing through the plasma impact on the substrate and deform into "pancake"-shaped splats. Because of the higher particle velocity in the LPPS process, the molten droplets flatten and spread to a much higher degree than in the case of APS (Fig. 6.34) (Sampath and Herman, 1996). The high-temperature exposure, unique to the LPPS process, induces several phenomena different from APS. Thermodynamic and kinetic equilibrium are achieved, which allow multiple phases to form, conforming to the phase diagram. For MCrAlY coatings, this generally means that a predominantly $\beta + \gamma$ phase structure forms (Fig. 6.35a), with the presence of other minor phases. The lack of texture in Fig. 6.35a should be compared with Fig. 6.35b. The latter exhibits elongated texture. Such a texture is the result of the vapor deposition process at molecular levels by which the coating of Fig. 6.35b was produced. The process is described in a later section. Extensive diffusion in the LPPS coating during

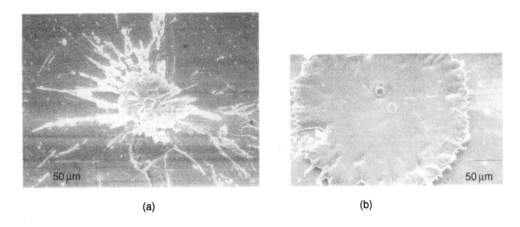

(a)   (b)

Figure 6.34 SEM micrograph comparing of splat size of nickel in (a) APS and (b) LPPS (S. Sampath and H. Herman, Rapid solidification and microstructure development during plasma spray deposition, *J. Therm. Spray Technol.*, 1996, 5(4), 445–456). Reprinted with permission from ASM International.

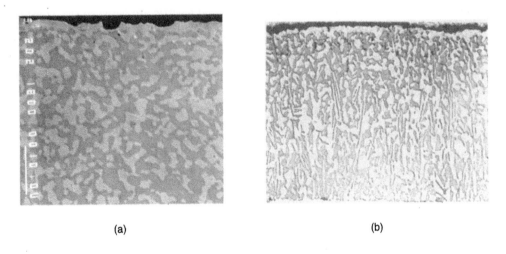

(a)   (b)

Figure 6.35 Microstructure of (a) NiCoCrAlY coating deposited by LPPS process, (b) CoCrAlY coating deposited by EB-PVD process, discussed in a later section. Dark and light areas are the β and γ phases, respectively. (b in G. W. Goward, Protective coatings for high temperature alloys: state of technology, in *Source Book on Materials for Elevated-Temperature Applications,* Ed. Elihu F. Bradley, ASM, Metals Park, OH, 1979, pp. 369–386). Reprinted with permission from ASM International.

deposition eliminates the splat boundaries and porosity observed in APS coatings. Absence of air during the coating process also eliminates oxidation and oxide inclusions in the coating.

Arc Process

Electric Arc spray

The electric arc spray process is different from other thermal spray processes in that the melting of the coating material is achieved not by the application of heat from external sources, such

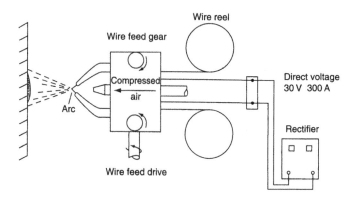

Figure 6.36 Schematic of electric arc spray process (H. D. Steffens, Spray and detonation gun technologies, laser assisted techniques, in *Coatings for High temperature Applications*, Ed. E. Lang, Applied Science Publishers, London, 1983, p. 128). Reprinted with permission from Elsevier.

as a flame or electric plasma, but by creating a controlled electric arc between two consumable electrode wires. The wires have composition close to that of the desired coating. The arcing melts the wire at the tip. A compressed air or inert gas stream fragments the molten material and propels fragments toward the substrate. A schematic of the electric spraying system is shown in Fig. 6.36 (Steffens, 1983). The arc electrode wires wound on reels pass through the gun and exit through flexible insulated conduits. The ends of the wires make contact at a small angle. Temperatures generated at the electrode tip depend on the current density. With iron wire electrodes, temperatures as high as 6000°C (10, 830°F) have been noted at electric current of 280 A (Marantz, 1974) with a voltage of a few tens of volts.

Arc sprayed coatings exhibit higher bond strengths compared with equivalent coatings by thermal spray process, 4.5 versus 3.0 ksi (31 MPa versus 21 MPa) for mild steel. The deposition rates for arc spray processes are generally high, about 15 g/s for Ni base alloys.

Microstructure of Arc Spray Coating

The microstructure of an arc sprayed coating is very similar to that of air plasma sprayed coating, as illustrated with Ni18.5Cr6Al in Fig. 6.37 (Zajchowski and Crapo, 1996). Both exhibit lamella structure with comparable levels of porosity and oxide content.

Electro–Spark Deposition (ESD)

This process was developed as a microwelding technique to deposit materials from an electrode. It has been adopted to deposit coatings (Li et al., 2000). In principle the process equipment consists of a resistance–capacitance power source. The substrate to be coated is electrically connected to the negative terminal of the power unit while the coating material, usually in the form of a pin, is connected to the positive terminal. When the two electrodes come near each other, a spark occurs between them that releases the energy stored in the capacitor. The discharge time is very short, 1 to 10,000 μs. The short pulse of energy produces very high temperatures in the range of 5000 to 10,000°C (9032 to 18,032°F). The substrate as well as the coating electrode locally melt, combine, and form a coating. The fast cooling rate associated

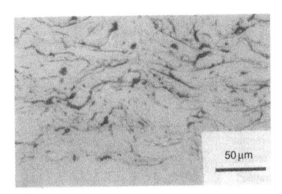

Figure 6.37 Microstructure of NiCrAl coating deposited by wire arc spray process (P. Zajchowski and H. B. Crapo III, Evaluation of three dual-wire electric arc-sprayed coatings: Industrial note, *J. Therm. Spray Technol.*, 1996, 5(4), 457–462). Reprinted with permission from ASM International.

with the process results in a fine-grained coating. Coatings deposited to date by this process include superalloy compositions and MCrAlYs.

Coating–Substrate Diffusion Effects

There are several mechanisms by which coatings degrade when exposed to a high-temperature environment for a prolonged period of time. These include oxidation, corrosion, erosion, and interdiffusion with the alloy substrate. The last involves outward diffusion of elements, such as Ni from the alloy and inward diffusional loss of the protective scale–forming elements, such as aluminum to the alloy substrate. For diffusion aluminides, the coating–substrate interaction has already been covered in an earlier section.

During oxidation, aluminum diffuses to the coating surface to form a protective alumina scale. This scale, under considerable residual stress at lower temperature due primarily to the difference in thermal expansion with the substrate, tends to spall on continued oxidation, particularly in cyclic thermal exposure. With the progressive loss of the oxide scale, more aluminum diffuses to the surface to reform the scale. This process continues until the aluminum concentration, particularly its activity in the coating, becomes insufficient to maintain a continuous alumina scale. At this stage, less protective oxides of other alloying elements begin to form. Because the scales formed are no longer protective, oxygen now easily diffuses into the coating to form internal oxides of the remaining aluminum. The aluminum concentration at which this transition from continuous alumina to nonprotective scale occurs depends on the alloy and coating composition, as well as their microstructures. For LPPS NiCoCrAlZr on simple Ni–Cr–Al based substrate, the transition in furnace oxidation test at 1150°C (2100°F) occurs at aluminum level of 0.5 to 1% (Nesbitt and Heckel, 1984). In a more severe cyclic burner rig test of LPPS NiCoCrAlY coating on nickel-base alloy U-700, the aluminum concentration at which the transition from continuous alumina scale to internal oxidation occurs is approximately 3.5% (Nesbitt et al., 1989). The loss of aluminum due to oxidation from the predominantly $\beta + \gamma$ structure of the coating manifests itself as a recession of the β phase from the coating–oxide scale interface to the interior of the coating. The β is slowly replaced by γ', which subsequently converts to γ.

Coating substrate interdiffusion, although not as extensive in MCrAlYs as in diffusion coatings, also involves loss of aluminum to the substrate. The coating, in turn, gains substrate

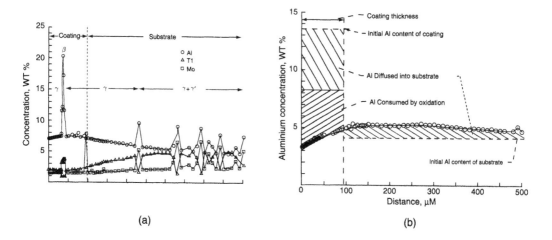

(a)   (b)

Figure 6.38 Concentration profiles as a function of depth in LPPS NiCoCrAlY coated U-700 exposed to cyclic oxidation in a Mach 0.3 burner rig at 1100°C (a) after 200 one-hour cycles, (b) after 535 one-hour cycle (J. A. Nesbitt, N. S. Jacobson, and R. A. Miller, Protective coatings for high temperature technology, in *Surface Modification Engineering*, Vol. II, *Technological Aspects*, Ed. Ram Kossowsky, CRC Press 1989). Reprinted with permission from Routledge/Taylor & Francis Group LLC.

elements with either detrimental or beneficial effects. Detrimental elements include Ti, Mo, and V. On the other hand, reactive elements such as Y and Hf provide beneficial effects. Figure 6.38 (Nesbitt et al., 1989) shows the concentration profiles in a LPPS NiCoCrAlY coated U-700 on exposure to burner rig flame at 1100°C (2012°F). Several features are clearly evident: (1) a positive concentration gradient of aluminum at the surface and a negative gradient in the γ phase at the coating–substrate interface indicate loss of aluminum down the gradient to the oxide scale as well as to the substrate alloy, respectively; (2) Ti and Mo have diffused from the alloy into the coating; (3) the β phase evident in (a) at 200 cycles has converted to γ' phase in (b) at 535 cycles; (4) loss of aluminum due to oxidation roughly equals that due to diffusion into the alloy; and (5) the depth of diffusion of aluminum into the alloy at 553 cycles is >4 times the coating thickness. Although some of the phases and features just described are specific to the particular coating–alloy system, the general processes involved are all applicable to MCrAlY coatings on nickel or cobalt base superalloys.

Commercial Overlay Coatings

A number of overlay coatings produced and used in the industry are listed in Table 6.8.

6.6 OVERLAY COATINGS BY PHYSICAL VAPOR DEPOSITION (PVD)

In physical vapor deposition, coating materials are vaporized in the form of atoms, molecules, or ions by physical process rather than through chemical reaction. The species so formed deposit on the substrate. The whole process is conducted in a vacuum or under low pressure. There are numerous PVD processes. They differ in the way the vapor species are generated. However,

Table 6.8 Commercial Overlay Coatings

Coating Designation	Producer	Composition	Comments
PWA 286	Pratt & Whitney	NiCoCrAlY + Hf, Si	LPPS
PWA 270	Pratt & Whitney	NiCoCrAlY	EB-PVD
BC21 (Marine)	GE	Co22.5Cr10.5Al0.3Y	PVD
BC22	GE	Co26Cr10.5Al2.5Hf	LPPS
BC23 (Marine) three-layer	GE	Co26Cr12Al1Hf5Pt	PVD + aluminize + Pt + diffusion heat treat
BC23 (Marine)	GE	Co26Cr12Al1Hf5Pt	LPPS
BC51	GE	Ni4Co9Cr6Al0.3Y	LPPS
BC52	GE	Ni10Co18Cr6.5Al0.3Y	LPPS
GT 29 (Plasmaguard)	GE	CoCrAlY	LPPS
GT 29+	GE	CoCrAlY + aluminized	LPPS + aluminized
GT 33	GE	NiCoCrAlY	LPPS
SICOAT 2231	Siemens AG	CoNiCrAlY	LPPS
SICOAT 2442	Siemens AG	NiCoCrAlY	LPPS
SICOAT 2453	Siemens AG	Ni10Co23Cr12Al13Re0.6Y	LPPS/HVOF
SICOAT 2412	Siemens AG	CoNiCrAlYRe	LPPS
SDP 2	N K Engines, Moscow	NiCrAlY	EB-PVD
SermAlcote	Sermatech	Overaluminized MCrAlY	LPPS + pack
SCC-103	Allied Signal	NiCrAlY	Overlay

all of these processes have some features in common. These include a coating chamber with a good vacuum system and methods to minimize contamination, the capability to introduce inert or reactive gases with controlled partial pressure, and suitable holders for both the coating material and the substrate to be coated. Prior to insertion of the substrate into the coating chamber, the surface is suitably cleaned and prepared for good adherence of the coating. Figure 6.39 lists some representative PVD processes.

Sputtering

The sputtering process involves a working gas, a "target" coating source, and the substrate to be coated. A high-energy beam of neutral atoms or charged ions of the working gas, created in a plasma by glow discharge, is directed to strike the target. Atoms of the target are physically ejected by momentum transfer on impact from the striking species. The momentum exchange extends into the target to within about 1 nm (10 angstroms, Å) of the surface. The ejected atoms deposit as thin films on the substrate that is held in their path. Sputtering is effectively a line-of-sight nonthermal evaporation process.

Typical sputtering equipment (Fig. 6.40) consists of a chamber evacuated to low pressure. The coating material constitutes the sputtering target. The substrate to be coated is positioned on a holder. Valves allow entry and exit of the flowing working gas. Various controls and

```
┌─────────────────────────────────────────────────┐
│                 PVD Processes                     │
├─────────────────────────────────────────────────┤
│  • Ion beam processes                             │
│       Sputtering                                  │
│           Planar diode sputtering                 │
│           Triode sputtering                       │
│           Magnetron sputtering                    │
│           Radio frequency sputtering              │
│       Ion implantation                            │
│  • Ion plating                                    │
│  • Evaporation                                    │
│       Electron beam physical vapor deposition     │
└─────────────────────────────────────────────────┘
```

Figure 6.39 Representative PVD processes.

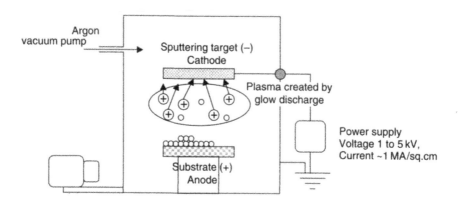

Figure 6.40 Schematic of planar diode sputtering process.

monitoring devices are attached to the equipment to control vacuum levels and gas flow rates, analyze residual gas, and monitor deposition rates. The target and the substrate are connected to a power supply in such a way that the target becomes a cathode relative to the grounded substrate. The working gas in the vicinity of the target ionizes to form the plasma, a collection of positive ions, electrons, and some neutral atoms. The positive ions of the plasma are accelerated toward and impinge on the substrate. The large momentum transfer on impact ejects atoms off the target surface. The ejected flux of target atoms deposit on the substrate, which is positioned to intercept the sputtered stream.

Roughly about 1% of the energy of impact is spent in ejecting particles from the target. A major fraction of the energy, roughly 75%, is spent in heating the target, the remainder, in producing secondary electron emission. The efficiency of the sputtering process is measured by the sputtering yield, which is defined as the number of target particles ejected per incident particle of the plasma. The sputtering yield depends on the target material, the angle of incidence of the impacting particles, and the nature of the target surface.

It is, however, independent of the charge on the incident particle. At particular ion energy, sputtering yield for different elements is found to be in the following order:

$$Ag > Cu > Ni > Al > Ti >> C$$

The sputtering rate is given by (Thornton, 1973), $\acute{R} = 6.23~IPM/\rho$. where $\acute{R}$ is the deposition rate in nm/min (nm = 10 Å), I is the ion current in mA/cm^2, P is the sputtering yield as atoms ejected per ion, M is the atomic weight of the target in grams, and ρ is the density of the target material in g/cm^3. The relationship clearly shows that the deposition rate is increased by higher ion current and for materials of high M/ρ. A number of variants of the general sputtering method have been developed to increased yield and address special requirements for microelectronic and semiconductor device applications.

Selected Sputtering Methods

Planar Diode Sputtering

One of the simplest of the sputtering processes is planar diode sputtering. In this process, the target and the substrate are held parallel to each other. The target is kept at a high negative bias of about -3 kV. The substrate and the chamber wall are grounded or the substrate is kept at a low negative bias (-100 V). The substrate acts as an anode relative to the target. The typical dimensions include target and substrate diameter 0.1 to 0.3 m with separation kept between 4 and 5 cm. The chamber is evacuated to a pressure in the range 10^{-7} to 10^{-5} torr and back filled with inert or reactive working gas to a pressure between 10^{-1} and 2×10^{-2} torr. The deposition rate achieved in this process is around 0.01 μm/min at cathode bias of -3 kV. The number and energy of the impacting ions in the plasma control the deposition rate. Higher gas pressure can increase the number of ions. However, with increased pressure, gas scattering effects increase, which in turn reduce ion current and deposition rate. For higher deposition rates, triode sputtering or magnetron sputtering is used.

Triode Sputtering

In this process, the target no longer has the function of the cathode. A hot filament acts as the cathode and generates electrons by thermionic emission. The electrons are attracted to an anode kept at a small positive potential relative to the cathode filament ($\sim$100 V). In addition to the ions formed in the glow discharge plasma, ions are also generated during the passage of the thermally emitted electrons through the working gas. The ion current, therefore, does not depend solely on volume ionization of the working gas. The ions are accelerated toward and strike the target held at a negative voltage. Thus, ion currents and deposition rates are increased even at reduced gas pressure.

Magnetron Sputtering

In triode sputtering, the number of ions is increased not by increasing the gas pressure, but by providing more electrons, which produce additional ions on collision with the working gas species. In magnetron sputtering, the ionization efficiency of the electrons is increased by trapping them in a combination of electric and magnetic field. The electron path lengths are increased significantly so that the electrons will continue to ionize the working gas. Figure 6.41 shows the principle of magnetron sputtering. A magnetic field H is created by permanent magnets. The combination of this magnetic field with the electric field E existing between the cathode and the anode determines the electron trajectory. The electrons drift along the vector v which is perpendicular to both E and H, and is given by the cross product of the E and

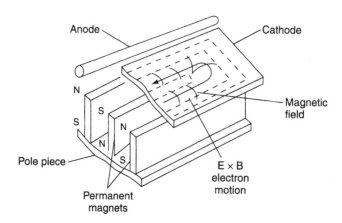

Figure 6.41 Schematic of magnetron sputtering showing the principle of trapping electrons to multiply ionization (R. F. Bunshah, in *Encyclopedia of Advanced Materials*, Eds. D. Bloor R. J. Brook, M. C. Flemmings, S. Mahajan, and R. W. Cahn., Pergamon Press, Oxford, 1994). Reprinted with permission from Elsevier.

H vectors, that is, $v = E \times H$ where $\times$ symbolizes a vector cross product. The combination of drift and the directional velocity of the electrons creates a helical path around the magnetic field lines. The electrons now are trapped in this helical path for travel over short, linear distances. As a result, the residence time of the electrons in the gas is increased, allowing continued ionizing collisions. The increased ionization in turn is responsible for a significantly higher sputter deposition rate, of the order of $1\,\mu m/min$. One of the disadvantages of the planar magnetron is the erosion of the cathode by the trapped electrons of the "plasma ring." This drawback is eliminated in cylindrical magnetrons.

DC magnetron sputtering has been demonstrated as a viable technique for deposition of thin aluminide coating containing reactive elements (Zhao et al., 2002; Ning and Weaver, 2004). NiAl containing 0.1% Hf has been deposited to a thickness of about $20\,\mu m$ on nickel-base superalloy CMSX-4. CoCrAlY coatings have also been successfully deposited on nickel-base superalloys by magnetron sputtering (Beregovsky et al., 1991).

Radio Frequency (RF) Sputtering

Sputtering from nonconductive targets is not possible by the use of the direct current (DC) sputtering processes just described. Because of the lack of an electrically conducting path through the target, electrical charges build up on the target surface, preventing the passage of positive ions, which would normally bombard the target to eject atoms. Use of a radio frequency AC power supply, capacitatively coupled between the nonconductive target and the substrate, alleviates the problem. During each AC cycle, the target surface alternately becomes positive and negative. When the surface is positive, electrons are attracted to it; positive ions are attracted to the target when the surface becomes negative. Being lighter, the electrons arrive faster and in large numbers, eventually charging the target surface negatively. The negative surface repels the arriving electrons. As a result, a dynamic equilibrium is established; creating a negative bias potential on the target. The negative potential, however, attracts the incoming positive ions, which are accelerated toward the target to collide with, and release, atoms. The typical RF frequency used is 13.56 MHz, which lies between the plasma electron and plasma ion natural frequencies.

Ion Plating

Ion plating is a plasma-assisted physical vapor deposition process, which provides highly adherent coatings. It complements the benefits of ion bombardment with the high deposition rate of physical vapor deposition. The coating material is evaporated from the source. However, prior to deposition, the vapors are bombarded with ions. The source of the ions defines the details of the process. In *glow discharge ion plating*, the ion source is a glow discharge in a low-pressure inert or reactive gas (typical pressures being 0.1 to 10 Pa; 0.001 to 0.1 torr). In *ion beam ion plating* the ion beam is generated in an external source and is allowed to bombard the vapor of the depositing material in high vacuum with pressure between 10^{-2} and 10^{-5} Pa (10^{-4} to 10^{-7} torr). The substrate is biased at a high negative potential in the range of 2 to 5 kV.

Ion Implantation

In this process (Smidt and Sartwell, 1985; Sinha, 2000), material surfaces are bombarded with accelerated ions of selected elements. The process is conducted in a moderate vacuum with ion energies between 50 and 100 keV. The implanted ions form a Gaussian distribution of ions versus depth, with the mean penetration depth less than 1 μm. Ion implantation is generally performed in an ion accelerator (Fig. 6.42) (Walters and Nastasi, 1995) with typical doses of implanting species between 1 and 5×10^{17} ions per square centimeter. The advantages of ion implantation are its capability to selectively dope metals, alloys, dielectrics, semiconductors, ceramics, and polymers to vary their surface properties. The process is used for surface modification rather than to deposit bulk coatings.

Electron Beam Physical Vapor Deposition (EB-PVD)

In this process, a vapor cloud is created in a highly evacuated chamber by melting ingots or granules of the coating material. The melting is achieved by the use of focused electron beams. The vapor species consist of atoms, molecules, ions, and cluster of atoms. The component to be coated is rotated in the vapor, which deposits to form the coating. The EB-PVD process will be discussed in detail in Chapter 7, which deals with thermal barrier coatings. Two important issues pertain to metallic coatings deposited by the EB-PVD process. First, the deposition rate is proportional to the vapor pressure of the coating constituent. In order to achieve a particular coating chemistry, the raw material composition, therefore, has to be adjusted, keeping the difference in vapor pressure of the constituents in mind. Second, the microstructure of the deposited coating generally exhibits a texture with intercolumnar grain boundaries perpendicular to the substrate surface, as was shown in Fig. 6.35b. A number of coatings have been deposited by this process, including Pratt & Whitney's PWA 270 with composition containing NiCoCrAlY. Recently, NiAl coatings (thickness ~20 μm) for TBC bond coat application have been deposited at about 1000°C (1832°F) by a modified EB-PVD process called directed vapor deposition (Yu et al., 2005); see the chapter on TBC. Nickel and Aluminum are coevaporated on a Hastelloy X substrate by electron beams in the presence of an inert gas jet to collimate the vapor stream. The deposited coating exhibited β phase without the need for postcoating heat treatment. The resulting coating was highly textured, similar to other coatings deposited by EB-PVD. The microstructure exhibited elongated grains and grain boundaries perpendicular to the Hastelloy X substrate.

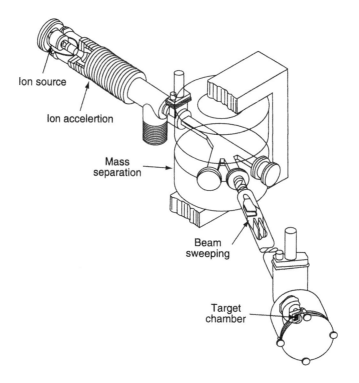

Figure 6.42   Schematic of an ion beam accelerator (Kevin C. Walters and Michael Nastasi, Ion implantation, in *Kirk-Othmer Encyclopedia of Chemical Technology*, Vol. 14, 4th ed., Wiley, 1995). Reprinted with permission from John Wiley & Sons, Inc.

Microstructure of Coatings

The coatings produced by the majority of the processes just described have unique microstructural characteristics. Some of the microstructural features and the mechanisms by which they are formed have already been discussed. The grain size, texture, porosity, phase distribution, and interaction with the substrate are controlled by the details of the process and the thermal history of the coating.

Mechanical Properties of Coatings and Coated Materials

The understanding of the deformation behavior of high-temperature metallic coatings and the mechanical and thermal mechanical properties of coated substrate alloys were initially limited to simple concepts of strain tolerance and load-bearing capability. These concepts were qualitatively assessed from coating ductility, or lack of it. The metallic coatings used at high temperatures belong to a class of materials, which are brittle at low temperatures and become increasingly ductile as the temperature is raised. For many of the coatings, the transition from ductile to brittle behavior is relatively well defined, the transition temperature being known as the ductile-to-brittle transition temperature (DBTT). Ductility, and the concept of DBTT, although helpful in qualitatively explaining the structural behavior of the coatings, are not adequate to develop models to predict the lives of coated articles in actual operation. A detailed understanding of the behavior of the coating under mechanical and thermal mechanical loading is essential.

Ductile-to-Brittle Transition Temperature

Both diffusion and overlay coatings exhibit low ductility below the DBTT, whereas above that temperature the ductility rapidly increases. Figure 6.43 (Boone, 1977) shows the variation of ductility of several overlay coatings as a function of temperature. The ductility is measured as strain to initiate cracking of the coating in tensile and often in bend tests. The transition from high-temperature ductile to lower temperature brittle behavior below the DBTT is very clear. The coating process, phase distribution, composition, heat treatment history, and microstructure are some of the factors influencing the DBTT. Diffusion aluminides with predominantly β phase, which is inherently brittle, exhibit higher DBTT (Fig. 6.44) (Meetham, 1986) than do overlay coatings, which have microstructures containing both β and the ductile γ phases. DBTT is related to the capability of the phases to plastically deform under load.

The strain these coatings experience in applications such as in gas turbine engines is imposed by the operating conditions. In such uses, the ductile-to-brittle transition behavior represented at the imposed strains helps in the selection of coatings, as well as their thickness for the structural application.

The effect of composition on DBTT of aluminides is very clear from the work of Goward (1970, 1976), who determined that the DBTT of NiAl is reduced by more than 100°C (180°F) when aluminum content is lowered from 32 to 25 wt %. Cobalt aluminide (CoAl) formed by aluminiding cobalt-based superalloys exhibits a DBTT somewhat higher than that of nickel aluminide (NiAl) (Boone, 1976). Also, platinum aluminide coating on nickel-base superalloys IN 100 and IN 738LC exhibits higher DBTT than that of plain aluminides because of the formation of a brittle $PtAl_2$ phase near the surface (Lowrie, 1952). Estimated DBTT values (Strang and Lang, 1982) of some of the aluminides are given in Table 6.9.

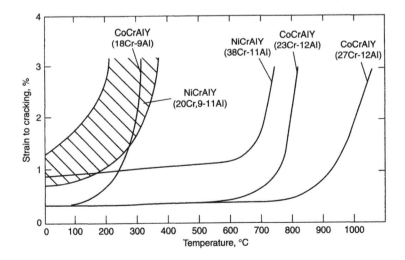

Figure 6.43 Temperature dependence of ductility of various MCrAlY coatings (D. H. Boone, Airco Temescal Data Sheets, Airco Inc., Berkeley, CA, USA, 1977). Courtesy of Donald H. Boone.

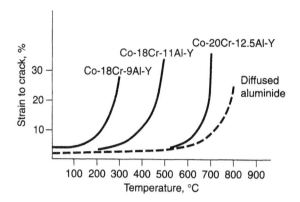

Figure 6.44 DBTT of CoCrAlY coatings compared with diffusion aluminide (G. W. Meetham, Use of protective coatings in aero gas turbine engines, *Mater. Sci. Technol.*, 1986, 2, 290–294). Reprinted with permission from Maney Publishing.

Table 6.9 Approximate Values of DBTT of Aluminides and MCrAlYs

Coating	Estimated DBTT, °C (°F)
NiAl	868–1060 (1594–1940)
(Ni,Pt)Al	> plain aluminide
CoAl	878–1070 (1612–1958)
Co18Cr9Al1Y	150–200 (302–392)
Co18Cr11Al1Y	250–300 (482–572)
Co20Cr12.5Al1Y	600–650 (1112–1202)
Co29Cr6Al1Y	700–800 (1292–1272)
Co27Cr12AlY	800–900 (1272–1652)
Ni20Cr9-11AlY	25–200 (77–392)
Ni38Cr11AlY	600–650 (1112–1202)
PtAl$_2$	870–1070 (1598–1958)
Commercial platinum aluminide (temperature at which fractures at 3% strain)	~930(1706)

DBTT also depends somewhat on the substrate on which the coating is deposited. Coating thickness is also known to have an effect (Nicholl and Hildebrandt, 1979) on the measured DBTT.

Tensile Properties

Freestanding Coatings

Data on tensile properties of freestanding coatings are scarce. The available data show that elastic constants of both diffusion aluminide and low-pressure plasma sprayed MCrAlY coatings are comparable at about 170 GPa (24.67 Msi) at room temperature (Wood, 1989). For comparison, polycrystalline superalloys have a higher modulus, typically around 200 to 220 GPa (29.02–31.92 Msi). The aluminide coatings may also exhibit strong texture (Wood, 1988a,b) based on the elastic anisotropy of the βNiAl phase. The grain sizes are often large relative to the coating thickness, particularly for the outward-growing aluminides, which experience higher processing temperatures. The elastic constants are, therefore, orientation dependent in line with the behavior of NiAl, which has modulus 100 GPa (14.51 Msi) along <100>, 190 GPa (27.57 Msi) along <110>, and 290 GPa (42.08 Msi) along <111>. Tensile strength, elastic modulus, and ductility of MCrAlY coatings have also been reported (Strangman, 1977; Smith, 1981). The MCrAlYs exhibit large strengths at low temperatures comparable to, and in some instances exceeding, those of many superalloys. However, unlike superalloys, the tensile strengths decline rapidly beyond 500 to 600°C (930 to 1110°F). In parallel with the loss of tensile strength, the ductility tends to increase rapidly, exhibiting the characteristic DBTT (Fig. 6.45 (Smith, 1981)).

The ductility and DBTT of coatings have also been estimated from measurement on coated substrates in cases where generation of the freestanding coating of appropriate composition and microstructure have not been feasible (Vogel et al., 1987). One such example

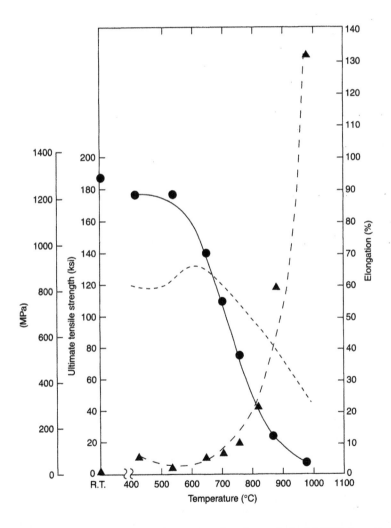

Figure 6.45 Tensile strength (filled circles) and ductility (triangles) of LPPS-deposited freestanding Co29Cr6AlY coating, trade name GT 29 from General Electric, and tensile strength of nickel base superalloy IN 738 (R. W. Smith, Mechanical properties of a low pressure plasma applied CoCrAlY coating, *Thin Solid Films*, 1981, 84, 59–72). Reprinted with permission from Elsevier.

is the platinum-containing aluminides, which exhibit several microstructural variants including single-phase NiAl with Pt in solid solution, two-phase PtAl$_2$–NiAl, and single-phase PtAl$_2$. Some of these microstructures can be generated by either low-temperature high-activity (LTHA) or high-temperature low-activity (HTLA) coating processes.

The tensile ductility plots of these coatings are shown in Fig. 6.46 (Vogel et al., 1987). One general observation can be made that the introduction of Pt increases the DBTT of the aluminides, and the temperature range appears to be dependent on aluminum and platinum content.

Effect of Coatings on Tensile Properties of Substrates

When coatings are used on superalloy substrates to protect against oxidation and corrosion, some mechanical properties of the coating as well as the alloy may be affected while others may not be influenced to any significant extent. The effects originate from several sources: (1) the postcoating heat treatment may alter the alloy microstructure, (2) interdiffusion between the alloy and the coating may deplete selected species and may also lead to the formation of brittle phases at the interface, (3) the coating may resist the environmental degradation of the alloy, which normally occurs during mechanical testing, (4) the coating may form a significant fraction of the load-bearing cross section, particularly for thin sections of substrates, and (5) because of the differences in the coefficients of thermal expansion and the modulus of elasticity, the coating as well as the alloy may have residual stresses that need to be accounted for in analysis of the data.

Tensile Strength and Ductility

Tensile strength and ductility of LPPS Co29Cr6Al1Y coating and of coated IN 738 have been measured by Smith (1981). The data (Fig. 6.47) show that there is no difference in tensile

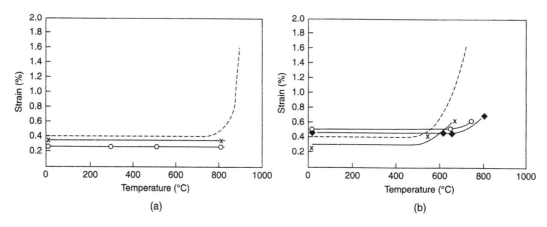

Figure 6.46 Ductility curves for (a) low temperature high activity coatings: - - -, coating without Pt; x, coating with Pt diffused for 0.5 hr at 870°C; o, coating with Pt diffused for 1 hr at 1052°C, (b) high temperature low activity coatings: - - -, coatings without Pt; x coating with Pt diffused for 4 hr at 1080°C; o, coating with Pt diffused for 2 hr at 980°C; ♦, coating with Pt diffused for 1 hr at 1052°C (D. Vogel, L. Newman, P. Deb, and D. H. Boone, Ductile-to-brittle transition temperature behavior of platinum-modified coatings, *Mater. Sci. Eng.*, 1987, 88, 227–231). Reprinted with permission from Elsevier.

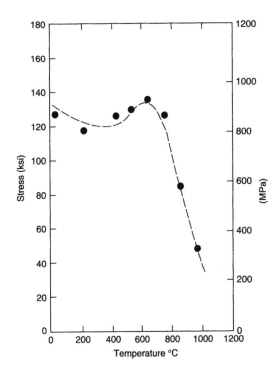

Figure 6.47 Tensile strength of IN 738 uncoated (---) and LPPS coated with Co29Cr6AllY (filled circle) (R. W. Smith, Mechanical properties of a low pressure-plasma-applied Co–Cr–Al–Y coating, *Thin Solid Films*, 1981, 84, 59–72). Reprinted with permission from Elsevier.

strength between uncoated and coated IN 738. Similar results have been found by Strang (1979) for uncoated, aluminide-coated, and NiCoCrAlY-coated nickel-based superalloy IN738LC and cobalt-based superalloy FSX414. Tensile ductility (Fig. 6.48) also follows a similar trend.

In those instances in which coating reduces the tensile properties of the substrate alloy, the effect is generally attributable to the formation of brittle phases at the coating–alloy interface, or to exposure of the alloy to the temperature of the coating process with attendant reduction in many of the critical properties.

Creep and Rupture Properties

Freestanding Coatings

The limited creep data available on MCrAlYs (Swanson et al., 1986; Hebsur and Miner, 1987; Wood and Restall, 1986) show that creep strength falls off very rapidly with temperature and follows power-law creep represented by the creep strain rate being proportional to some power of the stress (Fig. 6.49, based on Swanson et al., 1986). At higher temperatures, MCrAlYs exhibit superplastic behavior, and like other superplastic materials, the strain to rupture is strain-rate sensitive.

Whereas creep data on MCrAlYs are very limited, the data on aluminide coatings are practically nonexistent. A relative assessment of aluminide coatings can be made by studying NiAl and its ternary alloys. Wood (1989) has compared strain rate data of NiAl with that of

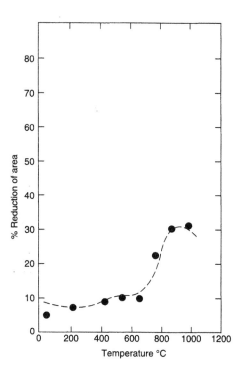

Figure 6.48 Tensile ductility of IN 738 uncoated (- - -) and LPPS coated with Co29Cr6AllY (filled circle) (R. W. Smith, Mechanical properties of a low pressure-plasma-applied Co–Cr–Al–Y Coating, *Thin Solid Films*, 1981, 84, 59–72). Reprinted with permission from Elsevier.

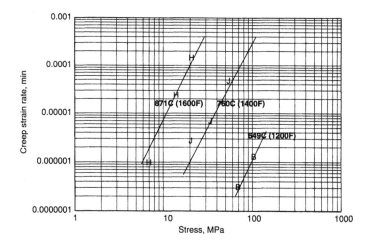

Figure 6.49 Creep behavior of LPPS NiCoCrAlY (based on data in G. A. Swanson, I. Linask, D. M. Nissley, P. P. Norris, T. J. Meyer, and K. P. Walker, *1st Annual Status Report, NASA-CR-174952*, 1986, National Aeronautics and Space Administration, Washington, DC). Used with permission.

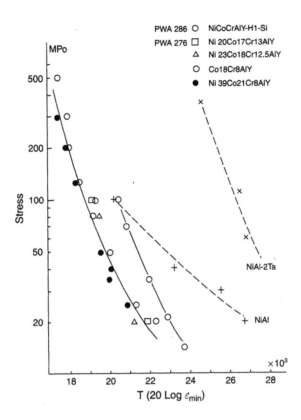

Figure 6.50 Strain rate Larson-Miller plot comparing NiAl with MCrAlYs stress rupture life (M. I. Wood, The mechanical properties of coatings and coated systems, *Mater. Sci. Eng.*, 1989, A121, 633–643). Reprinted with permission from Elsevier.

MCrAlYs in a Larson–Miller type plot using a Monkman–Grant relationship between rupture life and minimum strain rate (Fig. 6.50). It is evident that NiAl has higher creep strength, which can be further increased by alloying with small additions of such elements as Ta, Nb, Hf, and Mo. However, solubility of these elements in NiAl is very small.

Effect of Coatings on Creep Properties of Substrates

Coatings do not have a significant effect on the creep properties of coated substrates except through the influences of the following factors: (1) heat treatment during and after the coating process, (2) residual stresses related to the coating process and due to the thermal expansion and elastic modulus mismatch between the coating and the substrate, (3) diffusion of substrate alloy elements into the coating and coating elements into the substrate, and (4) cracking of the coating. Of all these factors, the heat treatment regimen associated with the coating process has the strongest effect; the remaining factors having only marginal influence, and only for thin sections of the substrate. Test data (Strang and Lang, 1982) running to 30,000 hours at 750 to 850°C (1382 to 1562°F) on uncoated IN738LC and FSX414 alloys as well as with plain aluminide, platinum aluminide, and NiCoCrAlY coatings show no influence of coatings as long as postcoating corrective heat treatment is followed. For thin sections of substrates, coatings occupy a significant fraction of the load-carrying cross section. Therefore, thickness

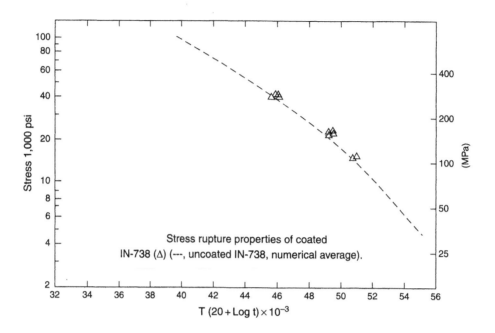

Figure 6.51 Larson Miller plot of rupture strengths of both coated and uncoated nickel base alloy IN-738 (R. W. Smith, Mechanical properties of a low-pressure-plasma-applied Co–Cr–Al–Y coating, *Thin Solid Films*, 1981, 84, 59–72). Reprinted with permission from Elsevier.

corrections should be important. Diffusion-related degradation could also be a prominent factor in influencing creep of the substrate. An example of the lack of influence of coating on rupture life is shown in Fig. 6.51 (Smith, 1981).

Low-Cycle and Thermal Fatigue

Free-standing Coatings

Low-cycle fatigue (LCF) data on freestanding coatings are rare. A study on LPPS-deposited NiCoCrAlY, trade name PWA 276, from Pratt & Whitney, is shown (Gayda et al., 1986) in Fig. 6.52. The LCF testing, conducted at frequency of 0.1 Hz, compares the behavior at 650°C (1202°F), at which the coating is strong with a yield stress of 490 MPa (71 ksi) with modest ductility (17% elongation), to that at 1050°C (1922°F), at which the coating is weak with yield stress less than 10 MPa (1.45 ksi), but good ductility. The fatigue life on a total strain range basis is almost an order of magnitude higher at 1050°C (1922°F) than at 850°C (1562°F). Also, the fatigue life of the coating exceeded that of the coated single-crystal alloy PWA 1480 at 1050°C (1922°F). But at 650°C (1202°F), the coated alloy fares much better.

Effect of Coatings on Low-Cycle Fatigue and Thermal Fatigue Properties of Coated Substrates

Three important factors influence the LCF of coated superalloys (Wood, 1987): (1) the resistance of the coating to crack initiation, (2) behavior of the cracks in terms of arrest or propagation, once they reach the interface between the coating and the substrate, and (3) nucleation

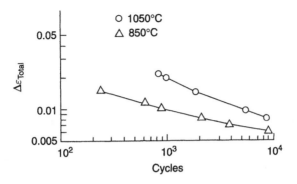

Figure 6.52 Effect of strain range on fatigue life of free standing NiCoCrAlY (J. Gayda, T. P. Gabb, and R. V. Milner, *Int. J. Fatigue*, 1986, 8(4), 217–223). Reprinted with permission from Elsevier.

and growth of cracks into the substrate as a response to (1) and (2). An overall review of data shows that the LCF properties of nickel and cobalt base alloys are not significantly affected by the presence of coatings at high temperatures. This is evident in 800°C (1472°F) data presented in Fig. 6.53b (Wood, 1988a,b, 1989). The coating, however, is found to crack long before the failure of the coated system occurs. The coated system does not exhibit the third factor described earlier, and the cracks are found to be arrested at the alloy–coating interface. Data generated at room temperature, where the coating is strong and brittle (Fig. 6.53a), on the other hand show that the coating indeed reduces the LCF life. The cracks in the coating in this case penetrated into the alloy. In some instances, coatings have been found to improve LCF life

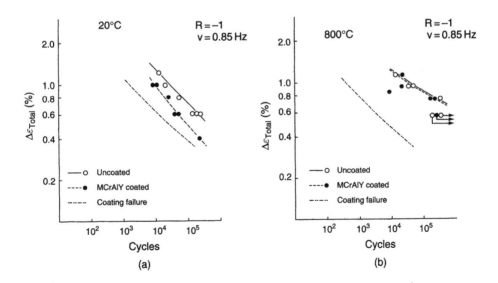

Figure 6.53 Effect of MCrAlY coating on the fatigue life of <100> oriented single crystal SX60A (a) at 20°C (68°F), (b) at 800°C (1472°F); figures based on Figs. 9a and b, M. I. Wood, The mechanical properties of coatings and coated systems, *Mater. Sci. Eng.*, 1989, A121, 633–643). Reprinted with permission from Elsevier.

of the alloy. This improvement can generally be attributed to the coatings' ability to impart oxidation resistance to the alloy during the testing.

Thermal Fatigue

Thermal fatigue, also known as thermomechanical fatigue (TMF), is a degradation mode, which involves simultaneous occurrence of both thermal and mechanical strain. Various combinations of mechanical strain (or stress) and temperature cycles are possible to generate thermal fatigue data (Fig. 6.54) (Wood, 1989). Unlike thermal fatigue, typical LCF testing is conducted with strain cycled at constant temperature. The most damaging cycle combination in thermal fatigue testing for coatings, which are brittle below the DBTT, is tensile strain at low temperature changing to compressive (or lower tensile) strain at high temperature. This is the traditional "out-of-phase" cycle, temperature being out of phase with tensile stress or strain. That the "out-of-phase" cycle is more severe than the "in-phase" cycle is clearly demonstrated by the fatigue behavior of LPPS NiCoCrAlY-coated nickel single-crystal superalloy PWA 1480 (Swanson and Bill, 1985). Out-of-phase cycle life is 500–1500 cycles compared with 10,500 cycles for in-phase cycle.

The stresses and strains involved in thermal fatigue should be corrected to include the contribution from the mismatch of coefficient of thermal expansion between the substrate and the coating. The magnitude of this contribution depends on the type of the coating and the stress free temperature, which is typically the coating deposition or postcoating diffusion heat treatment temperature. The fact that the coefficients of thermal expansions of aluminides, MCrAlYs, and the superalloys are strong functions of temperature and exhibit crossovers (Fig. 6.55) (Pint et al., 2000; Cheng et al., 1998) should be taken into consideration. The data on the magnitude and temperature dependence of coefficient of thermal expansion for Pt-containing aluminides are, however, somewhat limited. In the case of single-phase platinum aluminide, $(Pt, Ni)Al$ formed on Ni base superalloy René N5, measurement of thermal expansion by Pan et al. (2003), using a noncontact interferometric displacement gauge, gives an expansion

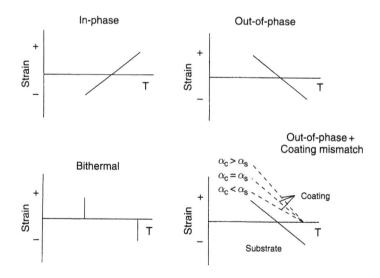

Figure 6.54 Various Thermal Fatigue cycles; α_c, and α_s are thermal expansion coefficients for coating and substrate, respectively (M. I. Wood, The mechanical properties of coatings and coated systems, *Mater. Sci. Eng*, 1989, A121, 633–643). Reprinted with permission from Elsevier.

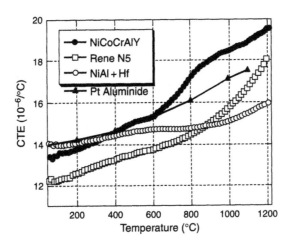

Figure 6.55 Coefficient of thermal expansion of nickel base superalloy single-crystal, nickel aluminide, and NiCoCrAlY (B. A. Pint, J. A. Haynes, K. L. More, I. G. Wright, and C. Layens, Compositional effects on aluminide oxidation performance: Objectives for improved bond coats, in *Superalloy 2000*, Eds. T. M. Pollack et al., TMS, 2000; Pt aluminide data from J. Cheng, E. H. Jordan, B. Barber, and M. Gell, Thermal/residual stress in thermal barrier coating system, *Acta Mater.*, 1998, 46, 5839–5850). Reprinted with permission from The Minerals, Metals & Materials Society, and Elsevier.

coefficient value of 15.5×10^{-6} per °C (7.6×10^{-6} per °F), independent of temperature between 400 and 850°C (752 and 1562°F). However, as shown in Fig. 6.55, Cheng et al. list temperature-dependent values, increasing from 13.6×10^{-6} per °C (7.6×10^{-6} per °F) at room temperature to 17.6×10^{-6} per °C (9.8×10^{-6} per °F) at 1100°C (2012°F). The relative trend seen in Fig. 6.55 among NiCoCrAlY, NiAl, and (Pt,Ni)Al agrees with more recent data from Haynes et al. (2004) on freestanding cast compositions. The latter data show that hyperstoichiometric (Pt,Ni)Al has a higher average thermal expansion coefficient than stoichiometric NiAl, whereas hypostoichiometric (Pt,Ni)Al has a lower thermal expansion coefficient.

The general trend of thermal fatigue data indicates that for conventionally cast polycrystalline superalloys, coatings improve fatigue life by protecting against grain boundary oxidation. Because single crystals do not have grain boundaries and are typically more oxidation-resistant than their polycrystalline counterpart, coatings tend to reduce thermal fatigue life because of their susceptibility to cracking. The latter case is illustrated in Fig. 6.56 (Bain, 1985; Fairbanks and Hecht, 1987) with data on single-crystal PWA 1480 coated with aluminide, PWA 273 and overlay NiCoCrAlY, PWA 276. The data also show a critical difference in behavior between aluminide and overlay NiCoCrAlY coatings: The thermal fatigue life is higher for aluminide coating at a low strain range, whereas NiCoCrAlY coating is superior at a higher strain range.

High-Cycle Fatigue

The very limited HCF data available on the effect of coatings on nickel and cobalt base alloys show mixed results. In the presence of coatings, the HCF life of the alloy could increase or decrease depending on the type of the test, the type of the coating and the coating process, the coating thickness, and heat treatment, particularly as it relates to microstructure and associated defects.

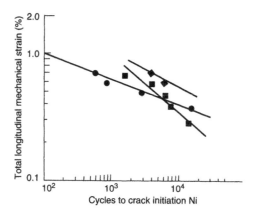

Figure 6.56 (Circles = aluminide;   squares = Co–Cr–Al–Y   overlay;   diamonds = uncoated; T_{max} 1038°C (1900°F), T_{min} 593°C (1100°F), ΔT 427°C (800°F)) Effects of coatings on the strain-controlled thermal fatigue life of PWA 1480 (Fig. 3 in K. R. Bain, The effect of coatings on thermomechanical fatigue life of a single crystal turbine blade material, Rep. AIAA-85-1366, 1985, American Institute for Astronautics and Aeronautics; John W. Fairbanks and Ralph J. Hecht, The durability and performance of coatings in gas turbine and diesel engines, *Mater. Sci. Eng.*, 1987, 88, 321–330. Reprinted with permission from Elsevier and from American Institute of Aeronautics and Astronautics).

6.7 RELATIVE OXIDATION AND CORROSION RESISTANCE OF COATINGS

Oxidation Resistance

The mechanism of oxidation of coatings used in high-temperature applications is in many ways similar to that of the superalloys. The key differences lie primarily in the aluminum content, the phase distribution, the presence of reactive elements, and the effect of interdiffusion with the substrate. Whereas superalloys have an aluminum level not exceeding about 6 wt %, above which it adversely affects strength and ductility, coatings may contain a higher aluminum content because they are not required for dedicated load-bearing functions. The microstructure, particularly the β phase content, fine grain size, and reactive elements such as Y, Zr, Hf, also play critical roles. The oxidation resistance of the two major coating types, diffusion aluminide and overlay coatings, is compared in Fig. 6.57 (Pennisi and Gupta, 1981; Fairbanks and Hecht, 1987). The coatings were deposited on Ni base single-crystal superalloy PWA 1480. It is clear that in spite of higher aluminum content, plain aluminide coatings are inferior in oxidation resistance to the MCrAlYs. The superior oxidation capability of the latter originates from increased Al activity due to the presence of Cr, and the improved adherence of the alumina scale because of the presence of reactive elements, such as Y and Hf. Presence of Si further enhances oxidation resistance. Comparison between electron beam physically vapor-deposited and LPPS-deposited NiCoCrAlY coatings shows the effect of microstructure and the deposition process on oxidation capability. The microstructure of the physically vapor-deposited coatings (Fig. 6.35b) contains significant numbers of "leaders," which are grain boundaries between columnar grains running from the surface to the coating–substrate interface. The leaders also contain open intergranular porosity providing access paths for oxygen. The negative effects

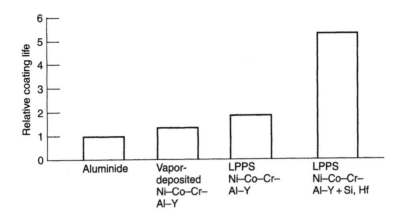

Figure 6.57 Comparative oxidation lives of some coatings at 1149°C (2100°F) in burner rig test (F. J. Pennisi and D. K. Gupta, Tailored plasma sprayed MCrAlY coatings for aircraft gas turbine application, *NASA Contract Rep CR 165234*, 1981; John W. Fairbanks and Ralph J. Hecht, The durability and performance of coatings in gas turbine and diesel engines, *Mater. Sci. Eng.*, 1987, 88, 321–330). Reprinted with permission from Elsevier.

of the column boundaries and porosity have generally been alleviated by shot-peening and appropriate heat treatment (Duret et al., 1982), both of which densify the coating.

A comparison among three major types of coatings, plain aluminide, platinum aluminide, and LPPS-deposited NiCoCrAlY is shown in Fig. 6.58 (Wood and Goldman, 1987). The trend supports the explanation put forth earlier. However, in recent years Pt aluminides containing reactive elements have been developed that exhibit improved oxidation resistance compared with traditional Pt aluminides. To the best of our knowledge a systematic analysis of the data on modified Pt aluminide has not been published.

Changes occurring within the microstructure of diffusion coatings as a result of oxidation have been discussed in Section 6.3. Continued loss of Al leads the Al-rich β to change to Ni-rich β, which slowly decomposes to $\beta + \gamma'$, then on to $\gamma + \gamma'$, and finally to all γ.

During oxidation, NiCoCrAlY loses Al to form the protective alumina scale. The β grains change phase to γ'. With continued loss of Al, the grains eventually convert to γ phase, which

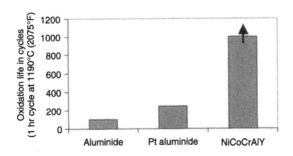

Figure 6.58 Comparison of oxidation resistance of coatings in cyclic thermal exposure in Mach 1.0 gas velocity in air at 1190°C (2075°F) (Based on data in John H. Wood and Edward H. Goldman, Protective coatings, pp. 359–384 in *Superalloy II*, Eds. C. T. Sims, N. S. Stoloff, and W. C. Hagel, Wiley, 1987.) With permission from John Wiley & Sons, Inc.

does not have adequate Al to form continuous alumina scale or repair spalled scale. The coating thus loses its resistance to oxidation.

Oxidation Resistance of HVOF coatings

For comparable composition, LPPS-deposited coatings have higher density and lower internal oxide content than coatings deposited by HVOF. This is a direct consequence of higher coating temperature and elimination of access of oxygen in the low-pressure plasma spray process. The internal oxides, except those involving reactive elements, tie up protective scale-forming elements. As a result, MCrAlYs deposited by the LPPS process exhibit better oxidation resistance than those deposited by HVOF.

Corrosion Resistance

The details of the process of hot corrosion have been covered in Chapter 5. Generally, there are two types of hot corrosion, type I occurring over a temperature range around 925°C (1700°F) and type II around 700°C (1300°F). The range of temperatures depends on the composition of the alloy and the coating, the corroding environment, and the salt chemistry. For example, the presence of vanadium reduces the hot corrosion temperature significantly.

One of the alloying elements with a significant effect in resisting hot corrosion is Cr. It is therefore not surprising that alloys containing higher levels of Cr perform very well in a corrosive environment. However, increased Cr in the substrate alloy comes at the expense of Al and refractory elements, adversely affecting mechanical properties and high-temperature oxidation resistance. As a result, alloys with load-carrying function are not adequately resistant to hot corrosion and are rarely used without specifically tailored coatings. The response of some of these coatings for protection against type I and type II hot corrosion is summarized next (Goward, 1986).

Diffusion Coatings

Diffusion coatings provide varying degrees of protection against hot corrosion depending on the composition. As has already been discussed in Section 6.3, plain aluminides offer practically no protection. Chromizing of aluminides, as well as siliconizing, protects against type II hot corrosion. The presence of Pt in a solid solution, or as a two-phase coating, provides resistance toward type I hot corrosion. When present as a continuous $PtAl_2$ phase, the coating also protects against type II hot corrosion.

Overlay Coatings

Coatings represented by MCrAlX (M being Ni, Co, Fe, and X standing for reactive elements) when deposited by some of the overlay processes such as LPPS or HVOF exhibit good oxidation resistance due to the presence of adequate levels of Al, Cr and the excellent oxide scale adherence afforded by the reactive elements. However, the coating composition for good hot corrosion resistance is predominantly CoCrAlX with a high Cr/Al ratio. A composition with 30% Cr, 5% Al has been found to have good performance in a coal combustion environment between 593 and 816°C (1100 and 1500°F). Siliconizing of MCrAlYs also improves resistance to hot corrosion.

Coatings for Marine Application

The marine environment poses a special challenge because of the severity of corrosion due to the presence of high concentrations of chlorides, sulfates, and seawater. The limited number of studies (Grisik et al., 1980; Grossklaus, Jr. et al., 1987; Driscoll et al., 2002) in this area shows that coatings have varied resistance to marine environments. The coatings evaluated in simulated, as well as in actual, environments include (a) siliconizing, (b) Pt aluminide, (c) CoCrAlYs, (d) pack-aluminized CoCrAlHf followed by platinizing, and (e) the coating in (d) followed by additional chromizing. Posttest analyses showed that the overchromizing or overplatinizing of CoCrAlHf improves performance. Coatings (d) and (e) were superior to (b) and (c), the latter two being similar in performance.

6.8 MODELING OF OXIDATION AND CORROSION LIFE

As we have seen in previous chapters, a typical component in high-temperature environments, such as in a gas turbine engine, consists of a load-bearing structural alloy. It requires an oxidation- or corrosion-resistant high-temperature metallic coating on which, depending on the performance and durability requirements, a low thermal conductivity ceramic layer may be deposited for thermal insulation. The useful life of the component is controlled by the degradation of the three individual constituents, the substrate alloy, the metallic coating, and the ceramic coating. The degradation mechanisms of the individual constituents (except for the ceramic coating) have already been discussed in the previous chapters. Here we focus on the methodology for predicting life of these constituents. The metallic coating life is controlled by one or more of the following degradation modes: (1) oxidation, (2) hot corrosion, and (3) thermal fatigue cracking. The principal modes for ceramic coating failure involve (1) spalling of the coating, (2) hot corrosion, and (3) erosion. Finally, the substrate itself may fail because of environmental interaction by (1) oxidation, (2) corrosion, and structurally by (3) creep rupture, high- or low-cycle fatigue, and thermal fatigue. The structural degradation of the alloy will not be considered here. Although there is significant interaction among the three constituents, we will deal with the coatings and only the environmental response of the alloy substrate.

Oxidation Life of Superalloys and Metallic Coatings

The basic principles of oxidation of superalloys and coatings were covered in Chapter 4. Oxidation results in loss of component cross section due to conversion of the metal to oxides. The oxides sporadically spall in a cyclic thermal environment. The component is overstressed and loses load-bearing capability, leading to its failure. The focus of the present section is to predict lifetimes of the components in an oxidizing environment. Several methodologies are available to frame a useful oxidation life prediction model.

Life Prediction Methodologies

An early oxidation life model known as COREST was developed by Barrett and Pressler (1976) of NASA for materials forming volatile oxides such as chromia-forming alloys or refractory metal alloys whose oxides tend to spall at oxidation temperature. The model assumes constant yet continuous loss of weight due to volatilization of the oxide formed. In most practical cases, however, the oxide does not vaporize but periodic spalling occurs. Examples of such cases include alumina forming alloys. For such system, a more general model called COSP (Cyclic Oxidation Spalling Program) was constructed by Lowell et al. (Lowell et al., 1991; Probst and Lowell, 1988). This model mathematically combines continued growth of oxide with a spall term, which is proportional to the pre-spall oxide weight. A statistical spalling model has been proposed by Poquillon and Monceau (2003), which introduces a spalling probability term in addition to the growth of the oxide in the COSP model. A multistage model has been proposed by Wright et al. (1988) in which the authors consider three stages of oxidation: transient, steady state, and modified steady state. The transient state is short lived for most alumina formers and may be neglected. The last stage, on the other hand, is governed by the reservoir of aluminum left after the steady-state stage. A somewhat different, but conceptually simple, model of life prediction was proposed by Lee et al. (1987), which is based on the depletion of aluminum of the βNiAl phase as a result of oxidation. The depletion occurs by two processes. The first involves the loss of aluminum to the coating surface to form, grow, and reform alumina scale if it spalls. The second is due to the inward diffusion of aluminum into the alloy and the diffusion of elements such as Ni and Co out of the alloy and into the coating. The resulting depletion zones at the coating surface as well as at the coating–substrate interface appear as well-defined γ phase regions devoid of any β phase. The width of the depleted zone X can be metallographically measured and related to the temperature T (K) and time of exposure through diffusion equations (Krukovsky et al., 2004). In a simple form, the parameters are related by the usual kinetic growth equation

$$X^n = kt, k = k_0 \exp(-\Delta H / RT),$$

where n, k_0 are constants, ΔH is the activation energy of diffusion, and R is the gas constant. The various parameters are experimentally measured for each coating–alloy system. Life is defined as the time after which the coating becomes completely depleted of β phase, that is, X equals the thickness of the coating. A model called COATLIFE has been developed jointly by the South West Research Institute and EPRI in the US (Chan et al., 1999), which treats the failure process consisting of the following elements: (1) oxidation kinetics, (2) fracture and spalling of the oxide scale, and (3) loss of aluminum in forming, as well as reforming the oxide scale after spall, and due to inward diffusion into the substrate. The end-of-life criterion has been set as the time or cycles to achieve a minimum aluminum level corresponding to complete depletion of the β phase in the coating. This assumption that only the volume fraction of β phase ultimately controls oxidation resistance does not, however, seem to explain the significant life remaining in coating systems that rely on the presence of reactive elements. A model for MCrAlY-coated substrate that not only addresses oxidation, but also includes one-dimensional transport associated with oxidation as well as interdiffusion with the substrate has been developed at NASA by Nesbitt (2000). The model, known as the Coating Oxidation and Substrate Interdiffusion Model (COSIM), involves a finite-difference computer program that is executable on a desktop PC.

Life Equation Formulation

For simplicity, we will focus on the elements of the COSP model, which is user friendly with Microsoft Windows compatible software available from the authors of the model (NASA Glenn Research Center, Mail Stop 49-1, 21000 Brookepark Road, Cleveland, OH 44135, USA). The model is based on a number of assumptions. The rate of oxide growth on a metal is treated as independent of spalling of part of the oxide and subsequent regeneration. It is assumed to be a function only of the thickness, the composition of the oxide, and the temperature. Spalling of the oxide depends monotonically on the oxide thickness. Both the growth and spalling of the oxide are modeled by using average thickness of the oxide.

The model can be formulated by utilizing Fig. 6.59 (Probst and Lowell, 1988). It schematically shows the growth of oxide scale during the heating portion, O to A, of the first cyclic exposure and a spall of the oxide during the cool-down, A to B, at the end of the first cycle. This is followed by oxide growth during the heating part of the second cycle. At this point, a few parameters as parts of the formulation need to be introduced. If W_r is the specific weight of the total oxide scale present before the cooling part of the cycle begins (A in Fig. 6.59) and W_{s1} is the specific weight of the oxide spalled on cooling (AB in Fig. 6.59), the fraction spalled is given by

$$f = W_{s1}/W_r.$$

Specific weight is defined here as weight divided by the area. The fraction spalled, f, is assumed to be proportional to the prespall specific weight. Probst and Lowell (1988) have produced supporting oxide spall data on TD-NiCrAl, an alumina scale forming alloy, and IN-601, a chromia scale former. Both of these systems exhibit a linear relationship between the fraction of oxide spalled on cooling and the specific weight of the oxide scale prior to cooling. Thus,

$$f = Q_0 W_r,$$

where Q_0 is a constant of proportionality with units of specific weight. Combining the two foregoing relationships leads to

$$W_{s1} = Q_0 (W_r)^2.$$

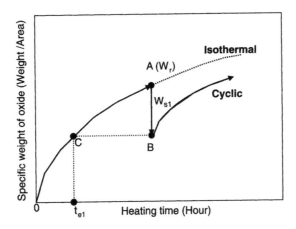

Figure 6.59 Weight of oxide formed on metal through heating and cooling cycle followed by the second heating cycle in a cyclic oxidation test (H. B. Probst and C. E. Lowell, Computer simulation of cyclic oxidation, *JOM,* Oct. 1988, pp. 18–21). Reproduced with permission from The Minerals, Metals & Materials Society.

Weight Change after the First Thermal Cycle

The specific *weight change of the sample* at the end of the first cycle is given by

$$\Delta W/A = W_r - W_{s1} = (k_p \delta t)^{1/2} - Q_0(W_r)^2.$$

In the foregoing equation, the first term corresponds to the growth of the oxide scale during the heat-up period δt of the first cycle. It is essentially the weight of oxygen picked up by the metal in the oxidation process. The growth is assumed to follow a parabolic rate with rate constant k_p. The second term corresponds to the oxide spall resulting from the cool-down, AB in Fig. 6.59. The spall weight contains both metal and oxygen. To calculate the contribution due only to oxygen, a constant α, which is the fraction of oxygen in one molecule of aluminum oxide, is defined by

$$\alpha = (3/2M_{O2})/(2M_{Al} + 3/2M_{O2}),$$

where M represents atomic weight. The specific weight of the remaining oxide scale is, therefore, given by

$$\Delta W/A = (k_p \delta t)^{1/2}[1 - \alpha^2 Q_0(k_p \delta t)^{1/2}].$$

Weight Change after the Second Thermal Cycle

At the beginning of the second thermal cycle (B in Fig. 6.59), the oxide scale is thinner than before the first cool down (A in Fig. 6.59) due to the loss of part of the oxide at the end of the first cycle. As a result, the oxide growth kinetics at the beginning of the second cycle is accelerated. The specific weight of the *oxide* at the beginning of the second cycle is given by

$$W_r' = \alpha(k_p \delta t)^{1/2}[1 - \alpha Q_0(k_p \delta t)^{1/2}].$$

Concept of Effective Time   The specific oxide weight W_r' at the beginning of the second cycle can effectively be generated at the end of an effective time t_{e1} (C in Fig. 6.59), during the first cycle following parabolic kinetics:

$$W_r' = \alpha(k_p t_{e1})^{1/2}.$$

Equating this relationship with the one before defines the "effective time" at the end of the first cycle as

$$t_{e1} = \delta t[1 - \alpha Q_0(k_p \delta t)^{1/2}]^2.$$

A similar argument provides the effective time at the end of the second cycle as

$$t_{e2} = (t_{e1} + \delta t)[1 - \alpha Q_0\{k_p(t_{e1} + \delta t)\}^{1/2}]^2.$$

Likewise, the effective time at the end of the nth cycle follows

$$t_{en} = (t_{e(n-1)} + \delta t)[1 - \alpha Q_0\{k_p(t_{e(n-1)} + \delta t)\}^{1/2}]^2.$$

Cumulative Specific Weight Change of the Sample and Metal Loss

The weight of the oxide which stays on the sample after n cycles (corrected for the weight fraction of oxygen) is given by $(1/\alpha)(W_r')_n$. Similarly, the weight of the spalled oxide (corrected for the weight fraction of the metal) is given by $(1 - 1/\alpha)(\Sigma_{1 \text{ to } n} W_{sn})$. The specific weight change of the sample at the end of n cycles is then given by

$$(\Delta W/A)_n = (1/\alpha)(W_r')_n - (1 - 1/\alpha)(\Sigma_{1 \text{ to } n} W_{sn}).$$

Similarly, the weight of the metal converted to oxide in n cycles is given by

$$(W_m)_n = (1 - 1/\alpha)(W_r')_n + \Sigma_{1 \text{ to } n} W_{sn}.$$

Life Prediction

The important input parameters for life estimation are k_p, Q_0, and δt, which control the quantities $(\Delta W/A)_n$, $(W_m)_n$, $(W_r')_n$, and t_{en}. The variation of these quantities for selected ranges of the input parameters can be determined using a computer program developed at NASA Glenn Research Center. A representative example of the model output in terms of $\Delta W/A$ fitted to data for cyclic oxidation of NiAl with 0.1% Zr is shown in Fig. 6.60 (Smialek and Auping, 2002). One can easily determine oxidation life from such plots with appropriate definition of the end-of-life criterion. Such definition is generally derived from the design requirements of components and is related to loss of performance beyond a certain minimum level. Eventually the performance loss is correlated with loss of metal, which in turn can be related to Fig. 6.60 to estimate life. Oxidation life so determined has to be correlated with field data, corrected for the difference in temperature between the test and actual field operations, such as gas turbine engines. This is accomplished by the use of appropriate temperature and engine factors.

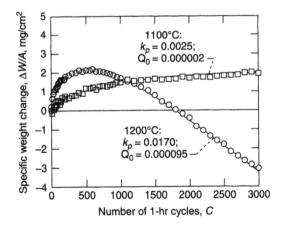

Figure 6.60 COSP model output for NiAl with 0.1% Zr (James. L. Smialek and Judith V. Auping, COSP for Windows: Strategies for rapid analyses of cyclic oxidation behavior, *NASA Report No. NASA TP-2002-211108*, 2002). Courtesy of James. L. Smialek, NASA Glenn Research Center.

Hot Corrosion Life of Superalloys and Coatings

Fundamentals of hot corrosion and the underlying mechanisms have been covered in Chapter 5. The process is illustrated schematically in Fig. 6.61 for propulsion, marine, and industrial gas turbine engines. Most Ni and Co base structural alloys, on exposure to the turbine environment in these engines, form protective alumina scales. Alloys with large concentration of chromium form a scale of chromia. These scales tend to dissolve to various extents in molten sulfate salt films that deposit on the components from external or internal sources. The solubility of the scales in molten sulfates is a strong function of the acidity (P_{SO3}) or basicity (a_{Na2O}) of the salt film, which is predominantly sulfates of Na, with varying amounts of sulfates of K, Ca, and Mg. Once the protective scale is defeated by dissolution, the bare alloy is exposed to the elements and rapid metal loss ensues. The hot corrosion process is generally limited to the temperature window between 700°C (1300°F) and 925°C (1700°F), except in the presence of impurities such as vanadium. Outside this temperature range, oxidation is the major mode of environmental attack, as shown in Fig. 6.62 (Strangman, 1990). Our interest in this section of the chapter is to understand the reduction in oxidation life due to the overlapping impact of hot corrosion, and to develop a procedure for the prediction of corrosion life. To illustrate the elements of the hot corrosion life prediction methodologies, a gas turbine engine is used as an example in the following section.

A number of methodologies are available for assessing hot corrosion life (Strangman, 1990; Hsu, 1987; Conde and McCreath, 1980; Rathnamma and Bonnell, 1986; Rhys-Jones et al., 1987) of engine components.

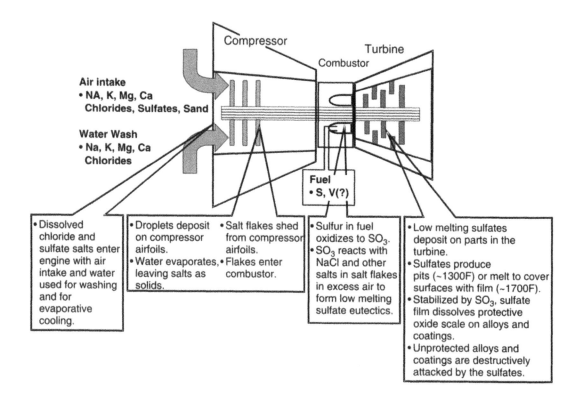

Figure 6.61 Hot corrosion process summary for gas turbine engine.

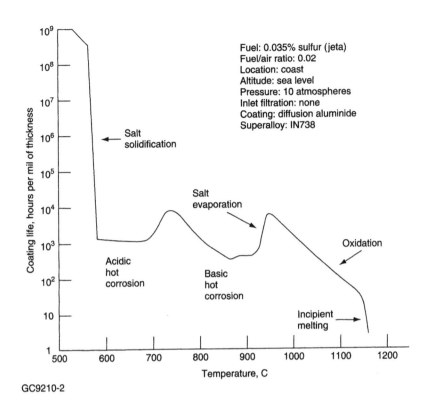

GC9210-2

Figure 6.62 Coating life in hours per mil loss of coating thickness as a function of temperature (T.E. Strangman, Turbine coating life prediction model, in *Proc. 1990 Coatings for Advanced Heat Engines Workshop, Department of Energy Report CONF-9008151*, 1990, Castine, ME, pp. II-35 to II-43). Courtesy of T. E. Strangman.

The first important requirement for life prediction is the definition of "end of life." It is generally defined as the total thickness loss of the metallic components due to hot corrosion, which results in the component being scrapped because of inability to meet load-carrying and performance requirements. If the maximum allowable thickness loss is L and the rate of thickness loss due to the combination of hot corrosion and oxidation is dX/dt, the projected life would be given by Life $= L/(dX/dt)$. The remainder of the exercise, therefore, consists of determining L from field experience, and expressing dX/dt in terms of various parameters (Strangman, 1990) related to the environment, temperature, burner rig (or other) corrosion test data, and engine experience.

Contributing Processes to the Corrosion Rate

The formulation of dX/dt can be easily understood from the schematic in Fig. 6.63 depicting the rate of metal loss as a function of temperature. The total metal loss has four contributing processes as shown: general oxidation, type I and type II hot corrosion due to molten sulfates, and hot corrosion due to the presence of vanadium in the fuel. The actual environment determines which of these processes would dominate.

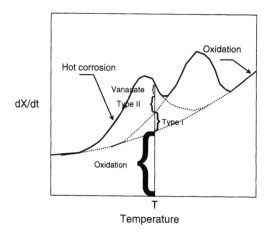

Figure 6.63 Schematic of corrosion rate measured as metal thickness loss as a function of temperature. Contributions to total metal loss from oxidation and from type I, II, and vanadic hot corrosion are indicated.

The rate of total metal loss dX/dt at any temperature T can be expressed as sum of losses due to oxidation $(dx/dt)_O$, sulfate hot corrosion type I $(dx/dt)_{typeI}$, type II $(dx/dt)_{typeII}$, and vanadic hot corrosion $(dx/dt)_V$. Thus,

$$dX/dt = (dx/dt)_O + m_s[(dx/dt)_{typeI} + (dx/dt)_{typeII} + (dx/dt)_V],$$

where m_s represents the effective molten salt concentration factor (Strangman, 1990).

Total Contaminant Concentration

The factor m_s is a composite of several terms representing contributions from many sources. These include contribution m_{fuel} from the fuel, which invariably contains sulfur and, in many instances, may contain vanadium, direct vapor m_{vap} from the inlet air, and solid salt flakes m_{shed} shed by the compressor. The salt content of air, which at sea level ranges between 0.001 and 0.100 ppm, is strongly dependent on wind velocity and the proximity to the ocean. Approximately 9% of sea salt consist of sulfates. Of the total salt entering the engine inlet, about 90% deposits on various compressor airfoils as solid after losing the water content through evaporation. Over time, this deposited salt sheds as flakes and enters the turbine. Only a fraction c_{sulf} of the total salt in the turbine is sulfate or gets converted to sulfate. Additionally, the corroding components retain some sulfate deposits m_{prev} from previous missions. In special cases, such as in power plants and engines using residual oil as fuel, there may be vanadium salt m_V. At temperatures characteristic of hot corrosion, a significant degree of evaporation m_{evap} of the deposited sulfate may occur. Once the total sulfate deposit is calculated, the fraction in the form of molten film is estimated by multiplying by the appropriate factor c_{melt},

$$m_s = [(m_{fuel} + m_{vap} + m_{shed})\ c_{sulf} + m_{prev} + m_V - m_{evap}]c_{melt}.$$

Each of the contributing terms can also be expanded to highlight contributions by various individual species of sulfates such as Na_2SO_4, K_2SO_4, $MgSO_4$, and $CaSO_4$. Contributions to the salt content from water used to wash or cool industrial gas turbines and power plants may also be included as required. When filters are used to remove contaminants, appropriate correction factors also need to be introduced.

Test Data Generation

The oxidation and corrosion rates dX/dt are usually measured in the laboratory for appropriate materials systems at a multitude of conditions. Generally, three categories of tests have been used: the crucible test, the furnace test, and the burner rig test (Saunders, 1986). In the crucible test, samples are immersed in molten sulfate salts held in crucibles. The salt mixtures consist of sulfates of alkali metals in such proportion as to form eutectics with melting points significantly lower than that of sodium sulfate. The advantage of the crucible test is the flexibility of using controlled atmosphere and adaptability of electrochemical procedures to control and monitor corrosion rates. The disadvantages include absence of high pressure, lack of exposure to fuel combustion products, absence of thermal transients, and aerodynamic characteristic of gas turbine engines. In the furnace test, the sulfate salts are deposited either continuously or intermittently on a specimen held in a furnace. The atmosphere is generally air or synthesized combustion gases. The disadvantages include the absence of transients and other characteristic of gas turbine engines. In the burner rig test, the basic principle of which has been described in Chapter 4, burning of a mixture of air and fuel oils creates combustion flames. Test samples in the form of pins are held in a rotating carousel and exposed to the flame. Engine cycles and transients are simulated by moving the samples in and out of the flames and by additional cooling with injection of cold air when the specimens are out of the flame. Salt solutions are introduced either in the fuel or the air stream. Although atmospheric-pressure burner rigs reproduce many aspects of hot corrosion, they do not duplicate all the features observed in gas turbine engines. This is due to the inability of the rigs to reproduce many critical parameters observed in actual gas turbine engines. Table 6.10 (Luthra, 1989) compares typical operating conditions in gas turbine engine (GTE) and atmospheric burner rig (ATB) tests. One aspect of GTE not reflected in the table is the contribution of the compressor and aerodynamic flow in the engine in deposition and shedding of salts. For the conditions in Table 6.10, the partial pressure ratio p^{SO3}/p^{SO2} has also been compared by Luthra for two temperatures, 977.4°C (1300°F) and 1172°C (1618°F).

Burner rigs operating at high velocities, and some additionally at high pressures providing better compromise between the laboratory test and the gas turbine engines, are now available. The corrosion rates are generally determined by measuring loss of diameter of specimen as function of time at selected exposure temperature (or as a function of temperature at fixed times) and salt concentration. Figure 6.64 (Conde and McCreath, 1980) shows an example of results of atmospheric burner rig tests. In this particular case, the rig used fuel with 1 wt %

Table 6.10 Comparison of Operating Conditions in Atmospheric Burner Rig (ATB) and Gas Turbine Engine (GTE) (Based on Luthra, 1989).

Parameters	ATB	GTE
Pressure (atm)	1	12
Air to fuel ratio (weight)	30	60
Oxygen partial pressure (atm)	0.1	1.89
Nitrogen partial pressure (atm)	0.76	9.34
CO_2 partial pressure (atm)	0.068	0.42
H_2O partial pressure (atm)	0.058	0.35
p^{SO3}/p^{SO2} at 1300°F (977.4°C) 1618°F (1172°C)	0.30/2.2 (for sulfur content of 9.6 and 16.8 wt % in fuel)	1.31/9.64 (for sulfur content of 1 wt % in fuel)

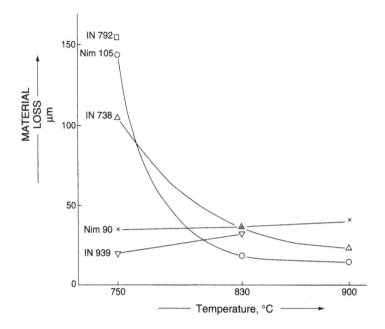

Figure 6.64 Corrosion of several alloys in 200 hours as a function of temperature in burner rig test (J. F. G. Conde and C. G. McCreath, The control of hot corrosion in marine gas turbines, *ASME 80-GT-126*). Reprinted with permission from American Society of Mechanical Engineers International.

sulfur, 0.1 ppm effective sea salt concentration in combustion air, air to fuel ratio in the range 40/1 to 50/1, and air flow velocity of 330 m/s.

The corrosion rates for various materials systems determined in similar burner rig tests are used in the life model equation.

Life Equation Formulation

The complete life model includes oxidation and corrosion rates obtained in cyclic burner rig tests with appropriate temperature and salt acidity (or basicity) factors, and correlated with engine experience. The individual terms of the corrosion rate equation are discussed next.

Oxidation

The details of the oxidation life model have already been covered in Section 6.8. In a simple form, the rate of metal loss can be expressed as follows:

$$(dx/dt)_O = (dx/dt)_{O \text{ b.rig}} f_O(T) \cdot g_O(\text{engine}).$$

Here, the rate of metal loss $(dx/dt)_{O\text{b.rig}}$ due to oxidation is measured in cyclic burner rig tests. The rig data are then corrected for appropriate temperature with the experimentally determined temperature factor $f_O(T)$. Finally, the data are correlated with field experience using the engine factor $g_O(\text{engine})$.

Type I Hot Corrosion

The metal loss due to type I hot corrosion may similarly be expressed as

$$(dx/dt)_{\text{type I}} = (dx/dt)_{\text{type I b.rig}} f_1(T) \cdot P_{1SO3} \cdot g_1(\text{engine})(1 - nV).$$

Just as in oxidation, metal loss is determined in a cyclic burner rig test, in which salt is introduced with the fuel or the air to induce type I hot corrosion. The rig data are then corrected for appropriate temperature with the temperature factor $f_1(T)$ determined experimentally. Additionally, the rig data need to be corrected for the acidity of the salt in the engine relative to that in the burner rig test. This is accomplished by multiplying with the factor P_{1SO3}. This correction factor is effectively the slope of the acidic branch of the solubility lines (see Fig. 5.7). The parameter nV is the mole fraction of sodium vanadate in the total salt. Finally, the metal loss data are correlated with field experience through the factor $g_1(\text{engine})$.

Type II Hot Corrosion

With arguments similar to the case of type I hot corrosion, the metal loss due to type II hot corrosion may be expressed as

$$(dx/dt)_{\text{type II}} = (dx/dt)_{\text{type II b.rig}} f_2(T) a_{2Na2O} g_2(\text{engine})(1 - nV).$$

The subscript refers to type II hot corrosion contribution. The basicity of the salt in the engine relative to that in the burner rig test is corrected by the factor a_{2Na2O}. This factor is effectively the slope of the basic branch of the solubility lines (see Fig. 5.7). Finally, the metal loss data are correlated with engine experience through the factor $g_2(\text{engine})$.

Vanadic Hot Corrosion

The metal loss due to vanadic hot corrosion may also be expressed, as in the case of type I and II hot corrosion, in the form

$$(dx/dt)_V = (dx/dt)_{V \text{ b.rig}} f_V(T) P_V g_V(\text{engine}) nV.$$

Here the nV is the mole fraction of sodium vanadate in the deposited sulfate salt, and $f_V(T)$, P_V, and g_V are the temperature, acidity, and engine correlation factors for vanadic hot corrosion.

Overall Corrosion Rate

The overall corrosion rate is obtained by adding the rates of each individual processes described above providing

$$
\begin{aligned}
dX/dt = {} & (dx/dt)_{O \text{ b.rig}} f_0(T) \, g_0(\text{engine}) + [\{m_s (dx/dt)_{\text{type I b.rig}} \, f_1(T) \, P_{1SO3} \\
& \times g_1(\text{engine}) + (dx/dt)_{\text{type II b.rig}} \, f_2(T) \, a_{2Na2O} g_2(\text{engine})\}(1 - nV) + (dx/dt)_{V \text{ b.rig}} \\
& \times nV f_V(T) P_V g_V(\text{engine})].
\end{aligned}
$$

The absolute corrosion life is given by Life $= L/(dX/dt)$.

Influence of Other Variables

Engine mission   Component temperature profile, air to fuel ratio, and time at temperature significantly influence hot corrosion rates. These parameters are affected by the engine mission profile. Short-hop missions of aircrafts involve more takeoffs and landings and, therefore, more fuel and air, each of which contributes to the deposition of salts.

Airport Locations   Engines operating among airports located in the coastal areas exhibit more hot corrosion than those limited to flying inland. This is directly attributable to higher salt content in the coastal air. As a corollary, flight at higher altitudes where the air is less laden with salt results in negligible hot corrosion.

Materials System   Some of the alloying constituents, such as tungsten and molybdenum, after oxidation, react with deposited sulfates to form low-melting complex oxides. WO_3 and MoO_3, for example, react with Na_2SO_4 and form low-melting reaction products:

$$WO_3 + Na_2SO_4 = Na_2WO_4 \text{ (melting point 696°C; 1285°F)} + SO_3.$$

The reaction increases the acidity of the melt and, depending on the amount of Na_2SO_4, enhances the dissolution of protective oxides and accelerates the corrosion rate.

6.9   INTERACTION OF EROSION–OXIDATION AND EROSION–CORROSION

We have already discussed the processes and ramifications of oxidation, as well as hot corrosion of alloys and metallic coatings. However, in industrial applications such as gas turbine engines and coal gasifiers, erosion by particulate materials is also present. In the presence of particulate impact, the protective oxide scale that forms on the metallic systems tends to erode away. Reduction or elimination of protective scale leads to increased oxidation and corrosion rates, which in turn increases erosion rate. The erosion profile of the scale, as well as the metal varies with the angle of impact, particulate size, and velocity, as we will see in Chapter 7. Erosion does not play a significant *direct* role in the removal of metal. Continued metal loss is the indirect result of formation of protective (alumina) as well as nonprotective scales (nickel oxide, sulfides) due to oxidation and corrosion, followed by the removal of products of oxidation and corrosion by the process of erosion. Pettit et al. (Chang et al., 1990; Rishel et al., 1992) have demonstrated that the total degradation can be represented by a combination of scale growth and a loss of scale by erosion,

$$dx/dt = [k_{(c,o,e)}/x] - k_{er},$$

where x is the scale thickness at time t, $k_{(c,o,e)}$ is the parabolic rate constant in the simultaneous presence of corrosion c, erosion e, and oxidation o, and k_{er} is the rate constant of erosion of the product of oxidation and corrosion. The increased rate of degradation is due to the fact that the rate constant $k_{(c,o,e)}$ in the presence of erosion is significantly higher than $k_{(c,o)}$ in the absence of erosion. For example, Pettit et al. have observed in the case of CoO, an increase in erosive particle velocity by a factor of four enhanced the rate constant roughly by about two orders of magnitude.

REFERENCES

Adesanya, O. A., K. Bouhanek, F. H. Stott, P. Skeldon, D. G. Lees, and G. C. Wood, Cyclic oxidation of two bond coats in thermal barrier coating systems on CMSX-4 substrates, *Mater. Sci. Forum*, 2001, 369–372, 639–646.

Albon, J. M., W. J. Davis, P. E. Skinner, and S. G. Warren, U.S. Patent No. 5,102,509, April 7, 1992.

Aldred, P., R. 125: Development and Application, S.A.E. Paper 751049, National Aerospace Engineering and Manufacturing Meeting, Los Angeles, 1975.

Angenete, J., Aluminide diffusion coatings for Ni based superalloys, Ph.D. Thesis, Chalmers University of Technology, Goteborg, Sweden, 2002.

Bain, K. R., The effect of coatings on thermomechanical fatigue life of a single crystal turbine blade material, *Rep. AIAA-85-1366*, 1985, American Institute for Astronautics and Aeronautics.

Barkalow, R. H., and F. S. Pettit, Degradation of coatings in simulated marine environments, *Report No. FR-10225*, U. S. Navy Contract N-00173-75-C0146, June 1978.

Barrett, C. A., and A. F. Pressler, COREST: a Fortran computer program to analyze paralinear oxidation behavior and its application to chromic oxide forming alloys, NASA Lewis Research Center, Cleveland OH, NASA—TN D-8132 E-8432, February 1976.

Basuki, E., A. Crosky, and B. Gleeson, Stages of interdiffusion behavior in an aluminide coated γ-Ni$+$ γ'-Ni$_3$Al alloy at 1150°C, in *Proc. International Symposium on High-Temperature Corrosion and Protection 2000*, pp. 315–322, Eds. Toshio Narita, Toshio Maruyama, and Shigeji Taniguchi, Hokkaido, Japan, 17–22 September 2000.

Beregovsky, V. V., G. G. Shergin, N. G. Shesterkin, and E. V. Kuznetsov, Magnetron sputter deposition of heat-resistant coatings on high temperature alloys, *Surf. Coat. Technol.*, 1991, 48, 13.

Blose, R. E., T. J. Roemer, A. J. Meyer, D. E. Beatty, and A. N. Papyrin, in *Thermal Spray 2003*, Vol. 1, pp. 103–111, Eds. C. Moreau and B. Marple, ASM International, Materials Park, OH.

Boone, D. H., Overlay coatings for high temperature applications, *Airco Temescal*, January 1976.

Boone, D. H., *Airco Temescal Data Sheets*, Airco Inc., Berkeley, CA, 1977.

Braithwaite, N., and B. Graham, Quest for perfect plasma, *New Scientist*, 27 Nov. 1993, pp. 34–38.

Bunshah R. F., in *Encyclopedia of Advanced Materials*, pp. 2008–2017, Eds. D. Bloor et al., Pergamon, Oxford, 1994; *IEEE Trans. Plasma Sci.*, 1990, 18(6), 846–854.

Cape, T., U.S. Patent No. 3,107,175, October 15, 1963.

Cashon, E. P., Wear resistant coatings applied by the detonation gun, *Tribol. Int.*, June 1975, pp. 111–115.

Chan, K. S., N. S. Cheruvu, and G. R. Leverant, Coating life prediction for combustion turbine blades, *J. Eng. Gas Turbines Power Trans ASME* 1999, 121, 484–488.

Chang, S. L., F. S. Pettit, and N. Birks, Some interactions in erosion-oxidation of alloys, *Oxid. Met.*, 1990, 34(1/2), 71–100.

Chen, J. H., and J. A. Little, Degradation of platinum aluminide coating on CMSX4 at 1100°C, *Surf. Coat. Technol.*, 1997, 92, 69–77.

Cheng, J., E. H. Jordan, B. Barber, and M. Gell, Thermal/residual stress in thermal barrier coating system, *Acta Mater.*, 1998, 46, 5839–5850.

Choquet, P., Doctoral Thesis, Orsay, 1987.

Conde, J. F. G., and C. G. McCreath, The control of hot corrosion in marine gas turbines, *ASME 80-GT-126*.

Czech, N., W. Eber, and F. Schmitz, Effect of environment on mechanical properties of coated superalloys and gas turbine blades, *Mater. Sci. Technol.*, 1986, 2(3), 244–249.

Deb, P., D. H. Boone, and R. Streiff, Effects of microstructural morphology on the performance of platinum aluminide coatings, in *Surface Modification and Coatings*, pp. 143–159, Ed. Richard D. Sisson, Jr., American Society of Metals, 1986.

Driscoll, M., E. McFetridge, and W. Arseneau, Evaluation of sea tested LM2500 Rainbow rotor blade coatings, in *Proc. ASME Turbo Expo 2002*, pp. 1–17, Amsterdam.

Duret C., and R. Pichior, Protective coatings for high temperature materials: Chemical vapor deposition and pack cementation processes, in *Coatings for High Temperature Applications*, pp. 33–78, Ed. E. Lang, Applied Science, London, 1983.

Duret, C., A. Davin, G. Marijnissen, and R. Pichor, Recent approaches to the development of corrosion resistant coatings, in *High Temperature Alloys for Gas Turbines 1982*, pp. 53–77, Eds. R. Brunetaud, D. Coutsouradis, T. B. Gibbons, Y. Lindblom, D. B. Meadowcroft, and R. Stickler, D. Reidel, Dordrecht, The Netherlands, 1982.

English, L. K., Coatings technology for wear protection, *Mater. Eng.*, September 1985, pp. 53–56.

Fairbanks, J. W., and R. J. Hecht, The durability and performance of coatings in gas turbine and diesel engines, *Mater. Sci. Eng.*, 1987, 88, 321–330.

Felten, E. J., and F. S. Pettit, Development, growth, and adhesion of Al_2O_3 on platinum–aluminum alloys, *Oxid. Met.*, 1976, 10(3), 189–223.

Fitzer, E., H. J. Maurer, W. Nowak and J. Schlichting, Aluminum and silicon base coatings for high temperature alloys-process development and comparison of properties, *Thin Solid Films*, 1979, 64, 305–319.

Fox, P., and G. J. Tatlock, Poster paper: Effect of tantalum additions on oxidation of overlay coated superalloys, *Mater. Sci. Technol.*, 1989, 5, 816–827.

Gayda, J., T. P. Gabb, and R. V. Milner, Low cycle fatigue behaviour of a plasma-sprayed coating material, *Int. J. Fatigue*, 1986, 8(4), 217–223.

Gleeson, B., W. Wang, S. Hayashi, and D. Sordelet, Effects of platinum on the interdiffusion and oxidation behavior of Ni–Al-based alloys, *Mater. Sci. Forum*, 2004, 461–464, 213–222.

Göbel, M., A. Rahmel, M. Schutze, M. Schorr, and W. T. Wu, Interdiffusion between the platinum-modified aluminide coating RT 22 and nickel-based single-crystal superalloys at 1000 and 1200°C, *Mater. High Temp.*, 1994, 12(4), 301–309.

Godlewska, E., and K. Godlewski, Chromaluminizing of nickel and its alloys, *Oxid. Metals*, 1984, 22(3/4), 117–131.

Goward, G. W., Current research on the surface protection of superalloys for gas turbine engines, *JOM*, October 1970, pp. 31–39.

Goward, G. W., in *Symposium on Properties of High Temperature Alloys*, Las Vegas, NV, USA, October 1976, pp. 806–823.

Goward, G. W., Recent Developments in High Temperature Coatings for Gas Turbine Airfoils, *High Temperature Corrosion*, Ed. Robert A. Rapp, NACE, Houston, Texas, 1981, *pp.* 553–567.

Goward, G. W., Low-temperature hot corrosion in gas turbines: A review of causes and coatings therefor, *J. Eng. Gas Turbines Power Trans. ASME*, 1986, 108, 421–425.

Goward, G. W., Progress in coatings for gas turbine airfoils, *Surf. Coat. Technol.*, 1998, 108–109, pp. 73–79.

Goward G. W., and D. H. Boone, Mechanisms of formation of diffusion aluminide coatings on nickel-base superalloys, *Oxid. Met.*, 1971, pp. 3, 475–495.

Goward, G. W., and L. L. Seigle, Diffusion coatings for gas turbine engine hot section parts, *ASM Handbook*, Vol. 5, *Surface Engineering*, pp. 611–620, ASM, 1994.

Grisik, J. J., R. G. Miner, and D. J. Wortman, Performance of second generation airfoil coatings in marine service, *Thin Solid Films*, 1980, 73, 397–406.

Grossklaus, W. D., Jr., G. B. Katz, and D. J. Wortman, Performance comparison of advanced airfoil coatings in marine service, in *High Temperature Coatings*, pp. 68–83, Eds. M. Khobaib and R. C. Krutenat, Metallurgical Society, Warrendale, PA, 1987.

Hancock, P., and J. R. Nicholls, in *Coatings for Heat Engines* (Workshop Proc.), pp. 31–58, Eds. R. L. Clark et al., U.S. Department of Energy, Washington, DC, 1984.

Haynes, J. A., B. A. Pint, W. D. Porter, and I. G. Wright, Comparison of thermal expansion and oxidation behavior of various high – temperature coating materials and superalloys, *Mater. High Temp.*, 2004, 21(2), 87–94.

Hebsur, M. G., and R. V. Miner, Stress rupture and creep behavior of a low pressure plasma-sprayed NiCoCrAlY coating in air and vaccum, *Thin Solid Films*, 1987, 147, 143–152.

Herman, H., Plasma-sprayed coatings, *Sci. Am.*, 1988, 259 (3), 112–117.

Hsu, L. L., Total corrosion control for industrial gas turbines: High temperature coatings and air, fuel and water management, *Surf. Coat. Technol.*, 1987, 32, 1–17.

Irons G., and V. Zanchuk, in *Proc. 1993 National Thermal Spray Conference*, pp. 191–197, Anaheim, CA, June 7–11, 1993.

Irving, B., R. Knight, and R. W. Smith, The HVOF process, the hottest topic in the thermal spray industry, *Welding J.*, July 1993, pp. 25–30.

Janssen, M. M. P., and G. D. Rieck, Reaction diffusion and Kirkendall-effect in the nickel–aluminum system, *Trans. Met. Soc. AIME*, 1967, 239, 1372–1385.

Kadyrov, E., Gas–particle interaction in detonation spraying system, *J. Therm. Spray. Technol*, 1996, 5(2), 185–195.

Kadyrov, E., and V. Kadyrov, Gas detonation gun, *Adv. Mater. Proc.*, 1995, 148(2), 21–24.

Kandasamy, N., L. L. Siegle, and F. J. Pennisi, The kinetics of gas transport in halide-activated aluminizing packs, *Thin Solid Films*, 1981, 84, 17–27.

Kircher, T. A., B. G. McMordie, and A. McCarter, Performance of a silicon-modified aluminide coating in high temperature hot corrosion test conditions, *Surf. Coat Technol.*, 1994, 68/69, 32–37.

Krukovsky, P., V. Kolarik, K. Tadlya, A. Rybnikov, I. Kryukov, N. Mojaiskaya, and M. Juez-Lorenzo, Lifetime prediction for MCrAlY coatings by means of inverse problem solution (IPS), *Surf. Coat. Technol.*, 2004, 177–178, 32–36.

Lee, E. Y., D. M. Chartier, R. R. Biederman, and R. D. Sisson, Jr., Modelling the microstructural evolution and degradation of MCrAlY coatings during high temperature oxidation, *Surf. Coat. Technol.*, 1987, 32, 19–39.

Lee, W. Y., and G. Y. Kim, Kinetic considerations for processing a diffusion NiAl coating uniformly doped with a reactive element by chemical vapor deposition, in *Elevated Temperature Coatings: Science and Technology III*, pp. 149–160, Eds. J. M. Hampikian and N. B. Dahotre, TMS, Warrendale, PA, 1999.

Lehnert G., and H. W. Meinhardt, *Electro-dep. Surface Treat*, 1972/73, 1, 189.

Levine, S. R., and R. M. Caves, Thermodynamics and kinetics of pack aluminide coating formation on IN-100, *J. Electrochem. Soc.*, 1974, 121, 1051–1064.

Li, Z., W. Gao, P. Kwok, S. Li, and Y. He, Electro-spark deposition coatings for high temperature oxidation resistance, *High Temp. Mater. Process.*, 2000, 19(6), 443–458.

Lowell, C. E., C. A. Barrett, R. W. Palmer, J. V. Auping, and H. B. Probst, COSP: A computer model of cyclic oxidation, *Oxid. Met.*, 1991, 36, 81–112.

Lowrie, R., Mechanical properties of intermetallic compounds at elevated temperature, *JOM*, 1952, 4, 1093.

Luthra, K. L., Simulation of gas turbine environment in small burner rigs, *High Temperature Technology*, 1989, 7(4), 187–192.

Marantz, D. R., The basic principles of electric arc spraying, in *Science and Technology of Surface Coating*, pp. 308–331, Eds. B. N. Chapman and J. C. Anderson, Academic Press, New York, 1974.

Marijnissen, G. H., Codeposition of chromium and aluminum during a pack process, in *High Temperature Protective Coatings*, pp. 27–35, Ed. Subhash C. Singhal, American Institute of Mining, Metallurgical, and Petroleum Engineers, New York, 1982.

Meetham, G. W., Use of protective coatings in aero gas turbine engines, *Mater. Sci. Technol.*, 1986, 2, 290–294.

Mevrel, R., State of the art on high-temperature corrosion-resistant coatings, *Mater. Sci. Eng.*, 1989, A120, 13–24.

Mevrel, R., C. Duret, and R. Pichoir, Pack cementation process, *Mater. Sci. Technol.*, 1986, 2, 201–206.

Montavon, G., C. C. Berndt, C. Coddet, S. Sampath, and H. Herman, Quality control of intrinsic deposition efficiency from controls of the splat morphologies and deposit microstructure, *J. Therm. Spray Technol.*, 1997, 6(2), 153–166.

Muehlberger, E., Method of forming uniform thin coatings on large substrates, U.S. Patent No. 5,853,815, December 1998.

Nesbitt, J. A., COSIM—a finite difference computer model to predict ternary concentration profiles associated with oxidation and interdiffusion of overlay-coated substrate, *NASA Technical Memorandum NASA/TM—2000—209271*, August 2000.

Nesbitt, J. A., and R. A. Heckel, Modeling degradation and failure of Ni–Cr–Al overlay coatings, *Thin Solid Films*, 1984, 119, 281–290.

Nesbitt, J. A., N. S. Jacobson, and R. A. Miller, Protective coatings for high temperature technology, in *Surface Modification Engineering*, Vol. II, *Technological Aspects*, Ed. Ram Kossowsky, CRC Press, Boca Raton, FL, 1989.

Nicholl A. R., and U. W. Hildebrandt, *Proc. Conf. on Acoustic Emission*, Bad Nauheim, Germany, April 1979, pp 217–223.

Nicholl, A. R., U. W. Hildebrandt, and G. Wahl, The properties of a chemical-vapor deposited silicon based coating for gas turbine blading, *Thin Solid Films*, 1979, 64, 321–326.

Nicholls, J. R., Designing oxidation resistant coatings, *JOM*, 2000, pp. 52(1), 28–35.

Nicholls, J. R., Advances in coating design for high-performance gas turbines, *MRS Bull.*, September 2003, pp. 659–670.

Nicholls J. R., and P. Hancock, Advanced high temperature coatings for gas turbines, *Ind. Corros.*, 1987, pp. 5(4), 8–18.

Ning, A., and M. I. Weaver, A preliminary study of DC magnetron sputtered NiAl–Hf coatings, *Surf. Coat. Technol.*, 2004, 177–178, 113–120.

Pan, D., M. W. Chen, P. K. Wright, and K. J. Hemker, Evolution of a diffusion aluminide bond coat for thermal barrier coatings during thermal cycling, *Acta Mater.*, 2003, 51, 2205–2217.

Papyrin, A., Cold spray technology, *Adv. Mater. Proc.*, 2001, 159(9), 49–51.

Papyrin, A. N., A. P. Alkimov, V. F. Kosarev, and N. I. Nesterovich, Method of applying coatings, Russian Patent No. 1,618,778, 1990.

Pennisi, F. J., and D. K. Gupta, Tailored plasma sprayed MCrAlY coatings for aircraft gas turbine application, *NASA Contract Rep CR 165234*, 1981.

Pichoir, R., Aluminide coatings on nickel and cobalt–base superalloys: Principal parameters determining their morphology and composition, in *High Temperature Alloys for Gas Turbines*, pp. 191–208, Eds. D. Coutsoradis, P. Felix, H. Fischmeister, L. Habraken, Y. Lindblom, and M. O. Speidel, Applied Science, London, 1978.

Pint, B. A., J. A. Haynes, K. L. More, I. G. Wright, and C. Leyens, Compositional effects on aluminide oxidation performance: Objectives for improved bond coats, in *Superalloy 2000*, pp.629–638, Eds. T. M. Pollock, R. D. Kissinger, R. R. Bowman, K. A. Green, M. McLean, S. L. Olson, and J. J. Schirra, TMS, 2000.

Poquillon, D., and D. Monceau, Prediction of high temperature cyclic oxidation kinetics with a simple statistical spalling model, in *Materials Lifetime Science and Engineering*, pp. 165–172, Eds. P. K. Liaw, R. A. Buchanon, D. L. Klastrom, R. P. Wei, D. G. Harlow, and P. F. Tortorelli, TMS Proceedings, 2003.

Probst, H. B., and C. E. Lowell, Computer simulation of cyclic oxidation, *JOM*, 1988, 40(10), 309–320.

Purvis, Andrew L., Bruce M. Warnes, The effects of platinum concentration on the oxidation resistance of superalloys coated with single phase platinum aluminide, *Surf. Coat. Technol.*, 2001, 146–147, 1–6.

Rathnamma, D. V., and D. W. Bonnell, Contaminated fuel combustion and material degradation life prediction model, in *High Temperature Alloys for Gas Turbines and Other Applications 1986*, pp. 1105–1116, Eds. W. Betz, R. Brunetaud, D. Coutsouradis, H. Fischmeister, T. B. Gibbons, I. Kvernes, Y. Lindblom, J. B. Marriott, and D. B. Meadowcroft, D. Reidel, Dordrecht, The Netherlands, 1986.

Reid, M., M. J. Pomeroy, and J. S Robinson, Microstructural stability of a Ni–Pt–Al coating on CMSX-10 alloy at 950 and 1100C, *Mater. High Temp.*, 2003, 20(4), 467–474.

Restall, J. E., B. J. Gill, C. Hayman, and N. J. Archer, A process for protecting gas turbine cooling passages against degradation, in *Superalloy 1980*, pp. 405–411, Eds. John K. Tien, Stanley T. Wlodek, Hugh Morrow III, Maurice Gell, and Gernant E. Maurer, American Society of Metals, Metals Park, OH, 1980.

Rhys-Jones, T. N., J. R. Nicholls, and P. Hancock, The prediction of contaminant effects on materials performance in residual oil-fired industrial gas turbine environments, in *Plant Corrosion: Prediction of Materials Performance*, pp. 289–311, Eds. J. E, Strutt and J. R. Nicholls, Ellis Horwood, London, 1987.

Rickerby, D. S., and R. G. Wing, U.S. Patent No. 5,942,337, August 1999.

Rickerby, D. S., S. R. Bell, and R. G. Wing, U.S. Patent No. 5,981,091, November 1999.

Rishel, D., F. Pettit, and N. Birks, Some principal mechanisms in the simultaneous erosion and corrosion attack of metals at high temperatures, in *High Temperature Corrosion of Advanced Materials and Coatings*, pp. 13–28, Eds. Y. Saito, B. Önay, and T. Maruyama, North-Holland, Amsterdam, 1992.

Ross, E. W., and C. T. Sims. Nickel-base alloys, in *Superalloys II*, pp. 97–134, Eds. C. T. Sims, N. S. Stoloff, and W. C. Hagel, Wiley, New York, 1987.

Sampath S., and H. Herman, Rapid solidification and microstructure development during plasma spray deposition, *J. Therm. Spray Technol.*, 1996, 5(4), 445–456.

Saunders, S. R. J., Correlation between laboratory corrosion rig testing and service experience, *Mater. Sci. Technol.*, 1986, 2, 282–289.

Saunders S. R. J., and J. R. Nicholls, Coatings and surface treatments for high temperature oxidation resistance, *Mater. Sci. Technol.*, 5(8), 1989, pp. 780–798.

Shankar, S., D. E. Koenig, and L. E. Dardi, Vacuum plasma sprayed metallic coatings, *JOM*, October 1981, pp. 13–20.

Shankar, S., and L. L. Seigle, Interdiffusion and intrinsic diffusion in the NiAl (δ) phase of the Al–Ni system, *Metall. Trans.*, 1978, 9A, 1468–1476.

Sinha, A. K. *Physical Metallurgy Handbook*, McGraw-Hill, New York, 2003.

Sivakumar, R., An evaluation study of aluminide and chromoaluminide coatings on IN-100, *Oxid. Metals*, 1982, pp. 17, 27–41.

Sivakumar, R., and B. L. Mordike, High temperature coatings for gas turbine blades, *Surf. Coat. Technol.*, 1989, pp. 37, 139–160.

Smialek, J. L., and J. V. Auping, COSP for Windows: Strategies for rapid analyses of cyclic oxidation behavior, *NASA Report No. NASA TP-2002-211108*, 2002.

Smialek, J. M., and C. E. Lowell, Effects of diffusion on aluminum depletion and degradation of NiAl coatings, *J. Electrochem. Soc.*, 1974, 121, 800–805.

Smidt, F. A., and B. D. Sartwell, Manufacturing technology program to develop a production implementation facility for processing bearings and tools, *Nucl. Instrum Methods Phys. Res.*, 1985, B6, 70–77.

Smith, J. S., and D. H. Boone, Platinum modified aluminides—present status, ASME No. 90-GT-319, 1990.

Smith, M. F., and R. C. Dykhuizen, Effect of chamber pressure on particle velocities in low pressure plasma spray deposition, *Surf. Coat. Technol.*, 1988, 34, 25–31.

Smith, R. W., Mechanical properties of a low pressure plasma applied CoCrAlY coating, *Thin Solid Films*, 1981, 84, 59–72.

Sodeoka, S., M. Suzuki, and K. Ueno, Effects of high-pressures plasma spraying for yttria-stabilized zirconia coatings, *J. Therm. Spray Technol.*, 1996, 5(3), 277–282.

Steffens, H-D., Spray and detonation gun technologies, Laser assisted technologies in *Coatings for High Temperature Applications*, pp. 121–138, Ed. E. Lang, Applied Science, 1983.

Steffens, H. D., H. M. Hohle, and E. Erturk, Low pressure plasma spraying of reactive materials, *Thin Solid Films*, 1980, 73, 19–29.

Strang, A., The effects of corrosion resistant coatings on the structure and properties of advanced gas turbine blading alloys, *CIMAC Congress*, Vienna, May 1979.

Strang, A., and E. Lang, Effects of coatings on the mechanical properties of superalloys, in *High Temperature Alloys for Gas Turbines 1982*, pp. 469–506, Eds. R. Brunetaud, D. Coutsouradis, T. B. Gibbons, Y. Lindblom, D. B. Meadowcroft, and R. Stickler, D. Reidel, Dordrecht, The Netherlands, 1982.

Strangman, T. E., Fatigue crack initiation and propagation in electron beam vapor-deposited coatings for gas turbine superalloys, *Thin Solid Films*, 1977, 45, 499–506.

Strangman, T. E., Turbine coating life prediction model, in *Proceedings of the 1990 Coatings for Advanced Heat Engines Workshop*, Department of Energy report CONF-9008151, pp. II–35 to II–43, Castine, ME, 1990.

Streiff, R., and D. H. Boone, Corrosion resistant modified aluminide coatings, in *Coatings and Bimetallics for Aggressive Environments*, Ed. R. D. Sisson, American Society of Metals, Metals Park, OH, 1984, pp. 159–169.

Streiff, R., and D. H. Boone, Corrosion resistant modified aluminide coatings, *J. Mater. Eng.*, 1988, 10(1), 15–26.

Sturgeon, A. J., M. D. F. Harvey, F. J. Blunt, and S. B. Dunkerton, in *Conference Proceedings: Thermal Spraying—Current Status and Future Trends*, pp. 101–106, Ed. A. Ohmori, High Temperature Society of Japan, Osaka, 1995.

Swanson, G. A., and R. C. Bill, Life prediction and constitutive models for engine hot section materials, *21st Joint Propulsion Conference*, Monterey, CA, July 1985, AIAA-SAE-ASME-ASEE, Paper AIAA-85-1421.

Swanson, G. A., I. Linask, D. M. Nissley, P. P. Norris, T. J. Meyer, and K. P. Walker, *1st Annual Status Report, NASA-CR-174952*, 1986, National Aeronautics and Space Administration, Washington, DC.

Takeda, Koichi, Michihisa Ito, and Sunao Takeuchi, Properties of coatings and application of low pressure plasma spray, *Pure Appl. Chem*, 1990, 62(8), 1773–1782.

Tawancy, H. M., N. M. Abbas, and T. N. Rhys-Jones, Role of platinum aluminide coatings, *Surf. Coat. Technol.*, 1991, 49, 1–7.

Tawancy, H. M., N. Sridhar, B. S. Tawabini, N. M. Abbas, and T. N. Rhys-Jones, Thermal stability of platinum aluminide coating on nickel-based superalloys, *J. Mater. Sci.*, 1992, 27(23), 6463–6474.

Tawancy, H. M., N. Sridhar, N. M. Abbas, and D. Rickerby, Comparative thermal stability characteristics and isothermal oxidation behavior of an aluminized and a Pt-aluminized Ni-base superalloy, *Scripta Met.*, 1995, 33(9), 1431–1438.

Taylor, A., and R. W. Floyd, The constitution of nickel rich alloys of the nickel–chromium–aluminum system, *J. Inst. Metals*, 1952–53, 81, 451–464.

Thornton, J. A., Sputter coating: Its principles and potential, *Trans. SAE*, 1973, 82, 1787–1805.

Todd, H. H., U. S. Patent No. 3,494,748, February 10, 1970.

Tryon, B., F. Cao, K. S. Murphy, C. G. Levi, and T. M. Pollock, Ruthenium-containing bond coats for thermal barrier coating systems, *JOM*, Jan 2006, pp. 53–59.

Vetter, J., O. Knotek, J. Brand, and W. Beele, MCrAlY coatings deposited by cathodic vacuum arc evaporation, *Surf. Coat. Technol.*, 1994, 68/69, 27–31.

Vogel, D., L. Newman, P. Deb, and D. H. Boone, Ductile-to-Brittle Transition Temperature Behavior of Platinum-modified Coatings, *Mater. Sci. Eng.*, 1987, 88, 227–231.

Voyer, J., T. Stoltenhoff, and H. Kreye, in *Thermal Spray 2003*, Vol. 1, pp. 71–78, Eds. C. Moreau and B. Marple, ASM International, Materials Park, OH.

Walston, W. S., J. C. Schaeffer, and W. H. Murphy, A new type of microstructural instability in superalloys—SRZ, *Superalloys 1996*, pp. 9–18, Eds. R. D. Kissinger, D. J. Deye, D. L. Anton, A. D. Cetel, M. V. Nathal, T. M. Pollock, and D. A. Woodford, The Minerals, Metals and Materials Society, Warrendale, PA, 1996.

Walters, K. C. and M. Nastasi, Ion implantation, *Kirk Othmer Encyclopedia of Chemical Technology*, Vol 14, 4th ed., Wiley, 1995.

Warnes, B. M., Reactive element modified chemical vapor deposition low activity platinum aluminide coatings, *Surf. Coat. Technol.*, 2001, 146–147, 7–12.

Warnes, Bruce M., and David C. Punola, Clean diffusion coatings by chemical vapor deposition, *Surf. Coat. Technol.*, 1997, 94–95, 1–6.

Wood, John H., and Edward H. Goldman, Protective coatings, in *Superalloy II*, pp. 359–384, Table 3, Eds. C. T. Sims, N. S. Stoloff, and W. C. Hagel, Wiley, New York, 1987.

Wood, M. I., The role of coatings in the fatigue of superalloy single crystals, *ASM Europe Conference*, Paris, France, September 1987, 179–188.

Wood, M. I., *RAE TR 88-021*, 1988a, Royal Aircraft Establishment, UK.

Wood, M. I., *RAE TR 88-057*, 1988b, Royal Aircraft Establishment, UK.

Wood, M. I., The mechanical properties of coatings and coated systems, *Mater. Sci. Eng.*, 1989, A121, 633–643.

Wood, M. I., and J. E. Restall, The mechanical properties of coated nickel based superalloy single crystals, in *High Temperature Alloys for Gas Turbines and Other Applications 1986*, Liege, Belgium, October 6–9, 1986, Eds. W. Betz, R. Brunetaud, D. Coutsouradis, H. Fischmeister, T. B. Gibbons, I. Kvernes, Y. Lindblom, J. B. Marriott, and D. B. Meadowcroft, Reidel, Dordrecht, The Netherlands.

Wright, I. G., B. A. Pint, L. M. Hall, and P. F. Tortorelli, *Oxidation Lifetimes: Experimental Results and Modeling*, Report of U.S. DOE contract DEA0596OR22725, 1988.

Young, S. G., and D. L. Deadmore, An experimental low cost silicon/aluminide high temperature coating for superalloys, *Thin Solid Films*, 1980, 73, 373–378.

Yu, Z., D. D. Haas, and H. N. G. Wadley, NiAl bond coats made by a directed vapor deposition approach, *Mater. Sci. Eng.*, 2005, A394, 43–52.

Zajchowski, P., and H. B. Crapo, III, Evaluation of three dual-wire electric arc-sprayed coatings: Industrial note, *J. Therm. Spray Technol.*, 1996, 5(4), 457–462.

Zhang, Y., J. A. Haynes, B. A. Pint, I. G. Wright, and W. Y. Lee, Martensitic transformation in CVD NiAl and (Ni,Pt)Al bond coatings, *Surf. Coat. Technol.*, 2003, 163–164, 19–24.

Zhao, J. C., M. R. Jackson, and R. Darolia, U.S. Patent No. 6,475,642, 2002.

Chapter 7

THERMAL BARRIER COATINGS (TBCs)

Metallic coatings for protection of structural alloys against oxidation and corrosion were covered in Chapter 6. There is another group of coatings called thermal barrier coatings (TBCs), whose function is to reduce component temperatures and thereby increase life. TBCs are generally a combination of multiple layers of coatings (Fig. 7.1), with each layer having a specific function and requirement. The topmost layer provides thermal insulation and consists of a ceramic, with low thermal conductivity, typically ZrO_2, known as zirconia. As will be discussed later, zirconia has to be stabilized against polymorphic phase transformation. The ceramic insulating layer is deposited on the substrate alloy with an intervening oxidation resistant metallic layer called the "bond coat." The metallic coating is either a diffusion aluminide, such as platinum aluminide, or an overlay coating of general composition NiCoCrAlY, conforming somewhat to the substrate alloy composition. During the ceramic coating deposition, a thermally grown oxide (TGO), predominantly Al_2O_3, forms on the bond coat surface at the ceramic–bond coat interface. In effect, the TGO binds the ceramic layer to the bond coat. Thus, TBCs are systems consisting of a ceramic coating, the TGO, and a metallic bond coat on the substrate alloy. The approximate thickness range of each layer is as follows: ceramic 125 to 1000 μm, bond coat 50 to 125 μm, and TGO 0.5 to about 10 μm, the last depending on the duration of exposure of the TBC in the high-temperature environment of the coating process and in operation.

7.1 TEMPERATURE REDUCTION BY TBCs

In order for TBCs to be effective in reducing surface temperature, the coated component needs to be actively cooled to remove heat conducted through the TBC. Turbine blades, vanes, seals, and combustor panels in gas turbine engines are good examples of systems effectively cooled by TBCs. The mechanism by which TBCs reduce component surface temperature is explained in Fig. 7.2, which is applicable to all the components just listed. Here, a turbine blade is given as an example because it is one of the most demanding of applications. The blade, coated with TBC, is actively cooled by cold air from the compressor of gas turbine engines. The cold air is routed through the internal cavity of the blade to remove the heat being conducted through the coatings and the blade walls. The cooling air needs exit passages, which may be provided by fine cooling holes drilled through the wall. The diameters of these holes are of the order of a few hundred microns, the length being proportional to the thickness of the blade wall. The holes are drilled either before the ceramic coating is deposited by the process of electrical discharge machining, or after the deposition of the ceramic by lasers. The left side of Fig. 7.2 depicts the concave outer wall of the blade, which sees higher pressure than the convex outer wall depicted on the right side. Further away from the left outer wall, the gas temperature is very high. As one approaches the coated external surface of the blade, the temperature is reduced somewhat by convection due to a film created by the relatively cooler air exiting the

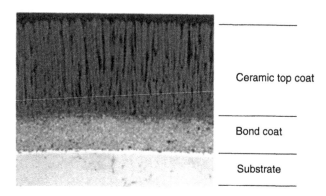

Figure 7.1 Thermal Barrier Coating consisting of metallic bond coat on the substrate and ceramic top coat on the bond coat.

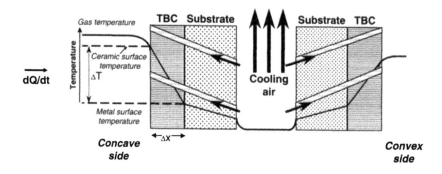

Figure 7.2 Mechanism by which TBCs reduce temperature of turbine blade.

holes drilled through the walls. Across the ceramic, the temperature drops significantly due to its lower thermal conductivity. Conductive heat flow in three dimensions is given by the vector equation

$$\partial(iQ_x + jQ_y + kQ_z)/\partial t = -KA(i\partial T/\partial x + j\partial T/\partial y + k\partial T/\partial z),$$

where K = thermal conductivity, (i, j, k) are the unit vectors in Cartesian x, y, and z directions, and $(\partial Q_x/\partial t)$, etc., are the respective heat fluxes passing across the cross section A. In one dimension, the heat transfer equation reduces to

$$dQ/dt = -KA \; dT/dx,$$

where dQ/dt is the heat flux and dT/dx is the local temperature gradient; the negative sign balances the negative slope of temperature or, in effect, indicates a temperature reduction. For thin coatings dT/dx can be approximated to $\Delta T/\Delta x$, where ΔT is the total temperature drop across the ceramic coating of thickness Δx. Rearranging the equation, the temperature reduction due to the ceramic coating is given by

$$\Delta T = (1/KA)\Delta x \, (dQ/dt).$$

Several important conclusions can be drawn from the foregoing equation:

- Because of the relatively high thermal conductivity, the metallic coatings have no significant impact in reducing surface temperature; that is, ΔT across the metallic coating is negligible.

- The temperature drop across the ceramic coating is inversely proportional to the thermal conductivity of the ceramic. Therefore, the lower the thermal conductivity of the ceramic layer, the cooler the metallic surface will be. This provides a strong incentive to identify and develop lower conductivity ceramic coatings.

- The temperature drop is proportional to the thickness of the coating. Thus, a thicker TBC is preferable. However, an unlimited increase of the thickness is not beneficial for three reasons. First, it adds parasitic weight to the part, which is detrimental to rotating components such as turbine blades. Second, thickness increases residual stress and total strain energy, making the ceramic coating prone to failure. Third, the more insulating the ceramic layer, the higher its surface temperature, leading to the possibility of sintering and associated structural and thermal problems.

- The temperature drop is proportional to the heat flux. In other words, components that experience more heat flux going through, effected by the cooling air, would have a larger drop in temperature for the same TBC thickness. This explains why a first-stage turbine blade may experience a larger temperature reduction than a second-stage blade for identical thickness of TBCs.

Magnitude of Temperature Reduction

The extent of the temperature reduction by TBCs has been assessed in gas turbine engines by conducting engine tests. Temperatures measured on turbine blades coated with TBC were compared with those measured on blades with metallic coatings only. The testing was done in Pratt & Whitney PW2000 gas turbine engine, which powers such aircraft as the Boeing 757. A map of the temperature reduction is shown in Fig. 7.3 (Meier and Gupta, 1994). On the concave side, at local hot spots, the surface temperature of the blade is reduced by as much as 139°C (250°F) by the use of 125-μm-thick zirconia-based TBC. The TBC was deposited by electron beam physical vapor deposition (EB-PVD).

This level of temperature reduction results in a several fold extension of the life of the component.

The Benefits of TBC

One of the dramatic uses of the systems benefit of TBC is provided by aircraft and land-based gas turbines. Examples of the former include Pratt & Whitney's JT9D, PW2000, V2500 (jointly with Rolls Royce and others), and PW4000 series engines, GE's CFM56 (with Snecma), CF6, and GE 90, as well as Rolls Royce's RB 211s. ABB's GT 24/GT29, GE's 7H, Siemens' V84/94.2, and MHI's 501G/701G are examples of the latter. The efficiency of these engines can be directly correlated with the temperature capability of turbine blades. From 1965 through 1985, the temperature capability gain had been approximately 80°C (144°F) (Fig. 7.4, DeMasi-Marcin and Gupta, 1994), accomplished by the development of several generations of nickel base single crystals of increasing creep and fatigue strength. Further improvements of similar magnitude through improvement of the alloys appear unlikely. However, since the introduction

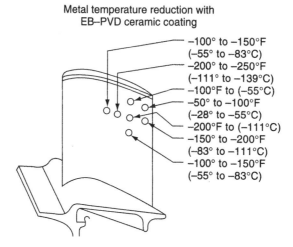

Metal temperature reduction with
EB–PVD ceramic coating

−100° to −150°F
(−55° to −83°C)
−200° to −250°F
(−111° to −139°C)
−100°F to (−55°C)
−50° to −100°F
(−28° to −55°C)
−200°F to (−111°C)
−150° to −200°F
(−83° to −111°C)
−100° to −150°F
(−55° to −83°C)

Figure 7.3 Local temperature reduction due to thermal barrier coatings applied by electron beam physical vapor deposition. The demonstration was done on first-stage turbine blades in Pratt & Whitney's PW2000 engine (S. M. Meier and D. K. Gupta, The evolution of thermal barrier coatings in gas turbine engine applications, *J. Eng. Gas Turbine Power Trans. ASME*, 1994, 116, 250–257). Reprinted with permission from American Society of Mechanical Engineers International.

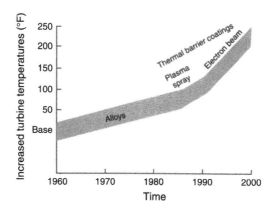

Figure 7.4 Schematic representation of temperature capability improvement over the years (J. T. DeMasi–Marcin and D. K. Gupta, *Surf. Coat. Technol.*, 1994, 68/69, 1–9). Reprinted with permission from Elsevier.

of TBC in the early 1980s on the turbine blades in Pratt & Whitney JT9D engines powering jumbo jets, the effective systems temperature capability has been improved by as much as 200°C (360°F). The direct benefits of the temperature reduction due to the TBC are reflected in reduction in oxidation. For rotating components, creep damage is reduced significantly. The substrate alloy retains a higher percentage of the room-temperature structural properties. Additionally, compressor air used to cool components may be rerouted to produce thrust, thus increasing efficiency of engines.

In addition to the direct benefits, there are a number of significant but indirect benefits. TBCs tend to reduce local variations in temperature, which results in reduced distortion and improved thermal fatigue life. Additionally, TBCs affect component response to transients in temperature. These transients originate from a sudden change in energy input to high-temperature components such as turbine blades during the takeoff, climb, and landing of an aircraft. The transients tend to adversely affect the thermal fatigue life of metallic components. Because of their lower thermal conductivity, the response of ceramic coatings to sudden temperature spikes is much slower than that of metals. As a result, TBCs smooth out the transients somewhat and improve thermal fatigue life.

7.2 MATERIALS REQUIREMENT FOR TBCs

The design of the TBC and the environment in which it operates impose restrictions on the materials of construction. Table 7.1 lists some of the property requirements of the ceramic coating of the TBC system. The requirements for the bond coat are very similar to those of the oxidation-resistant metallic coatings discussed in Chapter 6, which include the capability to form and maintain pure alumina scale with good scale adherence on exposure to an oxidizing environment at high temperature.

One of the most critical applications of TBC, as discussed in the earlier sections, is in gas turbine engines on turbine blades and vanes to reduce the temperature of the underlying metallic structure. Because the magnitude of temperature reduction, as shown earlier, is inversely proportional to the thermal conductivity, the ceramic coatings need to have low thermal conductivity. In gas turbines, the TBC surface may be exposed to hot gases with temperatures in excess of 3000°F (1649°C). In order to survive in such environments without melting, the ceramic top coat of the TBC should also have a high melting point. Additionally, it needs to have good oxidation resistance and a thermal expansion coefficient closer to that of metallic

Table 7.1 Materials Requirement for Ceramic Thermal Barrier Coating

Property	Requirement	Rationale
Melting point	High	Operating environment at high temperature
Thermal conductivity	Low	Temperature reduction inversely proportional to thermal conductivity
Coefficient of thermal expansion	High	Expansion should be close to that of superalloy substrate and bond coats on which coatings are deposited
Phase	Stable	Phase change in thermocyclic environment is structurally detrimental
Oxidation resistance	High	Operating environment highly oxidizing
Corrosion resistance	Moderate to high	Operating environment may be corrosive
Strain tolerance	High	Operating environment imposes large strain ranges

coatings. The last requirement reduces the thermal stresses arising due to a mismatch of the thermal expansion coefficient between the metallic coated substrate and the ceramic coating.

7.3 PARTIALLY STABILIZED ZIRCONIA

Not too many available materials satisfy all the requirements of TBC, one exception being zirconia. However, as shown by the phase diagram of Fig. 7.5, pure zirconia undergoes a polymorphic phase transformation (Scott, 1975) during heating and cooling.

The following characteristics of pure zirconia are evident from the phase diagram:

Melting point	2690°C (4874°F)
Cubic (C) to tetragonal (T) phase change	2370°C (4298°F)
Tetragonal (T) to monoclinic (M) phase change	1170°C (2138°F)

The tetragonal-to-monoclinic phase transformation is martensitic in nature because it does not require diffusion and occurs by small displacements of atoms in the structure. The transformation occurs on cooling zirconia from a high temperature and involves a 3–5% volume

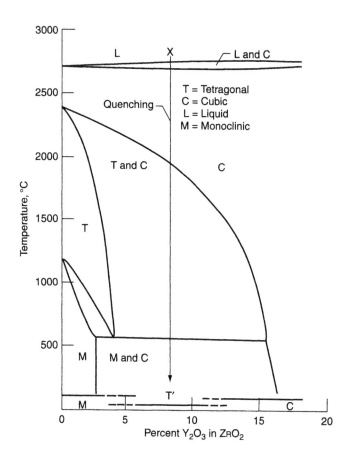

Figure 7.5 Phase diagram of ZrO_2–Y_2O_3 system (H. C. Scott, Phase relationships in zirconia–yttria system, *J. Mater. Sci*, 1975, 10, 1527–1535). Reprinted with permission from Springer.

increase. The volume change induces a significant shear strain, as high as 10%, in the structure, affecting the integrity of the coating. Alloying zirconia with other oxides such as CaO, MgO, Y_2O_3, CeO_2, Sc_2O_3, and In_2O_3 inhibits the phase transformation, stabilizes the high-temperature phase, and eliminates the volume change. Of these stabilizers, the most widely used is yttria, because of the long term stability of the resulting oxide alloy. Yttria and zirconia also have very similar vapor pressure, which is an advantage in processing by vapor deposition, to be covered in a later section. The optimum yttria stabilizer content is approximately 7 wt %. This has been determined by high-temperature durability rig testing at NASA (Fig. 7.6) of TBC coating as a function of yttria level (Stecura, 1985). Significantly lower yttria contents do not inhibit the transformation to monoclinic phase, while higher levels stabilize a cubic phase which lacks adequate strength and toughness. As a result, 7 wt % yttria-stabilized zirconia (written as 7YSZ) has become the industry standard.

In addition to the physical property constraints for ceramic material to function as TBC, there are microstructural requirements. Dense 7YSZ, for example, does not provide durable TBC, primarily because of its high modulus and poor tolerance to strain. Laminar and vertical microcracks with some levels of porosity improve strain tolerance and durability. Some of the thermodynamic parameters of zirconia are shown in Table 7.2.

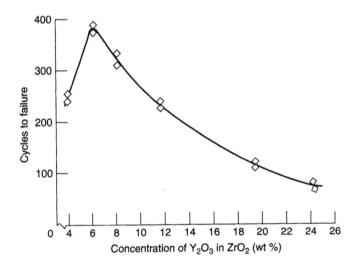

Figure 7.6 TBC composition selection is based on durability (S. Stecura, Optimization of NiCrAl–Y/ZrO$_2$–Y$_2$O$_3$ thermal barrier system, *NASA Tech. Memo. 86905*, 1985, NASA, Cleveland, OH). Used with permission.

Table 7.2   Thermal and Thermodynamic Parameters of Zirconia

Thermodynamic Parameter	Magnitude
Boiling point	4300°C (7772°F)
Enthalpy of tetragonal to monoclinic phase change	4.83×10^4 J/kg
Enthalpy of melting	7.08×10^5 J/kg
Enthalpy of evaporation	5.23×10^6 J/kg

Zirconia cannot be melted or vaporized by conventional heating processes because of its very high melting point, 2690°C (4874°F). In order to deposit coatings, some less conventional processes are, therefore, resorted to. These include air plasma spraying (APS) and electron beam physical vapor deposition (EB-PVD). Both of these processes deposit coatings along line of sight. Although the structures formed by these processes provide various levels of strain tolerance, the deposits, particularly by APS, seldom achieve thermal equilibrium during the deposition process. The phase diagram (Fig. 7.5) accommodates the formation of such nonequilibrium structures by introducing three discontinuous horizontal lines designated by M, T', and C at the bottom of the plot. For YSZ with less than about 4 wt % yttria, cooling from high temperature results in phase transformation from tetragonal T to monoclinic M, which is deleterious for the integrity of the TBC. With yttria content ranging between 4 and 10%, the phase formed is T', called nontransformable tetragonal, which remains stable on cooling without the formation of the monoclinic phase (Miller et al., 1981). For an yttrium level greater than 10%, the low-temperature phase is cubic C. This phase generally lacks the strength and durability characteristic of the T' phase. The ceramic composition of choice is therefore 7YSZ. The various processes by which the ceramic coating is deposited are discussed next.

7.4 PLASMA SPRAYED TBCs

Coatings based on zirconia can only be deposited by processes capable of adding enough energy to the raw materials to melt, evaporate, or chemically fragment to dimensions that can be deposited with adequate cohesive and adhesive strength. One of the processes that can deliver high energy is based on the phenomenon of plasma.

The Plasma Spray Process

In this process (Herman, 1988), a plasma jet melts the coating raw material in the form of powder. The plasma is created in a plasma gun. As explained in Section 6.5, because this process is conducted in air, it is called the air plasma spray (APS) process. A number of APS designs are commercially available.

A fair amount of the detail of the plasma spray process has already been given in the previous section on metallic coatings. At the heart of the process is the plasma gun described in Fig. 6.26. Additional details (Chagnon and Fauchais, 1984; Fauchais et al., 1998) are given next. While in some respect, the process for ceramic coatings is similar to that of the metallic coatings, higher melting points of ceramic materials require some modifications.

Using a carrier gas, the coating material, in the form of powders, is fed into the hottest zone of the plasma (Fig. 6.27), which is the recombination zone for the electron and the ions. The powder trajectory is shown in Fig. 6.29.

The velocity and point of injection of particles of the coating materials are critical parameters because the plasma behaves as a viscous fluid. Low injection velocity limits particles to the relatively colder periphery of the jet, resulting in insufficient melting. Higher velocity of injection forces particles to move past the hot zone into the colder periphery on the other side. The velocity, therefore, has to be adjusted carefully so that the particles are injected in the central hot zone.

The particle size of the powder is typically centered around 40 μm. Larger particles tend not to melt completely. Particles finer than 10 μm, on the other hand, do not penetrate the plasma and tend to get trapped in the colder periphery, and do not contribute to acceptable microstructure. The morphology of the powders is also important because it determines the ability to flow and melt. A few of the commercially available powders include:

Sintered and crushed (UCAR ZrO-137)

Hollow spherical (Metco HOSP 204B NS)

Spray dried and sintered (UCAR ZrO − 113)

The bonding between the air plasma sprayed coating and the substrate is considered to be predominantly mechanical. Interlocking of the depositing particles with the substrate surface roughness features provides the adhesion mechanism. Improved bonding, therefore, is achieved by having a rough bond coat surface prior to coating.

The bond coated substrate is generally preheated during the coating process. The preheat temperature plays a major role in determining the durability of the coating. It affects the residual stress generated in the coating as the molten droplets continually impact the substrate and freeze. The temperature also influences the bond strength of the ceramic coating with the bond coat. The typical preheat temperature lies between 127 and 227°C (260 and 440°F) (Fauchais et al., 1998).

The strength and density of the coating are a strong function of the angle of spray. This angle, defined as between the centerline of the spray stream and the substrate surface, is usually kept between 75 and 90 degrees to achieve good properties. The dimension of the plasma gun and the distance between the nozzle exit face and the substrate are also important parameters affecting coating properties.

In a production environment, the plasma spray process is conducted inside an enclosed booth. A schematic of a typical installation is shown in the sketch of Fig. 7.7. The major contents of the spray booth installation are the plasma spray gun with the gun manipulator, associated power supply capable of delivering currents up to 1000 A at a voltage of about 100 V, control panel, gas supply, powder feeder, and the holder for the component to be coated.

A variant of the plasma spray process is the solution precursor plasma spray (SPPS) process (Gell et al., 2004a; Bhatia et al., 2002), in which the raw material injected into the plasma is not in the form of powder but is a liquid, preferably an aqueous precursor, which undergoes physical and chemical changes of pyrolysis and sintering during flight through the plasma jet prior to deposition as 7YSZ coating. The microstructure of TBC deposited through the SPPS process is characterized by transverse microcracks and a splat structure of a finer scale compared with conventional plasma sprayed TBC. These features are discussed in detail in a later section. The microstructural features improve durability of SPPS-derived TBC and provide tolerance to increased thickness compared with the traditional APS-deposited TBC.

Microstructure of Plasma-Sprayed TBC

A typical microstructure of air plasma sprayed 7YSZ coating deposited on a NiCoCrAlY bond coat is shown in Fig. 7.8.

The characteristic features of the coating microstructure are summarized as follows:

- The ceramic layer contains 10 to 15 vol % porosity. Typically, finer powder particle size and closer spray distance result in lower porosity.

- The interface between the ceramic and the bond coat is rough.

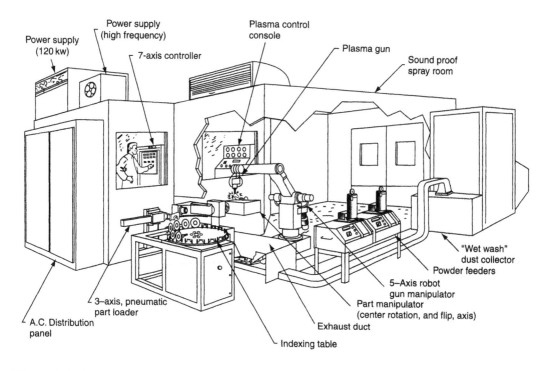

Figure 7.7 Schematic of a plasma spray manufacturing facility (P. Meyer and S. Muehlberger, Historical review and update to the state of the art of automation for plasma coating processes, *Thin Solid Films*, 1984, 118, 445–456). Reprinted with permission from Elsevier.

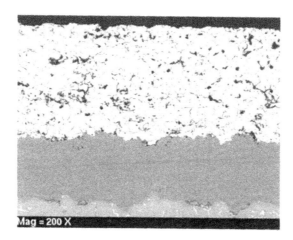

Figure 7.8 The microstructure of 7YSZ by APS process on a NiCoCrAlY bond coat deposited by LPPS.

- Minor amounts of monoclinic phase may be present in the ceramic in addition to major tetragonal phase.

- A higher magnification image shows splats (molten particles deformed on impact into a pancake shape) enclosing transverse microcracks, porosity, and occasional unmelted particles.

- A finer columnar structure forms within the individual splats.

- A thermally grown oxide (TGO) forms at the interface between the ceramic and the bond coat during the ceramic deposition. The TGO consists of aluminum oxide, at least initially. As the TBC continues exposure in oxidizing thermal environment during use, the TGO thickness increases. The growth generally occurs at the TGO–bond coat interface. With the continued loss of aluminum from the bond coat, the activity of aluminum falls below the level necessary to form alumina while the oxygen activity increases. The composition of the TGO below the alumina scale, therefore, tends to change to spinels, $(Ni,Co)Al_2O_4$, and for some low aluminum bond coats, it eventually changes to NiO. For high chromium overlay bond coats, the TGO changes to spinels of composition $(Ni,Co)(Al,Cr)_2O_4$ and possibly Cr_2O_3 (Shillington and Clarke, 1999). The scenario described earlier is valid for intact TBCs. However, if they fall off prematurely, as often happens in use, the oxidation process changes to that of a metallic coating exposed to oxidizing environment.

Microstructure Development and Structure–Property Relationship

Microstructure Formation

In order to understand the formation of microstructure of plasma sprayed TBCs, we need to follow the sequence of events leading to a molten droplet depositing on the substrate and freezing (Sampath and Herman, 1996) to a solid.

Zirconia particles injected into the plasma melt (larger particles may melt incompletely or may not melt at all) while transiting through the plasma plume. The molten particles impact the substrate surface, which is generally roughened to improve bonding. The depositing particles are flattened by the force of impact and solidify to form "splats" (Fig. 7.9) (Safai and Herman, 1977). The degree of flattening, or the flattening parameter, is given by the ratio between the diameter of the flattened pancake-shaped disk formed and its original diameter.

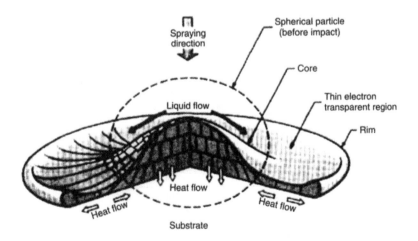

Figure 7.9 Schematic of deformation and heat transfer of a plasma sprayed droplet (Saed Safai and Herbert Herman, Microstructural investigation of plasma-sprayed aluminum coatings, *Thin Solid Films*, 45, 1977, pp. 295–307). Reprinted with permission from Elsevier.

This parameter depends on the velocity of the particles and of the plasma, the density of the coating material, and the kinetic viscosity of the melt (Madejski, 1976). The flattening parameter for a typical depositing zirconia particle is about 7. Because of rapid heat transfer (Fig. 7.9), the molten particles freeze at an extremely high cooling rate $\sim 10^6$ K/s ($1.8 \times 10^{6\circ}$F/s). The roughness of the surface provides irregularities, which the initial droplets anchor onto.

The deposited particles exhibit high residual stresses. There are two sources of the stress: the first arising from contraction of the particles as they are quenched from the molten state to solid splats at room temperature, and the second from the differential thermal contraction of zirconia relative to the underlying metallic substrate. In order to relieve these high stresses, the brittle zirconia splats, which exhibit low creep ductility below 1400°C (2552°F) (Firestone et al., 1982), form cracks. The cracks typically run perpendicular to the inter-splat boundaries. Calculated equilibrium crack distances lie between 25 and 50 μm, depending on the deposition temperature, see Fig. 7.10. Thus, two to four microcracks would be present every 100 μm lateral distance. These vertical cracks are independent of the positions of the underlying splats.

Molten particles continue depositing on already piled-up solidified splats. The adhesion among the splats is significantly different from that with the metallic substrate. Horizontal delamination cracks form at the splat boundaries because of relatively weak adhesion. The splats adhere to each other better if they are sprayed warm. The horizontal delaminations are, therefore, a function of deposition temperature also, as shown by the dependence of crack density on the substrate temperature (Fig. 7.11).

The propensity of the horizontal delamination cracks has significant effect on the TBC thermal conductivity and thermal shock resistance (Fig. 7.12) (Wigren et al., 1996).

The structure within each splat of deposited zirconia is crystalline with well-defined columnar grains (Fig. 7.13) (Bengtsson and Johannesson, 1995). The heights of the columns typically equal the thickness of the splats. The direction of heat transfer as the splats freeze determines the orientation of the columns within the splats.

Cross sections of the columnar grains are shown in Fig. 7.14 (DeMasi-Marcin et al., 1989). The grain sizes are of the order of 0.25 μm. Figure 7.14 also shows an intergranular crack formed during thermal exposure of the coating.

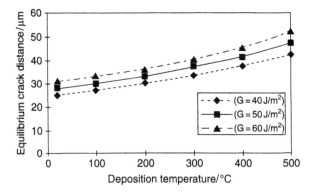

Figure 7.10 Theoretical calculation of distance between vertical cracks of a splat (J. F. deVries quoted in J. Wigren and L. Pejryd, Thermal barrier coatings—why, how, where and where to, in *Proc. 15th. International Thermal Spray Conf.*, 1998, ASM International, Ed. C. Coddet, pp. 1531–1542). Reprinted with permission from ASM International.

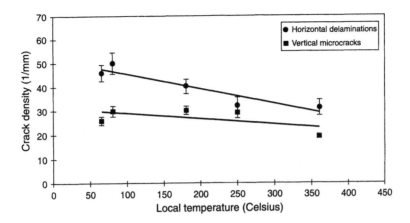

Figure 7.11 Vertical and horizontal crack density as a function of deposition temperature (J. F. deVries quoted in J. Wigren and L. Pejryd, Thermal barrier coatings—why, how, where and where to, *Proc. 15th. International Thermal Spray Conf.*, 1998, ASM International, Ed. C. Coddet, pp. 1531–1542). Reprinted with permission from ASM International.

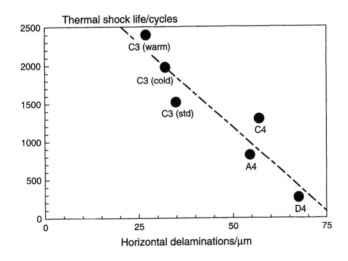

Figure 7.12 Thermal shock resistance as a function of horizontal delamination (J. Wigren, D. Grevin, and J.-F. DeVries, Effects of powder morphology, microstructure and residual stresses on the thermal shock life of thin thermal barrier coatings, *Proc. NTSC'96 in Cincinnati*, pp. 855–861). Reprinted with permission from ASM International.

Segmented TBCs

Through-thickness cracks running from the coating surface toward the bond coat are called segmentation cracks. Such cracks, studied first by Ruckle (1980), relieve stresses within the coatings. The propensity of these cracks together with transverse microcracks (Fig. 7.15) (Schwingel et al., 1998) can be controlled by variation of plasma spray parameters (Strangman, 1992) as well as by controlling the stabilizer content. For example, Ruckle introduced segmentation cracks by using 6% Y_2O_3 to stabilize zirconia. The origin of the cracks was attributed to strains associated with phase transformation–related volume change. The segmentation cracks

Figure 7.13 Columnar grains within the splats (P. Bengtsson and T. Johannesson, Characterization of microstructural defects in plasma sprayed thermal barrier coatings, *J. Therm. Spray. Technol*, 1995, 4(3), 245–251). Reprinted with permission from ASM International.

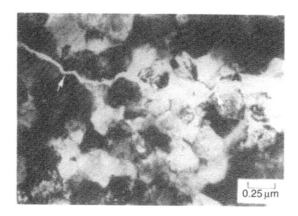

Figure 7.14 TEM micrograph of grain cross sections parallel to plane of coating (Jeanine T. DeMasi-Marcin, Keith D. Sheffler, and Sudhangshu Bose, Mechanisms of degradation and failure in plasma deposited thermal barrier coating, *ASME 89-GT-132*, 1989). Reprinted with permission from American Society of Mechanical Engineers International.

reduce effective modulus of elasticity and improve the thermocyclic durability of the TBC without affecting thermal conductivity. The laminar microcracks, on the other hand, reduce thermal conductivity because they are positioned transverse to the heat flow. The segmented microstructure is particularly attractive for thicker coatings because of higher residual stress and frequency of coating spallation in the absence of such cracks.

Phase Identification in the Ceramic Coating

Because the process of formation involves extremely fast quenching of molten particles, air plasma sprayed 7YSZ exhibits a nonequilibrium microstructure. The phases present in the

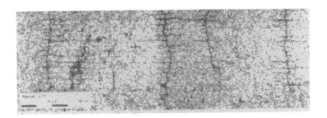

Figure 7.15 Plasma-sprayed segmented TBCs (D. Schwingel, R. Taylor, T. Haubold, J. Wigren, and C. Gualco, Mechanical and thermophysical properties of thick PYSZ thermal barrier coatings: Correlation with microstructure and spraying parameters. *Surf. Coat. Technol.*, 1998, 108–109, 99–106). Reprinted with permission from Elsevier.

coating, therefore, do not conform to the equilibrium phase diagram. The three horizontal lines at the bottom of the phase diagram (Fig. 7.5), identified as M, T′, and C, show the range of composition for the formation of monoclinic, nontransformable tetragonal, and cubic phases, respectively, for plasma-sprayed material. The actual phases and their content in the ceramic layer of TBC depend strongly on the process parameters, the characteristics of the spray powder used, and the thermal exposure history. Phase content is generally determined by the use of x-ray diffraction (XRD). However, the identification process is somewhat complex and requires resolution of specific diffraction lines into their individual components because the lines are composites of overlapping diffraction peaks from more than one phase. Figure 7.16a (Scardi et al., 1994) shows the XRD profile of plasma sprayed 7YSZ. A shoulder on the (111) line (Fig. 7.16b) (Scardi et al., 1995) usually identifies the presence of monoclinic phase.

Role of Substrate Surface Roughness

For TBCs based on air plasma sprayed zirconia, the bond between the ceramic and the metallic coating is presumed to be "mechanical." The bond is established early in the deposition process

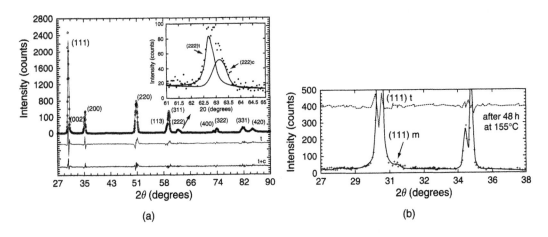

(a) (b)

Figure 7.16 XRD pattern of air plasma sprayed 7YSZ (a) (P. Scardi, E. Galvanetto, A. Tomasi, and L. Bertamini, Thermal stability of stabilized zirconia thermal barrier coatings prepared by atmosphere and temperature–controlled spraying, *Surf. Coat. Technol.*, 1994, 68–69, 106–112), (b) (P. Scardi, M. Leoni, and L. Bertamini, *Surf. Coat. Technol.*, 1995, 76–77, 106–112). Reprinted with permission from Elsevier.

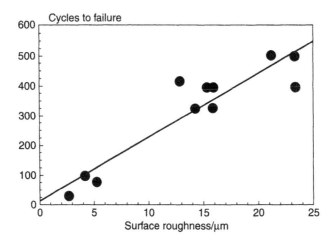

Figure 7.17 Effect of bond coat surface roughness on TBC life (J. Wigren and L. Pejryd, Thermal barrier coatings—why, how, where and where to, *Proc. 15th. International Thermal Spray Conf.,* 1998, ASM International, Ed. C. Coddet, pp. 1531–1542). Reprinted with permission from ASM International.

by the interlocking of the initial splats with the topological features of the bond coat surface. Many of the surface features are in the form of peninsular intrusion of the bond coat into the ceramic. The surface roughness of the bond coat, therefore, plays a critical role in controlling the durability of the TBC system. This is clearly demonstrated in the cyclic durability plot of Fig. 7.17 (Wigren and Pejryd, 1998).

Thick TBC

For many applications, thick (~1000 μm) TBCs are required. For air plasma sprayed TBCs, experimental data generated in cyclic burner rig testing show that the failure life decreases with increased TBC thickness (Fig. 7.18) (Bose and DeMasi-Marcin, 1997), as long as the

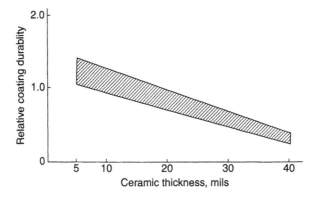

Figure 7.18 Reduction in TBC life with increase in ceramic layer thickness. (S. Bose and J. DeMasi-Marcin, Thermal barrier coatings experience in gas turbine engines at Pratt & Whitney, *J. Therm. Spray Technol.,* 1997, 6(1), 99–104). Reprinted with permission from ASM International.

TBCs of various thickness are tested at the same temperature. Failure modes also change from laminar compressive failure near the interface for thin coatings (<250 μm) to "mud flat" tensile cracking and loss of discrete segments for thicker coatings. In actual use, increased thickness results in higher ceramic surface temperature during operation, which induces sintering effects resulting in the mud flat cracks to relieve associated tensile stresses. It should be kept in mind, however, that in the field, thicker TBCs reduce substrate temperature by a larger amount and, therefore, prolong substrate life. Thus, with increased thickness there is a trade-off between higher TBC distress and lower substrate distress.

Thermal Properties and Consequences

As discussed in Section 7.1, the heat transfer properties of TBC can be illustrated by the simple one-dimensional equation

$$(dQ/dt) = -KA(dT/dx),$$

where dQ/dt is the heat flux across area A, dT/dx is the temperature gradient, and K is the thermal conductivity. The critical materials property controlling heat transfer is the thermal conductivity. Experimental techniques, however, seldom directly measure conductivity with the exceptions of the steady-state laser technique and the guarded hot plate technique. The parameter typically measured experimentally is thermal diffusivity, given by $\alpha = K/\rho C_p$, wherein ρ is the density and C_p is the heat capacity at constant pressure.

There are two widely used methods available to determine thermal conductivity. The traditional method utilizes steady state to measure heat input and the resulting temperature rise. For samples of small volume and low conductivity, a more attractive method is the laser flash technique, which measures thermal diffusivity (Taylor, 1979). In this method, a laser pulse (for example, a ruby laser with pulse duration 8×10^{-4} s) is shined on the surface of a sample of TBC coating in the form of a circular disk. The temperature increase of the opposite surface is measured with a thermocouple as a function of time. The thermal diffusivity is evaluated from the temperature profile using the equation $\alpha = 0.1388 L^2/t_{1/2}$, where L is the thickness of the disk and $t_{1/2}$ is the time for the back surface to reach half the maximum temperature it would eventually experience. Thermal conductivity is derived from the diffusivity by substituting values for density and heat capacity. When the diffusivity of plasma-sprayed YSZ is measured as a function of temperature both during heating up as well as cooling down, presence of a hysteresis (Fig. 7.19) (Schwingel et al., 1998) is observed, that is, the profile on heating does not retrace on cooling. This behavior is due to irreversible physical changes, such as sintering, that occurs in the ceramic on heating above some critical temperature, which in this case appears to be >780°C (1436°F).

The typical value of K for plasma sprayed 7YSZ without any sintering lies between 0.77 and 1.1 W/mK depending on porosity (Fig. 7.20) (Wigren and Pejryd, 1998) and details of microstructure. The horizontal crack density, which is related to deposition temperature, has a strong effect on the thermal diffusivity (Fig. 7.21).

Thermal Conductivity

Thermal conductivity of dense zirconia depends on the type of stabilizer used and the crystalline phase content. Attempts to derive thermal conductivity values from theoretical considerations

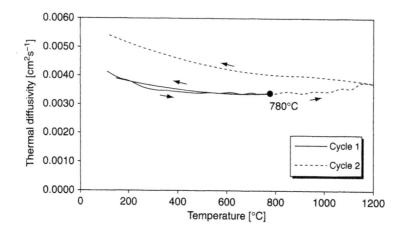

Figure 7.19 Thermal diffusivity as a function of measurement temperature. (Cycle 1) Lack of hysteresis when cycled between 100 and 780°C (212 and 1436°F). (Cycle 2) Clear hysteresis when cycled between 100 and 1200C (212 and 2190°F). (D. Schwingel, R. Taylor, T. Haubold, J. Wigren, and C. Gualco, Mechanical and thermophysical properties of thick PYSZ thermal barrier coatings: correlation with microstructure and spraying parameters. *Surf. Coat. Technol.*, 1998, 108–109, 99–106). Reprinted with permission from Elsevier.

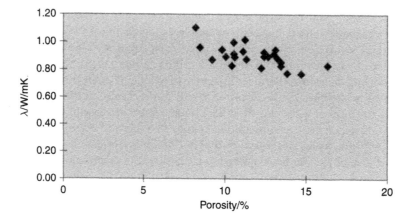

Figure 7.20 Effect of porosity on thermal diffusivity (J. Wigren and L. Pejryd, Thermal Barrier coatings—why, how, where and where to, *Proc. 15th. International Thermal Spray Conf.*, 1998, ASM International, Ed. C. Coddet, pp. 1531–1542). Reprinted with permission from ASM International.

have generally been less than successful. Thermal conductivity of solids can be expressed in the general form

$$K = 1/3 \; Cvl,$$

where C is heat capacity per unit volume, v is the velocity of the heat carriers, which for nonmetals are atomic vibrations called phonons, and l is the scattering mean free path. Because

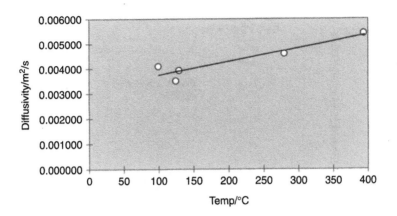

Figure 7.21 Effect of deposition temperature (indirectly horizontal cracking) on thermal diffusivity. (J. Wigren and L. Pejryd, Thermal barrier coatings—why, how, where and where to, *Proc. 15th. International Thermal Spray Conf.*, 1998, ASM International, Ed. C. Coddet, pp. 1531–1542). Reprinted with permission from ASM International.

the heat carriers are phonons with a range of frequency f from 0 to f_m, the equation should be written as (Klemens and Gell, 1998)

$$K = 1/3 \int C(f)\, v\, l(f)\, df, \text{ the limits of integration being from 0 to } f_m,$$

where f_m is the upper limit of the frequency. The carrier velocity has weak temperature dependence and, therefore, can be treated as constant. Debye theory of lattice waves restricts the upper limit of the frequency spectrum. The upper limit f_m is related to the Debye frequency f_D through $f_D = f_m N^{1/3}$, where N is the number of atoms per primitive unit cell ($N = 3$ for ZrO_2) of the crystal. Thermal conductivity at high temperatures is easy to understand with the concept of Debye temperature θ defined by $h f_D = k_B \theta$ where h and k_B are Planck's and Boltzmann's constants, respectively ($\theta \sim 380\,\text{K}$, 225°F, for ZrO_2). For temperature $T > \theta$ $C(f) \sim f^2$. Also for scattering of a phonon by other phonons $l(f) \sim (\mu a^3 v f_m) f^{-2} T^{-1}$, where μ is the shear modulus and a^3 is the atomic volume. These simplifications provide

$$K = (3/4) N^{-2/3} \mu v^2 / (f_D T).$$

This relationship describes the thermal conductivity of defect-free dielectric crystals at temperatures near and above the Debye temperature. Inserting the values $N = 3$, $\mu = 5.31 \times 10^4\,\text{J/cm}^3$, $v = (\mu/\rho)^{1/2} = 2.3 \times 10^5\,\text{cm/s}$, ρ is the density, $f_D = 5.2 \times 10^{13}\,\text{/s}$, we get $K \sim 20/T$, temperature T being in kelvins. This expression gives too small a value for the thermal conductivity. It is, therefore, surmised that there are additional heat transport mechanisms operating in ZrO_2. Also, at lower temperatures, defects such as vacancies, substitutional atoms, and grain boundaries contribute to scattering. Measured values of thermal conductivity from zirconia partially stabilized with 7% yttria (7YSZ) are shown in Fig. 7.22 as a function of measurement temperature (Meier et al., 1992).

The difference in the thermal conductivity between APS and EB-PVD TBC (discussed in Section 7.5) is significant and has a strong effect on their cooling efficiency when used to reduce component temperature. The difference arises from the microstructural defects, particularly the microcracks in APS TBC, which are transverse to the heat flow, and column boundaries in

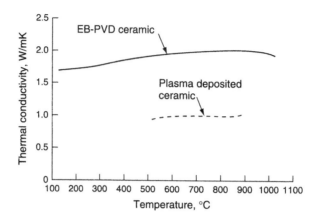

Figure 7.22 Thermal conductivity of seven percent yttria partially stabilized zirconia as a function of measurement temperature (S. M. Meier, D. M. Nissley, K. D. Sheffler, and T. A. Cruise, Thermal barrier coating life prediction model development, *J . Eng. Gas Turbine Power Trans. ASME*, 1992, 114, 258–263). Reprinted with permission from American Society of Mechanical Engineers International.

EB-PVD, which are parallel to the heat flow. Varying amounts and shapes of porosity in each case also make an important contribution to the thermal conductivity reduction. In addition to the phonon or lattice wave contribution to the conductivity (Klemens and Gell, 1998), there is a radiative or photon component. At high temperatures, the photon conductivity is given by $K = 4\beta T^3 n^3 L$, where β is the Stefan Boltzmann constant, n is the index of refraction, and L is the ceramic layer thickness. For a 0.25-μm-thick layer of zirconia, and $n = 2.7$, K comes out to be ~ 0.9 W/mK at 1000°C (1832°F). This is comparable to the phonon conductivity. One important point to note is that the functional dependence on temperature is different for phonon and photon conductivity.

Residual Stresses

The TBC consisting of the ceramic coating, the TGO, and the bond coat exhibits varying levels of residual stress in each layer. There are two primary sources of the stress within the ceramic coating, assuming no sintering. The first source is the deposition process itself (rapid quenching, solidification, thermal contraction, constraint due to underlying material, stress relief due to microcracking, and creep relaxation of the bond coat). The deposition stress, estimated at 30–40 MPa (4.3–5.8 ksi), is low and roughly independent of temperature (Elsing et al., 1990). The second source of residual stress is the mismatch between the coefficient of thermal expansion (CTE) of the ceramic and the bond coat. The CTE mismatch stress is generated on cooling from the deposition temperature. The TGO layer exhibits residual stress originating from conversion of high-density metal to low-density oxide in a constrained volume and the CTE mismatch between the TGO and the bond coat. A possible additional source of residual stress in the TGO is due to phase transformation and associated volume change that occurs when metastable alumina such as θ-Al_2O_3 changes to more stable α-Al_2O_3 (Lipkin et al., 1997). The stresses in the ceramic can be measured by standard x-ray diffraction methods. However, the measurement is limited only to close to the surface because of low penetration depths for conventional x-rays. More energetic and tunable x-rays of higher intensity from synchrotron light sources are used to probe deeper regions. For thin coatings with flat surfaces, elastic analysis indicates the biaxial CTE mismatch stress in the plane of the coating to be (Tien and Davidson, 1975):

$$\sigma = -[E_c(\alpha_m - \alpha_c)\Delta T]/[\{(1 - v_c) + 2\{h_c E_c(1 - v_c)/h_m E_m\}],$$

where subscripts m and c indicate metal and ceramic coating, respectively, E is the modulus of elasticity, v is the Poisson's ratio, h is the thickness, and ΔT is the difference between the stress free (usually the deposition) and measurement temperature. For thin ceramic coatings, $h_c << h_m$, the stress relationship simplifies to

$$\sigma = -E_c(\alpha_m - \alpha_c)\Delta T/(1 - v_c).$$

The stress-free temperature for APS TBC can be as high as 400°C (752°F) (Sevcik and Stoner, 1978). The associated in-plane strain is

$$\varepsilon = -(\alpha_m - \alpha_c)\Delta T.$$

The elastic stored energy in a volume is given by (Thornton et al., 1999),

$$J = \tfrac{1}{2}\sigma\varepsilon V \approx E_c t_c (\alpha_m - \alpha_c)^2 \Delta T^2/(1 - v_c),$$

for a unit area of cross section of the ceramic layer of thickness t_c, and volume V. The ceramic layer of the TBC will delaminate only if J exceeds the energy required to create the surfaces formed by the delamination process. This is a necessary, but not sufficient condition, as will be discussed in more detail in a later section.

For wavy surfaces there is a radial component to the stress. This component is alternately tensile and compressive, depending on the site being a "hill" or "valley," respectively. The stored energy drives the process of failure of the ceramic coating during the use of the TBC. The foregoing equations indicate that to improve TBC failure lives, the elastic stored energy has to be minimized. This is achieved by the reduction of the modulus of elasticity through microstructural optimization such as by introduction of segmentation cracks, microcracks, and porosity. An estimate of the residual stress (Miller and Lowell, 1982) from CTE mismatch for a plasma sprayed 7YSZ with a stress-free temperature of 375°C (707°F) gives a tensile stress value of $-120\,\mathrm{MPa}$ ($-17.4\,\mathrm{ksi}$) using $(\alpha_m - \alpha_c) = 5 \times 10^{-6}/°C$, $E = 48\,\mathrm{GPa}$ (69.6 Msi), and $v = 0.25$. This stress agrees well with $-100 \pm 70\,\mathrm{MPa}$ ($14.5 \pm 10.1\,\mathrm{ksi}$) measured by Thornton et al. (1999) for as-sprayed 7YSZ on NiCoCrAlY bond coat. This level of residual stress is obviously not enough to cause TBC failure.

Role of Thermally Grown Oxide (TGO)

The TGO plays the critical role of binding the ceramic layer to the metallic bond coat deposited on the substrate. The composition of the TGO, as discussed earlier, is predominantly α-Al_2O_3 for alumina-forming bond coats. For NiCoCrAlY bond coats, minor amounts of spinels $(Ni,Co)(Al, Cr)_2O_4$, and occasionally NiO, are also found, particularly in the region of the TGO closer to the bond coat. The growth rate of the TGO affects the eventual spallation of the TBC when it is cycled to high temperature in an oxidizing environment. Experimental and field data from aircraft gas turbine engines suggest that typical TGO thickness at TBC spallation is in the 6 to 7 μm range and certainly below 10 μm. Based on electron microscopic

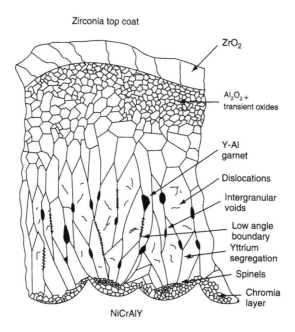

Figure 7.23 Microstructure model of TGO on Marcy bond coat (L. Lelait, S. Algerian, and R. Mevrel, Alumina scale growth at zirconia–MCrAlY interface: A microstructural study, *J. Mater. Sci.*, 1992, 27, 5–12). Reproduced with permission from Springer.

characterization, the general microstructural features of the TGO formed on overlay coatings of MCrAlY composition have been modeled by Lelait et al. (1992) (Fig. 7.23).

As can be seen from the model, the alumina grains at the TGO metal interface are large and elongated. The morphology is such that oxygen diffusion in the alumina grains is faster than aluminum diffusion. Yttrium segregates at the oxide grain boundaries either as garnet, $Y_3Al_5O_{12}$, or as finer precipitates. Zirconium is also found within the TGO, leading to the speculation that some of the alumina has formed at the expense of the reduction of zirconia at high temperature ($\sim 1200°C$, $2192°F$), where such reduction is thermodynamically feasible.

Structural Properties

Most ceramic materials are brittle, and at low to moderate temperatures have linear elastic stress–strain behavior with little or no plastic deformation. Plasma sprayed YSZ, however, exhibits nonlinear deformation characteristics from room temperature all the way to $1200°C$ ($2192°F$). This is demonstrated in Fig. 7.24a (DeMasi-Marcin et al., 1989) for samples machined from bulk plasma deposits. In the uniaxial tensile and compressive tests, the loads were applied in the plane of the deposit. It is evident from the plots that even at room temperature and low stresses the tensile deformation behavior of plasma sprayed 7YSZ is nonlinear, unlike other bulk ceramic materials. The ultimate tensile strengths and failure strains are low, of the order of $20\,MPa$ ($3\,ksi$) and 0.25%, respectively. The strength decreases only slightly between room temperature and $980°C$ ($1796°F$). Beyond $1100°C$ ($2012°F$), the strength falls off rapidly. The uniaxial compressive behavior of 7YSZ is initially linear at lower loads, transitioning to nonlinear behavior at higher loads (Fig. 7.24b). The compressive strength is significantly

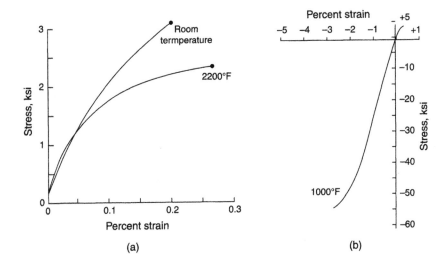

Figure 7.24 Actual stress–stain behavior (a) in tension and (b) in compression for freestanding plasma sprayed TBC (Jeanine T. DeMasi-Marcin, Keith D. Sheffler, and Sudhangshu Bose, Mechanisms of degradation and failure in a plasma deposited thermal barrier coating, *ASME 89-GT-132*, 1–8). Reprinted with permission from American Society of Mechanical Engineers International.

higher than the tensile strength. In addition to the nonlinear deformation behavior, plasma deposited 7YSZ also exhibits significant hysteresis when the load is reduced or reversed. Above 980°C (1796°F), the material exhibits significant creep. Stress-sensitive fatigue behavior is also observed.

The structural behavior of plasma sprayed 7YSZ just described is due to the details of the microstructure formed from the depositing splats. DeMasi-Marcin et al (1989) have proposed a stick-slip model to rationalize the metal-like behavior. The existing microcracks in the coating have associated mating surfaces, which are rough on a fine scale. Under load, they slide past each other. Initially the surfaces with finer surface roughness features move past each other under low stresses. With progressively higher stresses, surfaces with coarser features slide against each other to newer positions where they become locked. The shape of the stress–strain curves exhibits this progressive sliding and locking. In compression, higher loads are required because the surfaces are more restricted in movement because of closer contact under normal load.

Plasma TBC Durability

TBCs are used on components ranging from automotive valves and combustion chambers of industrial power generators to turbine blades in jet engines to extend the component lives and increase efficiency. In order to meet these requirements, the TBCs have to survive in the application environments for the required duration. The effect of such environments on durability of TBCs is, therefore, a major concern.

Effects of Thermal Cycling in Oxidizing Environment

Cyclic high-temperature exposure, the combination of thermal and mechanical stresses, and the presence of an oxidizing and often corroding environment characterize most of the applications of TBCs. Of all these factors, the effects of cyclic variation of temperature combined with bond coat oxidation make the greatest contribution to the total strain in the plasma sprayed TBC. The cyclic component contributes to the CTE mismatch strain while bond coat oxidation provides strain related to the continued growth of the TGO. In order to understand and simulate the effects of the various parameters in the laboratory, a number of cyclic tests have been devised. These include cycling in and out of a furnace followed by forced-air cooling and cyclic exposure to combustion flames in burner rigs. For lower thermal mass, the latter test simulates the thermal transients observed in gas turbine engines better than the furnace test. These tests have already been discussed in some detail in the Section 4.2 on oxidation. The critical parameters in the tests are the temperature, the dwell time at temperature, the total length of each cycle, the rates of heat-up and cool-down, and the microstructural details of the TBC, including the composition of the substrate alloy and the bond coat, and the thickness of the ceramic layer. In laboratory tests, the use of known standard baseline TBCs is always recommended to ensure that the tests are run correctly.

The standards selected should have known and predictable behavior. Failure of TBC is generally defined as the loss of significant sizes (about $1\,cm^2$ across or a constant percentage of the total area) of the ceramic coating in the hot zone (Fig. 7.25a) (DeMasi et al., 1989). Figures 7.25b (Miller, 1989) and 7.25c (Wortman et al., 1989) show the number of cycles to TBC failure as a function of the length of the heating cycle with maximum temperature of

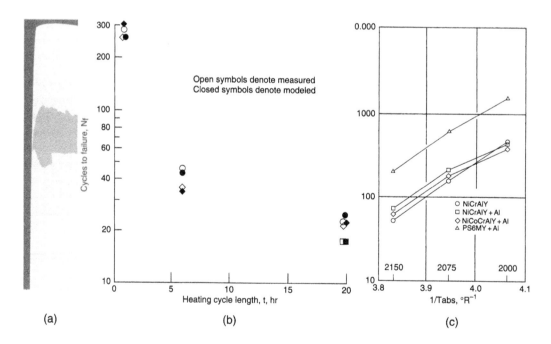

(a)   (b)   (c)

Figure 7.25 (a) A typical TBC failure (J. T. DeMasi, K. D. Sheffler and M. Ortiz, Thermal barrier coating life prediction model development, *NASA Contractor Report 182230*, 1989, p. 47. Courtesy of Robert A. Miller, NASA Glenn Research Center); (b) TBC lives in burner rig test, test temperature 1100°C (R. A. Miller, Life modeling of thermal barrier coatings for aircraft gas turbine engines, *J. Eng. Gas Turbines Power Trans. ASME*, 1989, 111, 301–305. Reprinted with permission from American Society of Mechanical Engineers International); (c) Furnace cycle lives of TBC with several bond coats on René N5 substrate disk samples (D. J. Wortman, B. A. Nagaraj and E. C. Duderstadt, Thermal barrier coatings for gas turbine use, *Mater. Sci Eng.*, 1989, A121, 433–440. Reprinted with permission from Elsevier).

1100°C (2012°F). It is evident that the TBC life is a function of both the nature of the cycle and the temperature. The bond coat composition also plays a very important role as seen in Fig. 7.25c.

TBC Degradation Modes and Locations

Continued exposure of TBC in a real operational environment results in incremental damage, which accumulates with time and cycles. Although thermal cycling drives a significant part of the damage, mechanical and chemical processes also play important roles. The TBC eventually fails by the local spalling of the ceramic coating. The details of individual modes involved in the degradation of TBC are covered in the section on EB-PVD ceramic. The modes observed for plasma sprayed ceramic coatings, particularly YSZ, are briefly summarized here:

(1) *Infant mortality*: This mode involves premature failure of TBC generally attributable to processing aberrations such as unclean surfaces on which ceramic is deposited, drifting of process parameters outside the prescribed envelope, and use of raw materials not meeting specifications. Adherence to good manufacturing practices generally eliminates infant mortality.

(2) *Thermocyclically induced degradation*: This is the predominant mature failure mode controlling TBC life. It involves simultaneous imposition of cyclic thermal and mechanical strains (or stresses).

(3) *Erosion-induced degradation*: This process occurs as a result of the slow loss of ceramic by the impact by fine particulate materials. Although it may affect performance, erosion seldom leads to TBC failure.

(4) *Environmental deposit–induced degradation*: The deposits chemically and mechanically affect the integrity of the ceramic and the bond coat. Salt deposits responsible for hot corrosion affect the TBC within a temperature range characteristic of the process as shown in Fig. 7.26 (Miller, 1989; Strangman, 1990). This mode of failure is strongly dependent on the stabilizer used. YSZ has been found to be resistant to hot corrosion.

In the mature failure of APS TBC indicated in 2 above, lifting of the ceramic layer, called spallation, occurs close to the ceramic–TGO interface but within the ceramic.

Failure Mechanism of Plasma-Sprayed TBCs

As discussed earlier, plasma-sprayed TBCs fail by spallation of the ceramic coating. For TBC thickness <250 μm, it involves local separation close to the bond coat but within the ceramic coating. The driving force for the spallation is the combination of cyclic thermal strain due to CTE mismatch, continued oxidation of the bond coat leading to TGO growth, and externally imposed cyclic mechanical strain. The failure sequence of the TBC may be described as follows:

- Formation of subcritical cracks commencing early (<25% of exposure life) in the ceramic.

- Progressive link-up of adjacent subcritical cracks.

- Link-up resulting in dominant in-plane crack within the ceramic but close to the bond coat–ceramic interface (Fig. 7.27) (DeMasi et al., 1989).

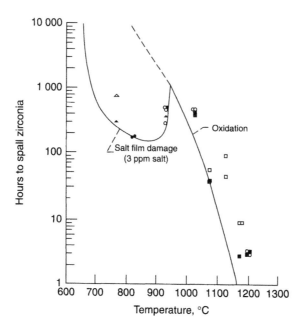

Figure 7.26 TBC life as a function of temperature and salt deposit (T. E. Strangman, Turbine coating life prediction model, *Proc. 1990 Coatings for Advanced Heat Engines Workshop, Department of Energy Report CONF-9008151*, Castine, ME, 1990, pp. II35–II43; R. A. Miller, Life modeling of thermal barrier coatings for aircraft gas turbine engines, *J. Eng. Gas Turbines Power, Trans. ASME*, 1989, 111, 301–305). Reprinted with permission from American Society of Mechanical Engineers International.

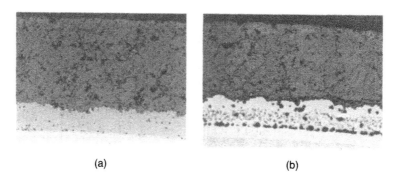

(a)   (b)

Figure 7.27 Microstructure of (a) pretest, (b) posttest plasma sprayed TBC (J. T. DeMasi, K. D. Sheffler, and M. Ortiz, Thermal barrier coating life prediction model development, *NASA Contractor Report 182230*, 1989, p. 56), showing large crack in the ceramic near the interface. Courtesy of Robert A. Miller, NASA Glenn Research Center.

• The separation of ceramic at the dominant crack leading to failure.

• For thickness $>> 250\,\mu$m, failure generally occurs within the ceramic layer, away from the interface.

    Residual stress within the TGO, surface roughness of the interface in the form of undulations, aluminum depletion of the bond coat due to oxidation and TGO growth, and bond coat creep play critical roles in inducing TBC spall. Cracking of the ceramic and crack link-up have been semiquantitatively analyzed by Schlichting et al. (2005) using a fracture mechanics approach. Although this is not a life model, the analysis validates the advantages of TBC design modifications provided by segmentation cracks and columnar growth structure.

Design Capable Phenomenological Life Model

A number of quantitative models to describe the failure mechanism of plasma sprayed TBCs have been proposed (Chang et al., 1987; Freborg et al.,1998; Cheng et al.,1998) focusing on finite element analysis and mostly assuming linear elastic behavior of the ceramic. However, as discussed earlier in Section 7.4, plasma sprayed TBC exhibits nonlinear rate-dependent behavior (DeMasi et al.,1989). In order to incorporate more realistic deformation characteristics, viscoplastic model, initially developed for monolithic ceramic, has also been applied to plasma sprayed TBC (Xie et al., 2003). Here we will discuss a user-friendly fatigue model based on the following elements:

- The failure results from fatigue induced by thermal cycling.

- Whereas elastic strain is benign to the TBC structure, every inelastic strain cycle induces a finite amount of damage.

- The fatigue life is calculated using the Manson–Coffin equation (Dieter, 1986) relating life to inelastic strain range.

- The temperature/time cycle of thermal exposure in the gas turbine environment, translates into a stress or strain–time cycle in the ceramic layer of the TBC.

Because most of the laboratory data are generated in burner rigs, the starting point in the life model is the burner rig temperature/time cycle to which the TBC is routinely exposed. Next, the corresponding stress/strain cycle is composed both as an idealized cycle followed by a calculated cycle based on the actual temperature/time profile of the burner rig test.

When the TBC coated hardware is exposed to a burner rig flame, the surface temperature time profile typically follows the trend shown in Fig. 7.28a. It takes a short time to heat up. The hold time may vary from a few minutes to several hours. The cool-down takes a few minutes. The thermal cycle induces stresses and strains in the system. The stress–strain cycle for an idealized linear elastic–ideal plastic ceramic layer is shown in Fig. 7.28b.

Figure 7.28c is a calculated stress–strain cycle based on a burner rig test in which the emphasis was on thermal strain rather than oxidation. For the idealized case of Fig. 7.28b, the assumed stress–strain curve is shown in Fig. 7.29.

The nonlinear stress–strain behavior has been explained by a proposed stick-slip model. Surfaces associated with microcrack having very small asperities slide past each other under smaller initial stresses. As stresses increase, surfaces with progressively larger asperities move past each other.

For the idealized model behavior, the contributions to the total inelastic strain range $\Delta\varepsilon_i$ of Fig. 7.28b are as follows:

0 to 1: The as-coated ceramic layer is typically under small compression due to processing effects.

1 to 2: As the TBC layer heats up because of flame impingement in the burner rig, the ceramic tries to expand, but the cold substrate constrains it into *compression, which is initially elastic.*

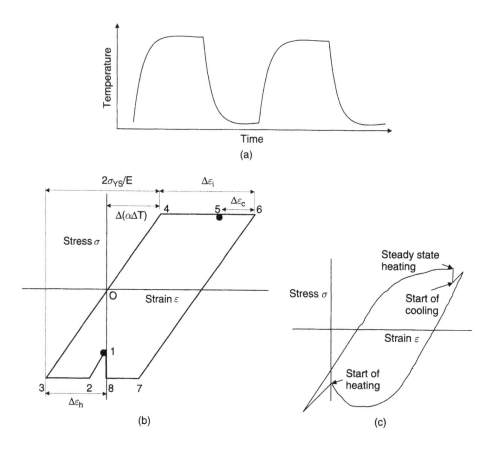

Figure 7.28 (a) Qualitative burner rig test temperature time profile. (b) Idealized stress–strain curve for the ceramic layer. (c) Calculated stress strain cycle for ceramic layer (J. T. DeMasi, K. D. Sheffler, and M. Ortiz, Thermal barrier coating life prediction model development, *NASA Report 182230*, Dec. 1989). Courtesy of Robert A. Miller, NASA Glenn Research Center.

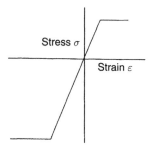

Figure 7.29 Idealized linear elastic–plastic stress–strain behavior. Uniaxial tensile and compressive stress–strain behavior of actual plasma–deposited ceramic is shown in Fig. 7.24.

2 to 3: Ceramic finally yields under *compressive stress, exhibiting inelastic behavior.*

3 to 4: The underlying metallic substrate including the bond coat heats up. The differential thermal expansion reverses the stress state and forces the ceramic into *tension, which is initially in the elastic range.*

4 to 5: Ceramic finally yields under *tensile stress, exhibiting inelastic characteristic.*

5 to 6: When the flame is turned off and cold air is injected on the surface of the sample, the ceramic tends to cool down faster but is restrained by the hot substrate. This generates additional *inelastic tensile strains in the ceramic.*

6 to 7: When the underlying metal finally cools down, the differential contraction introduces *compressive stress in the ceramic, initially in the elastic range.*

7 to 8: The ceramic finally yields under *compressive stress, exhibiting inelastic behavior.*

8: The system reaches equilibrium at the minimum exposure temperature.

The total inelastic strain range is, therefore, given by

$$\Delta\varepsilon_i = \Delta(\alpha\Delta T) + \Delta\varepsilon_c + \Delta\varepsilon_h - 2\sigma_{YS}/E,$$

where $\Delta\varepsilon_i$ is the total inelastic stain range; $\Delta\varepsilon_c$ and $\Delta\varepsilon_h$ are the inelastic strain ranges associated with the heating and the cooling transients, respectively; σ_{YS} is the yield strength of the ceramic; E is the modulus of elasticity; $\Delta\alpha$ is the difference in thermal expansion coefficients between the metal and ceramic layers; and ΔT is the temperature excursion range in the burner rig.

Life Equations

Using the Manson–Coffin fatigue model, cyclic life N in the burner rig is given by

$$N = (\Delta\varepsilon_i/\Delta\varepsilon_f)^b,$$

where, as explained above, $\Delta\varepsilon_i$ is the total inelastic strain range during thermal cycling, $\Delta\varepsilon_f$ is the inelastic strain range that causes failure in one cycle (failure strain range), and b is a constant.

Impact of Oxidation

Thermal cycling at high temperature in air results in the growth of the TGO that has formed during the plasma spray process. The cyclic inelastic strain ranges increase as the TGO grows. The contribution of TGO thickness change to the failure strain range is assumed to follow the relationships depicted in Fig. 7.30,

$$\Delta\varepsilon_f = \Delta\varepsilon_{fo}(1 - \delta/\delta_c) + \Delta\varepsilon_i(\delta/\delta_c),$$

where δ_c is the critical TGO thickness which induces TBC spallation in one cycle, and $\Delta\varepsilon_{fo}$ is the strain range in absence of oxidation, which causes failure in one cycle. Because the TGO thickness growth is a thermally activated process, the thickness follows the relationship

$$\delta = \delta_o \exp(-\Delta H/RT),$$

where ΔH is the activation energy of the process, δ_o is a constant, and R is the gas constant. Combining all the relationships, we get the TBC life in cycles:

$$N = (\Delta\varepsilon_i/\Delta\varepsilon_f)^b = [\{\Delta\alpha\Delta T + \Delta\varepsilon_c + \Delta\varepsilon_h - 2\sigma_{YS}/E\}/\{\Delta\varepsilon_{fo}(1 - \delta/\delta_c) + \Delta\varepsilon_i(\delta/\delta_c)\}]^b.$$

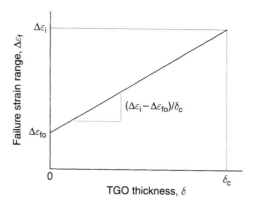

Figure 7.30 Assumed relationships between effective strain range and oxide thickness.

Using typical ceramic and bond coat parameters and specific burner rig cycles, life of plasma sprayed TBCs has been evaluated under various thermocyclic conditions. The calculated and observed TBC lives agree to within $\pm 2\times$ of each other. This is generally acceptable for APS TBC considering the usual scatter in the data obtained from the field as well as in burner rig tests. Typical values of the various parameters are $b = -30$ to -50, $\Delta\varepsilon_{fo} = 0.01222$ (1.22%), $\delta_c = 7.5\,\mu m$. The critical TGO thickness 7.5 μm for ceramic spallation is close to the value observed in the field from data on a number of gas turbine engines.

7.5 ELECTRON BEAM PHYSICAL VAPOR DEPOSITED (EB-PVD) TBCs

EB-PVD is another process by which the ceramic layer of TBCs can be deposited. It is a modification of the high-rate vapor deposition used for metallic coatings (Demaray, 1982; Strangman, 1985).

Why Electron Beam

Zirconia is highly refractory with a melting point of 2690°C (4875°F). As discussed before, because of its high melting point, traditional methods of heating do not melt the oxide. One, therefore, has to resort to special methods. One such method, the plasma spray process, has already been described in detail. A second method of delivering localized high energy to zirconia is by utilizing high power of focused electrons in the electron beam physical vapor deposition process. A beam of electrons generated by electron guns is focused on appropriately stabilized zirconia in the form of either granules or sintered cylindrical ingots.

General Principle

The basic principle of the EB-PVD process consists of creating a melt pool of the raw material in an evacuated chamber by heating 7YSZ with the focused high-energy electron beam. The pool generates vapor. The part to be coated is held over the pool. The coating on the surface of the part forms by the deposition of the molecules in the vapor as opposed to the deposition of large molten particles formed in the plasma spray process. The complete process is conducted inside an EB-PVD coating chamber.

Processing

The EB-PVD coater consists of the following components and functions:

- *An evacuated coating chamber*: Typical chamber pressure is about 10^{-4} torr. This is set by the maximum pressure that electron beam (EB) guns can tolerate before automatic shutoff to avoid filament burnout. At the start of the coating campaign, the chamber pressure is kept lower than the steady-state value. As the coating process progresses and additional oxygen (discussed later) flows in, the pressure increases to a steady-state level. Chamber geometry and dimensions vary with the manufacturers. In order to maintain quality of the coating, the leak rate of the chamber has to be maintained below prescribed levels.

- *Docking and preheating chambers*: There are two distinct coater designs available from the manufacturers. In one, the required preheating, explained later, is done in the same chamber in which the coating process takes place. In the other design, preheating is done in separate chambers. In these independently evacuated stations provided with load locks, the fixtured parts are preheated to required temperature prior to the coating process.

- *Raw materials*: Coating raw material is in the form of either granules or cylindrical ingots, the latter being more convenient for shipment, storage, and handling. The ingots are made first by cold pressing powders in cylindrical shape followed by sintering to improve structural integrity. When ingots are used, they are held in water-cooled crucibles.

- *EB guns*: EB guns provide a narrow electron beam of high energy. The design of the guns and their power ratings vary with manufactures and models. A typical power rating is 45 kW per gun, although guns with power ratings > 200 kW are not uncommon. The electron acceleration voltage is typically of the order of 40 kV. The number of guns used to preheat parts and melt raw materials also depends on the coater design. For example, the Pennsylvania State University coater (Fig. 7.33) (Singh, 2006), built by Sciaky, Inc., has six guns, whereas the Leybold designs made by Leybold-Heraus (Fig. 7.34) have two guns.

- *Electron beam deflection system*: Depending on the coater designs, electron beam paths are either linear, as in the Pennsylvania State University coater, or bent magnetically, as in the Leybold-Heraus coaters, by an angle as much as 270 degrees.

- *Melting*: Kinetic energy exchange between the electron beam (~ 40 kW) and the coating material results in heating of the raw material to melting and evaporation.

- *Vapor deposition*: Components to be coated are appropriately fixtured and held over the molten pool in the path of the vaporizing molecules. Because of the extremely low pressure of the EB-PVD chamber, the vapor molecules have large mean free paths (average distance traveled by a molecule between collisions with other molecules). Following ideal gas laws and the kinetic theory of gases, the mean free path λ at absolute temperature T is given by

$$\lambda = RT/(\sqrt{2}\pi d^2 N_A P),$$

where R is the gas constant, d is the diameter of the vapor molecule, N_A is Avogadro's number, and P is the ambient pressure. Using typical values of the chamber pressure and temperature, one can easily show that the mean free path is of the order of the chamber dimension, that is, the vapor molecules experience very minimal collisions among themselves. While colliding with the component, the molecules deposit to form the coating.

- *Ingot manipulation*: The ingots are manipulated by automatic feeder mechanisms, which raise the ingots to maintain the molten surface at constant distance from the parts being coated to compensate for the raw material consumed due to vapor deposition.

- *Electron beam pattern*: Evaporation from a large area of the ingot is achieved by rastering the focused electron beam over the surface of the coating source. The sweep patterns are a critical aspect of the process parameters.

- *Temperature control*: Controlled defocusing of the electron beam is used to reduce the temperature at the surface of the molten pool.

- *Multiple ingots*: Deposition from multiple ingots to vary coating composition is accomplished by using the jumping beam technique. The focused electron beam is made to jump alternately on multiple ingots, which may have different compositions, maintaining multiple molten pools.

- *Preheat*: In order to establish an adherent coating, the parts, held over the ingots at a safe distance to avoid overheating from the molten pool, need to be preheated to the desired temperature and maintained at that temperature during the coating process. Preheat provides thermal energy for surface diffusion of the deposited species leading to nucleation and growth of EB-PVD structure. Preheating and maintaining substrate temperature during the coating process is accomplished either from an electrically heated radiant heating source or by using an auxiliary over-source EB gun. For zirconia-based TBC, typical substrate temperature range is between 1000 and 1050°C (1832 to 1922°F) (Schulz et al., 1997). Figure 7.31 (Reinhold et al., 1999) shows the preheating temperature profile of a turbine blade to be coated for an industrial gas turbine engine.

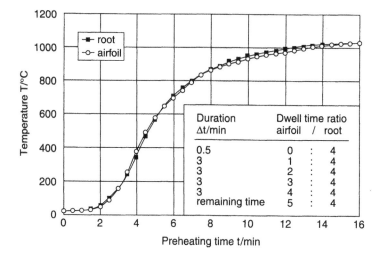

Figure 7.31 Preheat temperature profile of a turbine blade for TBC coating (E. Reinhold, P. Boltzer, and C. Deus, EB-PVD process management for highly productive zirconia thermal barrier coating of turbine blades, *Surf. Coat. Technol.*, 1999, 120–121, 77–83). Reprinted with permission from Elsevier.

- Parts to be coated are fixtured on a part manipulator attached to a rotating shaft called the sting. Production coaters generally have two stings. The regions of the part that should not require coating, such as the root of a turbine blade, are appropriately masked so that the coating vapors do not find access to these areas. Typical part movements require rotation as well as tilting (also known as pivoting) of the parts (Fig. 7.32) to achieve uniformity of coating thickness as well as favorable microstructural orientation for all exposed areas. The rotation of the parts is usually planetary. For turbine blades with platforms to be coated, tilt angles as high as ~30 degrees are used. The rate of rotation is determined from design of experiments. A rotational speed of 30 rpm is very common. Zirconia-based coatings tend to lose oxygen and become nonstoichiometric in the low-pressure, high-temperature environment of the coater. In order to maintain stoichiometry of zirconia during the coating process, oxygen is bled at the proximity of the parts at prescribed flow rates. The flow is monitored with an electronic mass flow control system.

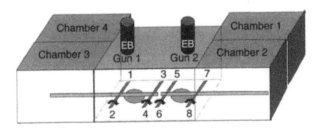

Figure 7.32 Schematic of the positioning of parts (1 through 8), part manipulation (rotation around part axis; options for parts tilting during rotation is also available in many designs), EB guns, ingots (cross section shown between parts), and various chambers (for loading, preheat, postcoat) in an EB-PVD coater.

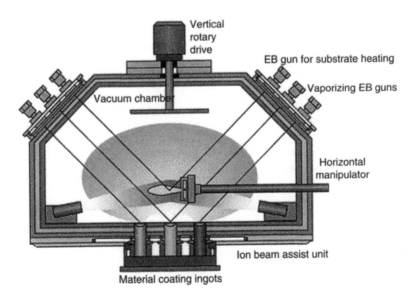

Figure 7.33 EB Coater designed and built by Sciaky, Inc., for the Pennsylvania State University. Courtesy of Prof. Jogender Singh, jxs46@psu.edu, The Applied Research Laboratory, The Pennsylvania State University.

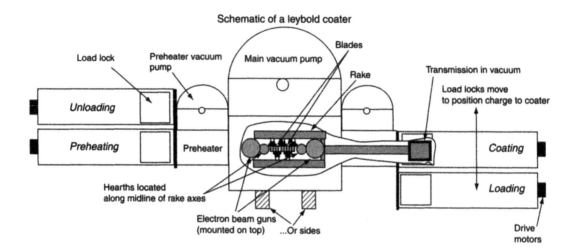

Figure 7.34 Schematic of a Leybold EB-PVD Coater (D. V. Rigney, R. Viguie, D. J. Wortman, and D. W. Skelly, PVD thermal barrier coating application and process development for aircraft engines, *J. Therm. Spray Technol.*, 1997, 6(2), 167–175). Reprinted with permission from ASM International.

- During the preheat phase, the bleeding of oxygen helps the formation of the TGO on the metallic bond coated hardware. The TGO acts as the bond between the ceramic layer and the metallic bond coat. It continues to grow in thickness during the coating process.

- Residual gas analyzers are used to monitor the quality of the ambient atmosphere within the coating chamber.

- Coating condensation rates vary depending on part position relative to the ingots, as well as the shape and size, with 5 to 7 μm/min being typical (Schulz et al., 1997).

- Many of the EB-PVD coaters have options to attach ion guns to perform ion beam assisted deposition (IBAD). IBAD essentially provides high-energy ($\sim$500 eV) argon ions to bombard substrates during selected periods of deposition, generally to improve bond strengths (Youchison et al., 2004).

Microstructure Formation

The ceramic deposition rate can be estimated from the Hertz–Knudsen equation, which is applicable to materials undergoing evaporation:

$$W = n'(P^* - P)(m/2\pi k_B T)^{1/2},$$

where n' is the dimensionless evaporation coefficient (varies between 0 and 1), k_B is the Boltzmann constant, m is the molecular weight of the depositing species, T is the absolute temperature, P^* is the vapor pressure of the depositing material, and P is ambient pressure in the coater chamber. Three important conclusions can be drawn from the deposition equation:

(1) Materials with higher vapor pressure (P^*) would deposit at faster rates. Temperature dependence of the vapor pressure of several oxides is given in Fig. 7.35 (Schulz et al., 2004).

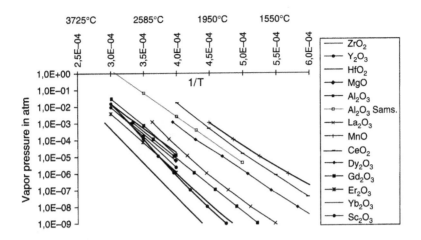

Figure 7.35 Temperature dependence of vapor pressure (Uwe Schulz, Bilge Saruhan, Klaus Fritscher, and Christoph Leyens, Review on advanced EB-PVD ceramic topcoats for TBC applications, *Int. J. Appl. Ceram. Technol.*, 2004, 1(4), 302–315). Reprinted with permission from Blackwell Publishing.

(2) The composition of the vapor has been subject of many studies. Belov and Semenov (1985) found that when ZrO_2, HfO_2, and Y_2O_3 are evaporated in a vacuum, the vapor species contained the oxide molecules, gaseous monoxides, and atomic oxygen. When evaporating from a single ingot containing multiple materials or from multiple ingots, the vapor composition generally reflects the composition of the depositing coating and is dominated by the molecular species of the ingot having the highest vapor pressure. Thus, if the ingot contains species of widely differing vapor pressure, the resulting coating would have a composition significantly different from that of the ingot, enriched with the highest vapor pressure component. The ingot composition, therefore, needs to be adjusted to achieve the desired coating chemistry. The deposition of YSZ does not suffer from such compositional issues because, as Fig. 7.35 shows, the vapor pressures of ZrO_2 and Y_2O_3 are reasonably close.

(3) A faster deposition rate can be achieved at a higher vacuum level (lower P).

Typical deposition rates are obtained with only 2 to 5% of the vapor flux depositing as coating. As in plasma spray, EB-PVD deposition is a "line of sight" process. In other words, the vapor molecules having large mean free paths travel in a straight line and deposit on any surface in the path at a point intersected by the trajectory. The model that explains the deposition profile is known as the Knudsen cosine law, shown in Fig. 7.36. In the equation, M is the mass of ingot evaporated, ρ is the density of the coating, and θ and φ are respectively the angles between the radial direction r and the normals to the vapor-emitting, as well as the receiving surfaces. It is clearly evident from the equation that the deposition rate is:

- Maximum right above the ingot ($\theta = 0$)

- Lower at higher θ angles

- Progressively decreases as the back of the part is approached (lower φ)

- Inversely proportional to the square of the distance

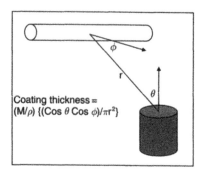

Coating thickness =
$(M/\rho) \{(\cos \theta \cos \phi)/\pi r^2\}$

Figure 7.36 Knudsen cosine law of vapor deposition.

The limitation of the line-of-sight process is that no coating would deposit at a point exactly opposite of the point of direct "hit" of a vapor molecule ($\varphi = 180$ degrees). However, some scattering among the vapor molecules themselves, and by the molecules of the residual gas in the coater chamber, still occurs and results in a very thin deposit at this location. A part held stationary during deposition will, therefore, have a nonuniform coating with maximum thickness at $\varphi = 0$ and minimum at the opposite point, $\varphi = 180$ degrees. In order to overcome this problem and achieve uniform coating thickness, the parts are rotated in the vapor. The complexity of rotation is unique to each coater design and the geometry of the parts.

EB-PVD coating is characterized by a columnar microstructure (Fig. 7.37). Neighboring columns have weak intercolumnar bonds. This allows the structure to provide strain tolerance during use. Several geometrical as well as thermally induced processes influence the formation of the microstructure. These include the adsorption of the vapor molecule (or atom), desorption of some of the molecules during deposition, the vapor impingement angle on the substrate, shadowing of the incoming molecules by the existing deposits, surface diffusion, and volume diffusion.

As the vapor molecules (or atoms) approach the surface of the part to be coated, some bounce off while others are adsorbed, releasing energy of adsorption. With the help of their thermal energy, the adsorbed molecules move on the substrate or partially coated surface by surface diffusion to seek low-energy sites. Once they become attached to the surface at these sites, nuclei tend to form. Whether the subsequent depositing molecules will form new nuclei

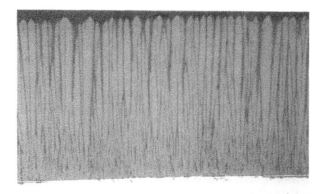

Figure 7.37 EB-PVD columnar microstructure.

or contribute to the growth of existing nuclei depends on the process parameters, including the molecular flux and energy, chamber pressure, and substrate temperature. Molecules obviously continue to deposit, resulting in growth of the structures. The deposition of a fresh layer of molecules, followed by diffusion and the ensuing energy exchange, leads to a columnar growth with the column axis always oriented toward the vapor source. The growing column shadows the deposition of fresh molecules from certain directions, which results in the restriction of column width. The combination of substrate rotation and the homologous temperature (T/T_m, where T is ambient temperature at the deposition surface and T_m is the molten pool temperature) exerts a significant influence on the texture of the columns, the intercolumnar porosity that controls bond strength between columns providing strain tolerance, and the durability of the coating (Kaysser et al., 1998). The formation of the general features of the microstructure has been explained by a zone model (Movchan and Demchishin, 1969; Thornton, 1977). Figure 7.38 shows the model, which maps the microstructure as a function of chamber pressure and homologous temperature. Three distinct zones are observed:

Zone 1 ($T/T_m < 0.3$, where T is the substrate temperature and T_m is the temperature of the molten pool) is characterized by tapered columns with domelike tops. The columns are separated by porosity. The structure and porosity result from insufficient energy of adatoms (condensed atoms) for surface diffusion. Adatoms, therefore, are unable to overcome the effects of shadowing of intercrystallite valleys by the crystallite peaks.

Zone 2 ($0.3 < T/T_m < 0.5$) is characterized by smooth, faceted, dense columnar grains. The structure results from dominance of surface diffusion of condensed atoms due to high thermal energy. This is the zone of interest in the deposition of TBC.

Zone T is a transition zone with structure somewhere between those of Zone 1 and Zone 2.

Zone 3 ($0.5 < T/T_m < 1$) is characterized by equiaxed grains formed due to dominance of bulk diffusion aided by high thermal energy available to the adatoms.

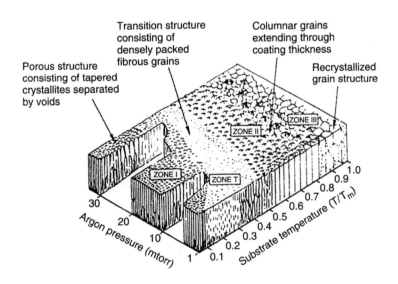

Figure 7.38 Thornton zone model of physically vapor deposited structure (J. A. Thornton, High rate thick film growth, *Ann. Rev. Mater. Sci.*, 1977, 7, 239–260). Reproduced with permission from Annual Reviews.

The columns constituting the microstructure of EB-PVD coatings exhibit strong texture or preferred orientation of crystal planes and directions. A simple description of texture is to identify the predominant plane of the zirconia crystal parallel to the substrate surface or the corresponding perpendicular direction. The texture is strongly dependent on process parameters, particularly the temperature of the substrate (or the coating chamber, which is typically 100 to 200°C (180 to 360°F) hotter than the substrate; Sohn et al., 1994). Sohn et al. report that in a Leybold coater design, the predominant texture varied from {111} through {311} to {200} as the coating chamber temperature was increased from 900°C (1652°F) through 1100°C (2012°F) to 1130°C (2065°F). The texture also varies with the angle at which the vapor molecules strike the substrate. For example, rotating turbine blades coated with airfoil surface perpendicular to the vapor incidence direction exhibit {200} texture, whereas a platform oriented at an angle of 45 degrees to the vapor incidence shows {220} texture.

Directed Vapor EB-PVD

Some of the limitations of the traditional EB-PVD process conducted in vacuum include low coating rates, difficulty in controlling and modifying microstructure, and deposition limited to the line of sight. The last limitation is alleviated by rotation of the substrate so that every point on the surface to be coated has a chance to receive the deposit. Many engineering components, however, have design features that shadow some areas of the components, resulting in vapor molecules being unable to arrive at those sites. Either the shadowed areas get no coating, or the deposited coating is thin with poor column orientation. Researchers at the University of Virginia in the United States and at the Fraunhofer Institute in Germany (Groves, 1998; Groves et al., 2000) have devised a modification of the EB-PVD process called Directed Vapor Deposition (DVD), which increases the deposition rate significantly with improved mixing of vapors from multiple ingot sources. Additionally, the process can modify the microstructure, vary porosity, and direct some of the vapor to the otherwise shadowed areas of the depositing surface. In one variant of the DVD process (Fig. 7.39) (Groves et al., 2000), the ingots are placed along the axis of a nozzle and a carrier gas such as helium is injected through an annulus surrounding the nozzle. Atoms in the carrier gas flow generate collisions with the vapor molecules. The collisions keep the vapor stream directed at the surface to be coated.

The collimating effect of the carrier gas has been found to increase deposition rate. Additionally, a plasma activation system to ionize the vapor and the gas stream combined with electrical biasing of the substrate further increases the deposition rate. In another variant of the process, the carrier gas is injected perpendicular to the vapor source. The gas scattering helps direct the vapor stream to otherwise shadowed regions. Manipulations of the carrier gas flow, vacuum level, and impingement angle have been used to modify the structure in the form of zigzag columns and with increased porosity. The carrier gas composition has also been varied by adding oxygen or nitrogen to the flow to achieve reactive deposition.

TGO

During the deposition of zirconia (7YSZ or other compositions) by the EB-PVD process in evacuated coating chamber, the zirconia molecules tend to loose oxygen and become nonstoichiometric. The loss of stoichiometry affects many of the properties of zirconia, including the color, which tends to become darker with continued loss of oxygen. To maintain stoichiometry,

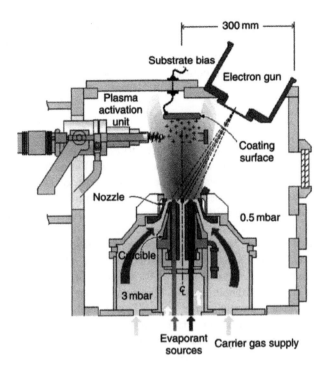

Figure 7.39 Directed vapor deposition in the EB-PVD process (J. F. Groves, G. Mattausch, H. Morgner, D. D. Haas, and H. N. G. Wadley, *Surf. Eng.*, 2000, 16(6), 461–464). Reprinted with permission from Maney Publishing.

oxygen is bled near the component to be coated. For typical metallic bond coats with high thermodynamic activity of aluminum, the coating chamber environment is sufficiently oxidizing. The bond coat, therefore, forms an alumina scale during the preheat stage. This scale at the ceramic metal interface is the thermally grown oxide (TGO) (Fig. 7.40), briefly described in Section 7.4.

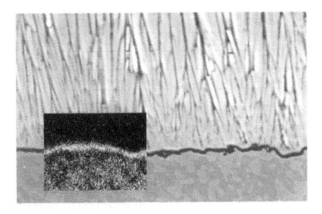

Figure 7.40 Thermally grown oxide (TGO) at the ceramic–bond coat interface of EB-PVD TBC. The inset is aluminum concentration map.

Being an ionic conductor, zirconia is essentially transparent to oxygen. With available access of oxygen through the zirconia layer, the TGO continues to grow on exposure to oxidizing environment at high temperature. The growth is a diffusion controlled process. The general trend of the increase of the TGO thickness, h_{ox}, can be expressed as a kinetic equation of the form

$$h_{ox} = kt^n,$$

where t is time, $k = k_0 \exp(-\Delta H/RT)$, k_0 is a constant, ΔH is the activation energy of the growth process, and n lies between 3 (Meier et al., 1992) and 3.33 (Schulz et al., 2001). Figure 7.41 (Schulz et al., 2001) shows the kinetics of growth of TGO thickness on Ni 22Co 12Al 0.1 to 0.2 Y on various substrates as a function of thermal cycle, each cycle consisting of 50 minutes at temperature and 10 minutes of forced air cool.

Limited available data show that TGO growth is faster in the presence of the ceramic layer of the TBC than on bare bond coat. Figure 7.42 (Stiger et al., 1999) compares the TGO growth at 1200°C (2192°F) with and without TBC on platinum aluminide bond coat. Electron microscopic analysis (Stiger et al., 1999) shows that the TGO grown in the presence of YSZ consists of two distinct zones. The zone closer to the bond coat consists of αAl_2O_3 of large grain size, whereas closer to YSZ, the TGO is a mixture of fine-grained alumina with dispersion of zirconia and yttria. In the absence of YSZ, the TGO is a single zone of αAl_2O_3. Increase in diffusivity in the presence of YSZ, is a possible explanation for the enhanced TGO growth. Similar consistent data on overlay bond coats have not been available.

It is generally believed that in the absence of TBC, the TGO growth occurs at the TGO–bond coat interface by diffusion of oxygen along grain boundaries and aluminum diffusion within the grains. In the presence of TBC the outward growth of the alumina increases, possibly due to the presence of transient aluminas (γ, θ phases).

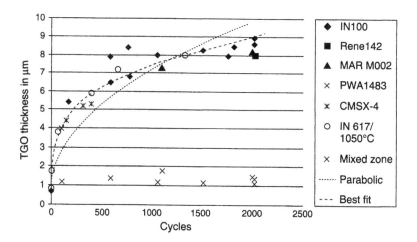

Figure 7.41 Growth TGO kinetics on NiCoCrAlY (Ni 22Co 12Al 0.1 to 0.2 Y) on various substrates (U. Schulz, M. Menzebach, C. Leyens, and Y. Q. Yang, Influence of substrate material on oxidation behavior and cyclic life time of EB-PVD TBC systems, *Surf. Coat. Technol.*, 2001, 146–147, 117–123). Reprinted with permission from Elsevier.

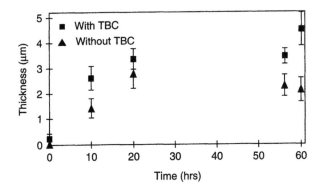

Figure 7.42 TGO growth at 1200°C (2192°F) with and without TBC on a platinum aluminide bond coat (M. J. Stiger, N. M. Yanar, F. S. Pettit, and G. H. Meier, Mechanism for the failure of electron beam physical vapor deposited thermal barrier coatings induced by high temperature oxidation, in *Elevated Temperature Coatings: Science and Technology III*, Eds. J. M. Hampikian and N. B. Dahotre, The Minerals and Materials Society, 1999, pp. 51–65). Reprinted with permission from The Minerals, Metals & Materials Society.

Role of Interface and Surface Roughness

In applications such as on turbine blades, thermocyclic durability is higher for EB-PVD TBC with a smoother rather than a rougher interface. Additionally, because the EB-PVD process reproduces the surface roughness of the underlying substrate, a smooth bond coat surface results in a smooth ceramic surface, which is aerodynamically beneficial and allows more efficient film cooling for coated gas-turbine components. On a microscopic level the interface topology can be idealized as undulations in the form of a sine wave. Whereas the TGO in valleys of the undulations will be in triaxial compression, the hills will be under in-plane biaxial compression and out-of-plane tension (Miller, 1987). The out-of-plane tension makes the TGO prone to delamination. The rough interface thus alters the local stress distribution detrimentally.

As the depositing species are of molecular dimension in the EB process, the surface of EB-PVD ceramic is much smoother. Typical surface roughness is 0.5 to 1.0 μm R_a. Because of the molecular dimensions of the depositing species, parts with cooling holes can be coated with very limited hole closure (5 to 10%) after the holes are drilled. Plasma sprayed coating, on the other hand, has rough surfaces. Typical value of surface roughness lies between 4 and 10 μm R_a. The higher roughness results because depositing fragments are larger in size. The large fragments of the depositing materials also force the complete closure of predrilled holes. Holes should, therefore, be drilled after the deposition of the APS ceramic.

EB-PVD TBC Degradation Modes and Locations

There are several modes by which TBCs degrade during their use. Plasma sprayed ceramic coating degradation was briefly discussed in Section 7.4. Although each degradation mode affects the performance of the TBCs to different extents, only a handful are life limiting. The individual degradation modes are considered next.

Infiltration by Environmental Deposits

Ingested sand and debris, which are solid below about 1100°C (2012°F), induce erosion and foreign object damage (FOD) when entrained in the air or gas stream in gas turbine engines. However, above 1200°C (2192°F), they melt and tend to adhere to the ceramic and slowly infiltrate the structure, making their way to the TGO–bond coat interface (Toriz et al., 1988; Siry et al., 2001). The composition and the melting points of the deposits vary according to the source and location of the sand and debris (Stott et al., 1994). The composition of the ingested ingredients is very important because it determines the effect on the TBC system. One form of the environmental deposits is called CMAS (calcium magnesium aluminosilicate, having the crystal structure of the mineral diopside, $Ca_5Mg_4Al_2Si_9O_{30}$). The composition of CMAS is somewhat independent of the geographic location of gas turbine engine operation. The deposits accumulate as benign solid particles on surfaces of components in the hot section of gas turbine engines. When the surface temperatures exceed the melting point of CMAS (> 1100°C, 2012°F), the deposits melt and infiltrate the structure of the TBC, leading to premature spallation. A subsurface delamination within the region of the TBC infiltrated with CMAS has been proposed as a failure mechanism (Mercer et al., 2005). Recent studies at UC Santa Barbara (Kraemer et al., 2006), using model synthetic CMAS, $35CaO$–$10MgO$–$7Al_2O_3$–$48SiO_2$ of melting point within the temperature range 1235–1240°C (2255–2264°F), found that for 7YSZ TBC deposited by EB-PVD on alumina substrate, the following sequence of processes occurs: (1) As soon as the CMAS melts, it rapidly infiltrates the TBC structure; (2) it chemically attacks the TBC; (3) the columnar morphology of the T' phase of YSZ is destroyed to a depth dependent on temperature, losing strain tolerance, and leaving on the surface monoclinic spherical particles of zirconia; (4) the infiltrating CMAS dissolves the alumina TGO; (5) the bottom part of the TBC in contact with alumina is attacked and converted from the T' phase of YSZ to monoclinic spherical particles; and (6) the intervening TBC remains intact. A number of solutions to the damaging effects of CMAS on TBC have been evaluated, including alternate compositions, modification of the microstructure, and deposition of physical barrier sealing against infiltration (Fig. 7.43) (Walston, 2004).

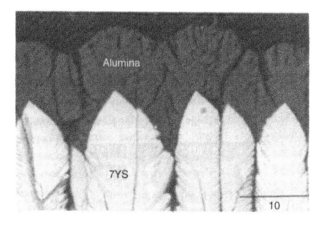

Figure 7.43 Alumina barrier on EB-PVD deposited 7YSZ (W. S. Walston, Coating and surface technologies for turbine airfoils, in *Superalloy 2004,* Eds. K. A. Green, T. M. Pollock, H. Harada, T. E. Howson, R. C. Reed, J. J. Schirra, and S. Walston, TMS, Warrendale, PA, 2004, pp. 579–588). Reprinted with permission from The Minerals, Metals & Materials Society.

Other types of deposits (Borom et. al, 1996) originate from concrete dust (di- and tricalcium silicate, tricalcium aluminate, and tetracalcium aluminoferrite) abundant around construction areas, fly ash that is inorganic residue of burning of coals in power stations, volcanic ash, and general dusts containing Fe_2O_3 and SiO_2. The last two oxides have been found to accelerate sintering of the ceramic coating (Trubelja et al., 1997). Sintering in turn increases the modulus of elasticity, adversely affecting strain tolerance.

A composite map of various failure modes of plasma sprayed TBCs used in combustors and shrouds is shown in Fig. 7.44 (Borom et al., 1996).

Hot Corrosion

Some of the environmental deposits with access to the TGO induce hot corrosion in the metallic bond coat. These deposits typically consist of sulfate salts of Na, K, Ca, and Mg. The salts become stabilized in the presence of SO_2 gas, which is a product of combustion of fuel in gas turbine engines. At appropriate temperatures, the salts melt, penetrate the ceramic structure if defects are present providing access, and reach the ceramic–bond coat interface. Through fluxing mechanisms, discussed in detail in Chapter 5, the deposits react with the protective alumina TGO and replace it with voluminous nonprotective oxides (Leyens et al., 2000). The phenomenon results in ceramic coating delamination. This type of TBC damage is seldom seen in aircraft engines because of the use of cleaner aviation fuel, but is more prevalent in industrial and marine gas turbine engines, which tend to use high-sulfur fuels, generating higher levels of SO_2 in the combustion product.

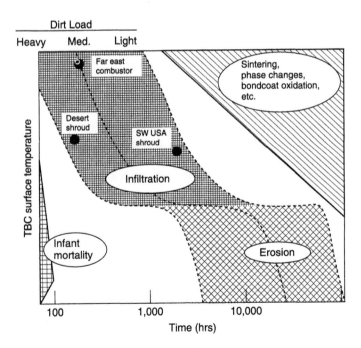

Figure 7.44 Temperature–time map of various failure modes of plasma sprayed TBC (Marcus P. Borom, Curtis A. Johnson, and Louis A. Peluso, Role of environmental deposits and operating surface temperature in spallation of air plasma sprayed thermal barrier coatings, *Surf. Coat. Technol.*, 1996, 86–87, 116–126). Reprinted with permission from Elsevier.

The ceramic coating may also be affected by the molten sulfates. The TBC stabilizers may be leached out of the coating by chemical reactions. In MgO-stabilized zirconia, for example, the stabilizer is leached out as $MgSO_4$, leaving ZrO_2 as destabilized (Singhal and Bratton, 1980), which now undergoes the detrimental phase transformation leading to failure. In the case of plasma sprayed Y_2O_3 stabilized zirconia (7YSZ), furnace testing (Barkalow and Pettit, 1979) of sulfate salt-coated samples in the presence of NaCl vapor exhibited hot corrosion of the samples. The stabilizer was released through acidic reaction either as yttrium sulfate, $Y_2(SO_4)_3$, or as hydrated sodium yttrium sulfate, $Na_2Y_2(SO_4)_4 2H_2O$ (Jones, 1991, 1992). However, a higher SO_3 partial pressure is required for reaction with YSZ. Y_2O_3 stabilized ZrO_2, therefore, tends to be more resistant to hot corrosion than zirconia stabilized with alkaline earth oxides. In cyclic burner rig tests by Nagaraj et al. (1992), hot corrosion of both plasma sprayed as well as EB-PVD deposited 8YSZ was evaluated. The systems consisted of René 80 substrate alloy, 8YSZ ceramic deposited for one set of samples by plasma spray on a bond coat consisting of BC52 of composition 65.2Ni10Co18Cr6.5Al0.3Y (with other elements or solid solution and grain boundary strengthening) with aluminide over coat, and for another set of samples by EB-PVD process on a bond coat comprising PBC22 of composition 62.5Co24.5Cr10.5Al2.5Hf. The salt environment was created in a burner rig, conducted at 1300° and 1700°F (704° and 927°C) using JP5 jet fuel doped with dibutyl sulfide. Synthetic seawater was injected into the combustor by atomized spray. Tests performed for up to 2000 hours showed no ceramic spallation and minimal bond coat attack. This test confirmed earlier engine test results reported by Wortman et al. (1989). The earlier test was conducted on turbine vanes by Fiat in Brindisi, Italy, on the coast of the Adriatic Sea where the air is laden with salt, an ideal environment for hot corrosion. The TBC system deposited on the outer platform of the vanes consisted of plasma sprayed 8YSZ and BC52 bond coat on René 80 substrate alloy. The rest of the vane airfoil surface was coated with an aluminide coating. After 687 hours/2750 cycles of testing, the aluminide coating showed severe hot corrosion attack with most of the coating consumed. However, the TBC and the bond coat underneath did not exhibit any signs of hot corrosion, although the ceramic coating was covered with salt containing about 17% Na_2SO_4, the remainder being $CaSO_4$.

Vanadate and phosphate deposits are known to leach out stabilizers including Y_2O_3, resulting in TBC destabilization and failure.

Recent furnace hot corrosion testing by Marple et al. (2006) showed that plasma spayed 7YSZ based TBC was resistant to sulfate salts but degraded rapidly in the presence of vanadium. $La_2Zr_2O_7$, an alternate ceramic for TBC, on the other hand, is susceptible to sulfate salt-induced hot corrosion but resistant to vanadate attack.

Erosion Damage

Erosion is the process by which material is gradually removed due to local impact by fine, abrasive particulate materials. Ceramic coatings are particularly prone to erosion, although it is not considered to be a life-limiting process for TBCs. The source of the particulates inducing erosion depends on the particular process and the details of the equipment design. Fly ashes generated in utility boilers using pulverized coal are the source of erosion for boiler tubes. In gas turbine engines, the particulates are the runway sand ingested with the air intake during taxi and takeoff as well as during thrust reversal during landing of the aircraft. Also, they are often generated within the combustor of gas turbine engines as hard carbon particles because of incomplete combustion of the fuel. Depending on the size, the particulates initially get entrained in the aerodynamic flow contour of the gas path. However, centrifugal load in the engine eventually forces them to deviate and impinge at various angles on engine parts in their path. The ultimate effect of erosion is gradual thinning of TBC, resulting in reduction of aerodynamic performance and loss of thermal insulation. The thinning may result in complete local removal of the ceramic coating, exposing the bond-coated alloy substrate to gas path temperatures, accelerating oxidation, and other degradation processes. The oxidation products of the metallic coating and substrate are also prone to erosion. In such cases, as the protective oxide scales are eroded away, a fresh metallic surface is exposed, and oxidation rates tend to increase significantly.

There are several factors that affect the rate of erosion. These include the microstructure of the ceramic coating, that is, plasma sprayed with splat boundaries versus columnar EB-PVD; porosity in the structure; the size, hardness, velocity, and flux of the particulates; the impingement angle; and the exposure temperature.

Erosion characteristics of TBCs are generally evaluated in the laboratory in modified burner rigs (Eaton and Novak, 1987; Bruce, 1998). However, to simulate engine-relevant conditions, and more particularly to incorporate corrosion processes simultaneously with erosion, the facilities become more complex (Nicholls, 1997; Nicholls et al., 1999).

A schematic of an erosion test facility (Tabakoff, 1989) is shown in Fig. 7.45. The particulate materials are held in a feeder (A) and injected into the heated gas stream through a preheater (D). The heat source is the combustor (C) in which a mixture of fuel and air (B) is combusted. The flowing particulates are accelerated by high-velocity air in the duct (F), which is cooled by steam. The specimens are held in the test section (G). Tilting of the sample relative to the gas stream controls the angle of impingement. Postimpingement particulates are mixed with the coolant and finally captured in the exhaust tank (H). The erodent medium generally consists of a measured quantity of abrasive particles, such as alumina, of the required particle size distribution. The airflow in the duct controls the particulate velocity. A laser Doppler velocimeter is used to measure particle velocity.

A representative range of test parameters used to simulate erosion in gas turbine engines is shown in Table 7.3 (based on Nicholls et al., 1999).

Dependence on Microstructure and Test Parameters

The impingement angle dependence of erosion rates measured for a nickel base superalloy Mar-M 002 and a number of coatings deposited both by air plasma spray (APS) and EB-PVD (PVD) are shown in Fig. 7.46 (Tabakoff, 1989). The coating compositions include 8% yttria partially stabilized zirconia (YSZ), zirconia titania yttria (ZTY), and alumina. Several important observations can be made from the data in Fig. 7.46:

(1) Maximum erosion occurs in ceramic coatings when the eroding particulates impinge at a 90-degree angle (perpendicular to the surface). This is significantly different from erosion of metallic materials, which exhibit maximum erosion at lower angles between 30 and 50 degrees. The difference relates to the mechanism of erosion. Metallic materials are ductile. Plastic deformation plays a critical role in their erosion. The maximum in erosion therefore follows the maximum in resolved shear stress. Ceramic coatings are brittle. Fracture, therefore, plays the key role in this case. As a result, the erosion profile follows maximum normal stress perpendicular to the impact surface.

(2) The columnar EB-PVD microstructure is more resistant to erosion by roughly a factor of 10 at normal impact compared with the plasma sprayed counterpart.

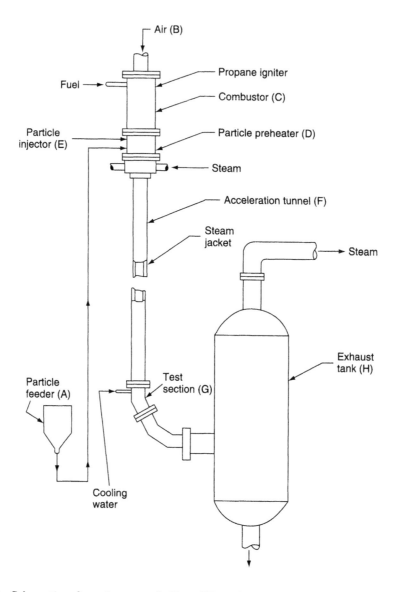

Figure 7.45 Schematic of erosion test facility (W. Tabakoff, Investigation of coatings at high temperatures for use in turbomachinery, *Surf. Coat. Technol.*, 1989, 39–40, 97–115). Reprinted with permission from Elsevier.

Table 7.3 Range of erosion test parameters to simulate engine relevant conditions

Parameters	Range
Temperature range (°C)	RT–920
Particulate velocity (m/s)	50–400
Particle size (μm)	20–1000
Impact angle (degrees)	20–90

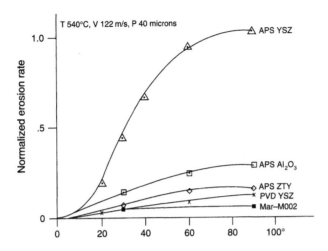

Figure 7.46 The dependence of erosion rate on impingement angle (W. Tabakoff, Investigation of coatings at high temperatures for use in turbomachinery, *Surf. Coat. Technol.*, 1989, 39–40, 97–115). Reprinted with permission from Elsevier.

(3) The benefit of the columnar EB-PVD microstructure over the APS counterpart at lower angles is relatively small, by about a factor of only 2. Thus, for combustors, and vane and blade platforms, which are oriented parallel to the airflow in gas turbine engines, EB-PVD structure provides less erosion benefit than for the segments of airfoil surfaces oriented at larger impact angles.

Erosion rates tend to increase with the velocity of the impinging particulates (Tabakoff, 1989; Nicholls et al., 1999), roughly following the relationship $e = cv^n$, where e is the erosion rate, v is the velocity, c is a constant, and n lies between 1 and 2.

The erosion rates have also been found to increase with the size of the impinging particulates up to about $100\,\mu$m (Nicholls et al., 1997). However, for large particulates, the extent of local damage, which is not registered as a mass loss, is expected to increase. This is where transition occurs from erosion into the realm of foreign object damage (FOD) by large particles.

The improved erosion behavior of EB-PVD ceramic over APS is due to the differences in failure modes for the two structures. On impact, the particulates generate cracks in the ceramic. In the APS ceramic, these cracks easily propagate by linking with favorably oriented weakly bonded splat boundaries and the network of microcracks, which are characteristics of the APS microstructure. As a consequence, erosion of APS ceramic involves relatively low energy. A larger amount of ceramic is removed by a small amount of particulates, resulting in higher erosion rates. The EB-PVD coating, on the other hand, has strong columns with weak vertical column boundaries. New cracks have to be created within the columns at the expense of large energy. On impact the columns undergo tensile cracking and fracture between column boundaries, thus limiting the amount of material released. The cracks in a column do not propagate to the neighboring column because of deflection at the weak boundaries. This is why the erosion rate is low for EB-PVD ceramic. As fracture is limited to the near-surface region, the surface finish of EB-PVD ceramic is not adversely affected by continued erosion, which is opposite of what happens to APS ceramic.

Sintering and densification of the ceramic coating occurs when it is exposed to high temperature for extended period of time. In APS ceramic, sintering results in reduction of microcracks, increase in strength, and elimination of splat boundaries. Consequently, a reduction in erosion rate takes place on sintering. However, in EB-PVD ceramic, sintering increases erosion rate (Wellman and Nicholls, 2004). This results from partial fusion of neighboring columns, which allows cracks to propagate across multiple columns, increasing the amount of mass loss. Thus, sintering tends to close the gap in erosion resistance between the APS and EB-PVD ceramic.

Foreign Object Damage (FOD)

FOD is created by a limited number of large particles impinging on ceramic coatings. Although the normalized rate of erosion by these large particles is lower than that by finer particles, $100\,\mu m$ or less in size, the degree of damage is greater and tends to increase with the particle size. For example, a $200\text{-}\mu m$ size alumina particle impacting on EB-PVD YSZ creates damage limited to the outer $40\,\mu m$ zone of the coating. In this zone, the column tips are fractured with loss of the tips and signs of some local plastic deformation. Impact by $1000\text{-}\mu m$ particles, on the other hand, creates extensive damage by compressing and densifying the ceramic. The local compaction results in increase in thermal conductivity of the ceramic and contributes to the formation of a hot spot on the bond coat below and eventual failure of the TBC. Extensive cracking occurs outside the compacted zone. The damage, in the form of cracks, extends to and run parallel with the TGO. With continued impact, the ceramic in the damaged area may be liberated. The sizes of the liberated areas are in the range of a few hundred microns to several millimeters.

Damage Due to Changes in the TGO

Extended life of TBC is generally associated with a protective, slow-growing, uniform, thin TGO, which is predominantly α alumina. However, with insufficient aluminum or after significant aluminum depletion, bond coats fail to maintain such a TGO (Shillington and Clarke, 1999). On most aluminum-containing bond coats, initially α alumina forms. However, with continued oxidation and growth of TGO, the aluminum activity in the bond coat decreases below the limit required to continue formation of alumina. The oxygen activity increases. At this stage, oxides of the other alloying elements start to form, beginning with spinel (Co, Ni) $(CrAl)_2O_4$, αCr_2O_3, and NiO. The bond strength of the interfaces of these oxides with the TBC as well as with the bond coat tend to be lower than that with α alumina. As a result, failure of the TBC by delamination occurs much sooner than for a TBC system with a bond coat having adequate aluminum activity.

Abrupt failure of TBCs on articles out of test or service, known as "desktop" failure, has been known to occur at room temperature. TBCs with significant prior thermal exposure history (generating residual stress, Section 7.5), which otherwise look fine to the naked eye, may spall without any sign of additional thermal exposure. Studies by Smialek et al., 2002; Smialek, 2006 indicate that this desktop failure involves moisture inducing hydrogen embrittlement, or more appropriately hydrogen charging at the TGO bond coat interface from the reaction Al (in bond coat) $+ 3H_2O$ (in air) $= Al(OH)_3{}^- + 3H^+$.

Role of Residual Stress

As in the case of plasma sprayed TBC, residual stress in the ceramic coating as well as within the TGO plays an important role in influencing the durability of the EB-PVD deposited TBC.

Stress measured within the ceramic layer gives values significantly lower than that within the TGO, which explains why failure seldom occurs within the ceramic for EB-PVD deposited TBC.

Stress within the TGO

- The TGO exhibits large residual biaxial compressive stress with a magnitude between 3 and 5 GPa at room temperature. There are two contributing components to this residual stress:

  1. Stress resulting from cooling to room temperature from the stress free state at high temperatures. This component arises as a result of the accommodation of mismatch of coefficient of thermal expansion (CTE) between the polycrystalline alumina TGO layer and the metallic bond coat. Elastic analysis, discussed earlier in Section 7.4 in relation to stresses in plasma-sprayed TBC, provides the CTE mismatch stress (Tien and Davidson, 1975),

  $$\sigma_{ox} = -[E_{ox}(\alpha_m - \alpha_{ox})\Delta T]/[(1 - \nu_{ox}) + 2\{h_{ox}\ E_{ox}(1 - \nu_m)/h_m E_m\}],$$

  where ox corresponds to the TGO oxide, m corresponds to the metallic bond coat, α is the CTE, E is the modulus of elasticity, ν is the Poisson ratio, h is the layer thickness, σ is the stress, and ΔT is the difference between oxidation temperature and room temperature. Since $h_{ox} << h_m$, the CTE mismatch residual stress reduces to

  $$\sigma_{ox} = -[E_{ox}(\alpha_m - \alpha_{ox})\Delta T]/(1 - \nu_{ox}).$$

  Introducing the variation of thermal expansion coefficients with temperature,

  $$\sigma_{ox} = -[E_{ox}/(1 - \nu_{ox})] \int (\alpha_m - \alpha_{ox}) dT,$$

  integrated between the oxidizing temperature T_0 and the room temperature (measurement temperature) T. The CTE mismatch stress has been estimated by Christensen et al. (1996) for aluminum oxide scale formed on Ni_3Al. Ni_3Al is the γ' phase that strengthens nickel base superalloys. Using appropriate α values for Ni_3Al and aluminum oxide, and room temperature elastic modulus of alumina ~380 GPa, the thermal expansion mismatch stress is calculated to be compressive ~3.5 GPa.

  2. Intrinsic stress associated with oxidation of metallic species to oxide of the TGO. There is an overall volume increase resulting from the conversion of the high-density alloy to the low-density oxide. The stress at which internal TGO growth is suppressed is given by (Cannon and Hou, 1998)

  $$\sigma = -k_B T[\ln (p^g_{O2}/p^{eq}_{O2)}][1 + 1/\{1 + 2(h_{ox}/G)\}]/24\Sigma_{Al2O3},$$

  where k_B is Boltzmann's constant, p is pressure, g is gas, eq represents equilibrium, G is the TGO alumina grain size, h_{ox} is the oxide thickness, and Σ is the molecular volume of alumina. Inserting typical values of the parameters gives a compressive stress value ~3 GPa (Cannon and Hou, 1998), which is significantly higher than the typical value of ~1 GPa estimated by subtracting calculated thermal expansion mismatch stress from the total stress measured experimentally. The discrepancy may be due to theoretical overestimation or to the fact that the stress exceeds the creep strength of the TGO, and plastic deformation of TGO occurs, leading to stress relaxation.

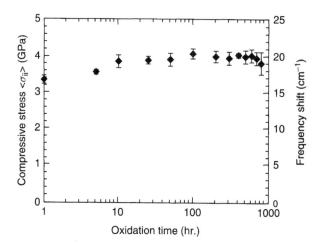

Figure 7.47 The evolution of room temperature residual stress in the TGO as a function of oxidation time at 1135°C (2075°F) (R. J. Christensen, D. M. Lipkin, D. R. Clarke, and K. Murphy, Nondestructive evaluation of the oxidation stresses through thermal barrier coatings using Cr^{3+} piezospectroscopy, *Appl. Phys. Lett.* 1999, 69(24), 3754–3756). Reproduced with permission from American Institute of Physics.

The residual stress in the TGO is measured by photoluminescence piezospectroscopy (PLPS). The basic elements of the technique are discussed in Chapter 8.

- The typical magnitude of the total residual stress lies between 3 and 4 GPa (Fig. 7.47) (Christensen et al., 1999), of which approximately 1 GPa is contributed by intrinsic oxide growth stress.

- There are several important features to the residual stress profile in the TGO (Fig. 7.47):

  (1) In most cases residual stress does not change significantly with time on continued oxidation.

  (2) At least in some systems, room temperature residual stress does not seem to depend on the oxidation temperature (Christensen et al., 1996).

  (3) There is a short transient period during which the stress increases. It is probably associated with transformation of transient alumina phases into stable α form.

  (4) Immediately prior to spallation of TBC, there is a decrease in stress possibly due to the onset of decohesion of the TGO from the metallic surface.

- Residual stress measured at various locations on turbine blades in coated as well as engine run conditions depends on the shape and size of the blade as well as the location of measurement. The TGO on the concave (pressure side) face is more stressed than on the convex face (Sohn et al., 2001) for aero engine blades, whereas for industrial gas turbine blades, the leading edge is more stressed, and the stresses in the concave and convex surfaces are approximately equal (Sohn et al., 2000). Engine exposure does not change the residual stress by any significant level, except in long exposures when stresses have been found to fall due to the initiation of microdelamination prior to spall.

- Except in the early transient period, stresses in the TGO are higher when the bond coat contains reactive elements such as yttrium (Christensen et al., 1997).

- Processing effects: There is significant difference in residual stress between samples grit blasted prior to ceramic deposition and those without grit blasting; the former is lower than the latter. The bond coat oxidation rate is also higher for samples grit blasted prior to the ceramic deposition. This is probably due to the incorporation of impurities at the bond coat surface by the grit blasting process and the blasting medium. TBC failure occurs early on grit blasted samples.

- Analyses of microstructure after isothermal and cyclic oxidation tests suggest that there is a critical TGO thickness at which failure occurs within the TGO.

- Some studies on specific bond coats show that the TGO growth rate is higher under the ceramic than the oxide scale growth rate on a bare bond coat (Fig. 7.41) (Stiger et al., 1999). As discussed before, the higher TGO growth rate under the TBC can often be due to the absence of the microspalling that occurs in the presence of TBC (R. A. Miller, personal communication, 2006). The other contributing factor is the presence of transient aluminas, γ, and θ.

- The stress contributions within the TGO at various stages of detachment of the elements of the TBC are shown in Fig. 7.48 (based on Peng and Clarke, 2000). The picture emerging from the limited data in Fig. 7.48 indicates that the TGO is under high compressive stress, about 70 to 80% of which is due to CTE mismatch between the TGO and the bond coat while the remainder, 30 to 20%, is due to the growth of the oxide of the TGO. The presence of the ceramic apparently increases the residual compressive stress in the TGO. Partial delamination reduces the CTE mismatch stress. However, even at full delamination, the TGO is under compressive stress contributed by the ceramic coating to which it is still adherent. A similar trend in reduction of TGO residual stress has also been observed in thermal cycling studies by Ferber and by Gell (Ferber, 1999; Gell et al., 2004).

- High residual stress in the TGO of the coating can initiate a number of degradation processes:

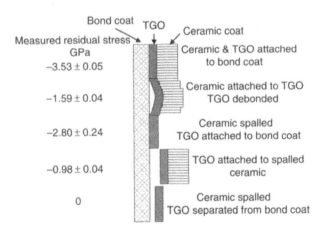

Figure 7.48 Room temperature residual stress within the TGO at various stages of TBC spallation (Xiao Peng and David R. Clarke, Piezospectroscopic analysis of interface debonding in thermal barrier coatings, *J. Am. Ceram. Soc.*, 2000, 83(5), 1165–1170). Reprinted with permission from Blackwell Publishing.

-Cracking of the oxide

-Separation of the oxide from the alloy substrate

-Plastic deformation of the bond coat and the substrate

-Plastic deformation of the oxide

-A variant of the deformation process: the penetration by the TGO into the bond coat. The TGO is stronger than the bond coat at high temperature. During continued growth of the TGO parallel to the interface, it penetrates into the creeping and weak bond coat, creating defect sites that significantly contribute to failure.

- The condition for fracture of the TGO is obtained by equating the elastic strain energy G_0 within the TGO with the fracture resistance of the interface (Wright and Evans, 1999),

$$G_0 = (1 - \nu_{ox})\sigma^2 h_{ox}/E_{ox} \geq \Gamma_0$$

where Γ_0 is the fracture toughness of the interface, and ν_{ox}, σ, h_{ox}, and E_{ox} have their usual meaning. This condition is necessary but not sufficient. Initiation of buckling of the TGO and the development of wedge cracks are some of the other processes consuming the elastic strain energy. The equation shows that spall within the TGO will occur if the stress is too high, the TGO is too thick, and/or the work of adhesion low (interfacial energy is high).

- Stress and strain distribution in the TGO and the bond coat at the interface (Tolpygo and Clarke, 1998): Although the ceramic coating does affect to some extent the residual stress within the TGO, as is evident from Fig. 7.48, a consistent quantitative measurement of the contribution has not been made. We will assume that the contribution is small and focus on the system without the ceramic layer. In the absence of ceramic coating, the compressive load within the TGO must be balanced by the tensile load in the metallic substrate so that the combined system is stress free. Therefore, force balance requires

$$\sigma_{ox} h_{ox} + \sigma_m h_m = 0.$$

The strain in the TGO has three contributors, which include elastic, creep, and growth components. The stress-free state is at the temperature at which the oxide forms and grows. The creep and elastic strains in the TGO are negative because it is deforming under compression during thermal cycling.

$$\varepsilon_{TO} = \varepsilon_{ox}{}^{elas} + \varepsilon_{ox}{}^{creep} + \varepsilon_{ox}{}^{growth}$$
$$= \sigma_{ox}(1 - \nu_{ox})/E_{ox} + (\alpha_m - \alpha_{ox})\Delta T + \varepsilon_{ox}{}^{creep} + \varepsilon_{ox}{}^{growth},$$

where ε is the strain, subscript TO indicates the total, ox is the TGO oxide, and m indicates metal. The strain in the metal has two contributing elements, which include elastic and creep components.

$$\varepsilon_{TO} = \varepsilon_m{}^{elas} + \varepsilon_m{}^{creep}$$
$$= \sigma_m(1 - \nu_m)/E_m + \varepsilon_m{}^{creep}$$
$$= -\sigma_{ox}(h_{ox}/h_m)(1 - \nu_m)/E_m + \varepsilon_m{}^{creep}$$

using the force balance equation just given.

From residual stress determination within the oxide and measurement of lateral growth of the oxide scale, Tolpygo et al. (1998) have estimated the values of the relaxation parameters ε_{ox}^{creep} and ε_{m}^{creep} for the model coating Fe22Cr4.8Al0.3Y. The values are dependent on temperature, time, and specimen thickness. The magnitude of ε_{m}^{elas} at room temperature is of the order of 1×10^{-5} to 4×10^{-4}. Length measurement after oxidation gives ε_{TO} to be an order of magnitude higher. ε_{m}^{creep}, the difference between ε_{TO} and ε_{m}^{elas}, is therefore large, indicating that the metallic coating deforms plastically during oxidation. Signs of such creep deformation have indeed been observed on many bond coats (see Chapter 10).

The residual stress within the TGO can be calculated from

$$\sigma_{ox} = -[E_{ox}(\alpha_m - \alpha_{ox})\Delta T]/[(1 - \nu_{ox}) + 2\{h_{ox}E_{ox}(1 - \nu_m)/h_m E_m\}].$$

Using the values (Tolpygo and Clarke, 1998; Watchman and Lam, 1959; Robertson and Manning, 1990; Touloukian, 1977) listed in Table 7.4, Tolpygo and Clarke calculated the value of thermally induced stress to be about -3.7 GPa for oxidation at 1100°C (2012°F) for 100 hours exposure. The growth stress of the oxide can be calculated by subtracting the thermal stress from the total stress measured by the PLPS technique (corrected for temperature using an appropriate value of elastic modulus). This gives the TGO oxide growth stress

$$\sigma_{ox}(\text{Growth})(E_{ox})_{RT}/(E_{ox})_T = \sigma_{total} - (\sigma_{ox}).$$

For oxidation at 1100°C (2012°F) for 100 hours, the value of $\sigma_{ox}(\text{Growth})$ comes out to be -1.2 GPa. Using the stress balance relationship, the stress in the metal after oxidation for 100 hours at 1100°C (2012°F) is estimated to be ~10 MPa. The measured growth stresses of alumina on various materials shown in Table 7.5 (Sarioglu et al., 2000) are within the range matching the value calculated above.

Table 7.4 Values of Parameters Contributing to Residual Stress

Temperature°(C)	h_{ox} μm	E_{ox} GPa	E_m GPa	$\alpha_{ox} \times 10^6$	$\alpha_m \times 10^6$
25	0	400	190		
1000	0.255	350		8.5	14.3
1100	0.655	340		8.6	14.6
1200	1.412	330		8.7	14.9
1300	2.63	320		8.8	15.1

Note: ox = Alumina, m = FeCrAlY, $\nu_{ox} = 0.25$, $\nu_m = 0.3$

Table 7.5 Growth Stress of Alumina Scale on Various Substrates

Alloy	T, °C (°F)	Growth Stress, GPa	Method
NiAl	1100 (2012)	~0	Hot stage (FIM, rocking)
Poly XL-LoS	1100 (2012)	~0	Estimated from residual stress
FeCrAlY	1100 (2012)	−1.3	Hot stage (FIM)
FeCrAlY	1000 (1832)	−1.7	Hot stage (FIM)
FeCrAl	1000 (1832)	−1.0	Hot stage (FIM)

Note: FIM = field ion microscope.

Growth stresses within the alumina scale formed on NiAl type alloys with varying Al content and other minor elements have also been experimentally measured by the x-ray diffraction technique at oxidation temperature to eliminate CTE-related stress. Synchrotron radiation sources have been used to exploit the high x-ray intensities available (Tortorelli et al., 2003; Hou et al., 2004). Generally, the measured stresses are dependent on the substrate, oxidation time and temperature, and are lower than the value predicted from PLPS data corrected for CTE mismatch stress.

Stress within the Ceramic Coating

The residual stress within the ceramic layer varies with processing conditions. Measurements made from change in substrate curvature upon coatings removal (Johnson et al., 1998) provide values between $-70\,MPa$ (compression) and $20\,MPa$ (tension). X-ray diffraction measurements on EB-PVD ceramic deposited on platinum aluminide bond coat (Jordan and Faber, 1993) give residual stress in the coated condition in the range 270 ± 9 and $304\pm15\,MPa$. On thermal cycling in air, this residual stress tends to decrease. The stresses in the ceramic obviously are significantly lower than those within the TGO and are unlikely to be contributing factors for failure of the YSZ-based TBC system.

Roles of Oxygen Reactive Elements

The addition of oxygen REs, such as yttrium, to the substrate alloys or the bond coat improves the oxide scale (TGO) adherence as discussed in Section 4.4. Residual stress measurements (Christensen et al., 1997) generally indicate that the TGO stresses are higher in the presence of yttrium. One possible explanation is that yttria dispersion within the TGO results in higher creep strength and elimination of stress relaxation. Other explanations have been covered in Section 4.4 in the chapter on oxidation. TBCs live on bond coats with REs are significantly higher than those on bond coats without these elements.

Bond Strength

In addition to the defects within each layer of TBC and at the interface, the bond strength of the interface between the ceramic coating and the TGO as well as between the TGO and the bond coat and their evolution with temperature and time play a key roll in controlling the durability of the TBC. A simple conceptual model for failure (Bose and DeMasi-Marcin, 1997) (Fig. 7.55) is that with continued thermal exposure of TBC, the residual stress increases, while the bond strength decreases, due to alterations in the composition of the interface region and the TGO, that changes from alpha alumina to spinels, and eventually to NiO. Failure occurs at a time, which corresponds with the intersection of the residual stress and bond strength curves. Details of residual stress profile in the TBCs have already been covered. We now focus on the out-of-plane bond strength at the ceramic-bond coat interface. The method generally used for bond strength measurement is the modified ASTM direct pull test (ASTM C 633-79) using an epoxy adhesive. The limited bond strength data reported in literature involve the following systems:

Alloy/bond coat: PWA 1484/NiCoCrAlY, where the alloy is a single-crystal Ni base alloy from Pratt & Whitney (Meier et al., 1992); Alloy/bond coat: N5-/Pt aluminide, N5+/Pt aluminide, where N5 is a single-crystal Ni base superalloy René N5 from General Electric,

+ indicates 100 ppm yttrium addition while – means no yttrium added (Rudd et al., 2001); Alloy/bond coat: N5, IN-939, CMSX-4, Mar-M 509/Pt aluminide or NiCoCrAlY, where IN-939 is a commercial polycrystalline Ni base alloy, CMSX-4 is a single-crystal Ni base superalloy from Canon–Muskegan, and Mar-M 509 is a commercial polycrystalline Co base superalloy (Gell et al., 1999) (Fig. 7.49).

The data exhibit the following general trend:

- The average interface bond strengths of the TBCs in the as-coated condition for most of the substrates are in excess of 60 MPa (8.6 ksi). For an EB-PVD coated PWA 1484/NiCoCrAlY system, the as-coated bond strength was estimated by Meier et al. to exceed the adhesive strength of the particular epoxy used in the pull test, 41.4 MPa (6 ksi), because the failure had occurred in the epoxy.

- With thermal cycling to $\sim$1130°C (2066°F), the interface bond strength decreases significantly, initially rapidly within the first 100 cycles.

- Similar values are obtained for both Pt aluminide and NiCoCrAlY bond coated samples.

- There is large scatter in the data. In a significant fraction of test samples, failure occurs at the ceramic–TGO interface.

- TBC on N5+ failed at roughly twice the number of cycles compared with TBC on N5−, showing a strong effect of the reactive element yttrium on the bond strength.

In order to predict life of the TBC and develop a failure model, the degradation of the bond strength needs to be complemented with evolution of residual stress, and understanding of initiation and growth of critical flaws.

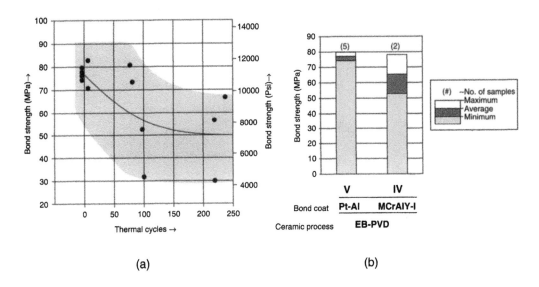

(a)   (b)

Figure 7.49 (a) Bond strength of TBC deposited by EB-PVD on Pt aluminide bond coat as a function of thermal cycling to 1121°C (2050°F). (b) Bond strength of as-coated samples with two different bond coats (Maurice Gell, Eric Jordan, Krishnakumar Vaidyanathan, Kathleen McCarron, Brent Barber, Yong-Ho Sohn, and Vladimir K. Tolpygo, Bond strength, bond stress and spallation mechanisms of thermal barrier coatings, *Surf. Coat. Technol.*, 1999, 120–121, 53–60). Reprinted with permission from Elsevier.

Structural Properties

Due to the columnar microstructure with weak inter-columnar cohesion, the mechanical properties, particularly stress–strain behavior in the deposition plane, are difficult, if not impossible, to measure for standalone EB-PVD ceramic coating. As a result, the properties are deduced from measurements made on TBC-coated and uncoated substrates, taking account of plastic behavior of the substrate at high temperature (Meier et al., 1992). Unlike plasma-sprayed ceramic, the tensile behavior of the EB-PVD ceramic is linear elastic up to fracture with tensile strength and elastic limit is less than 6.9 MPa (1.0 ksi). In compression, the response at room temperature remains linear elastic up to a strain of about 1%, increasing to 3% at 980°C (1796°F). Continued compression buckles the substrate with the ceramic remaining adherent. The calculated compressive stress–strain behavior of EB-PVD 7YSZ at 538°C (1000°F) is compared with that of the plasma-sprayed counterpart in Fig. 7.50. It is evident that EB-PVD ceramic exhibits higher compressive strength than the plasma-sprayed ceramic, the latter showing some plastic behavior even at 5 times higher strain rate. The elastic deformation of the EB-PVD ceramic, normalized for the same strain rate, is about 10% of that of the plasma-sprayed ceramic. The difference in the behavior is due to the widely different microstructure of the two systems. Prediction based on a viscoplastic model with assumptions similar to that for the plasma-sprayed TBC discussed earlier is also indicated in the Fig. 7.50

As expected for a columnar structure, the out-of-plane adhesive strength is significantly higher than the in-plane strength. Meier et al. report values in excess of 41.4 MPa (6 ksi).

The temperature dependence of in-plane Young's modulus, both in tension and compression (Fig. 7.51) (Meier et al., 1991), has also been determined from the stress–strain curves. Although the scatter in the tensile data is large, the general trend shows consistency between tension and compression.

Young's modulus in the plane of deposition as well as out-of-plane has also been measured by other methods. The data have been summarized by Schulz et al. (2000). Table 7.6

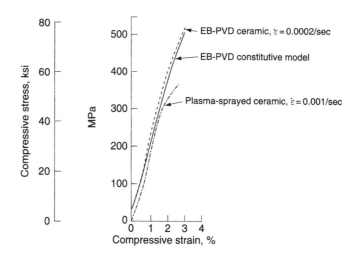

Figure 7.50 Compressive stress strain behavior of EB-PVD and plasma-sprayed 7YSZ at 538°C (1000°F) (Susan Manning Meier, David M. Nissley, Keith D. Sheffler, and Thomas A. Cruise, Thermal barrier coating life prediction model development, *J. Eng. Gas Turbine Power Trans. ASME*, 1992, 114, 258–263). Reprinted with permission from American Society of Mechanical Engineers International.

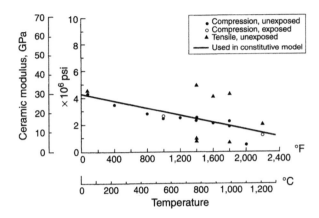

Figure 7.51 Temperature dependence of tensile and compressive Young's modulus (Susan Manning Meier, David M. Nissley, Keith D. Sheffler, and Thomas A. Cruise, Thermal barrier coating life prediction model development, *J. Eng. Gas Turbine Power, Trans. ASME*, 1992, 114, 258–263). Reprinted with permission from American Society of Mechanical Engineers International.

Table 7.6 Room Temperature Young's Modulus of EB-PVD

Orientation	Young's Modulus, Gpa (Room Temp.)	TBC Element	Test Method	Reference
Bulk	210			Schulz et., 2000
	42	Ceramic coating[a]	Tensile test	Meier et al., 1992
	100	TBC with bond coat	Compression laser ultrasonic	Morell et al., 1998
In-plane	15–77	TBC with bond coat	Flexural resonant frequency	Johnson et al., 1998
Out-of-plane	10–30	TBC with bond coat	Flexural resonant frequency	Johnson et al., 1998
	27	TBC with bond coat	Three point bend	Szücs 1998
In-plane	13–15	TBC	Three point bend	Szücs 1998
Out-of-plane	6	TBC	Three point bend	Szücs 1998

[a] Deducted from the TBC system and bond coal data.

shows the dependence of the modulus on the measurement technique. Other critical parameters include residual stress, microstructural detail, and texture. The effect of anisotropy in the EB-PVD structure is reflected in the flexural measurement, which includes out-of-plane contributions.

Oxidation and Thermocyclic Behavior

TGO growth during both cyclic as well as isothermal oxidation has been discussed in Section 7.5. On thermocycling at high temperature, TBCs deposited by the EB-PVD process eventually

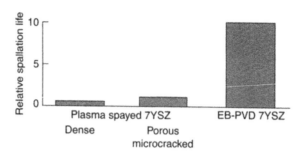

Figure 7.52 Comparison of life of TBCs deposited by EB-PVD with the plasma-sprayed counterpart (Susan Manning Meier and Dinesh K. Gupta, The evolution of thermal barrier coatings in gas turbine engine applications, *J. Eng. Gas Turbine Power, Trans. ASME,* 1994, 116, 250–257). Reprinted with permission from American Society of Mechanical Engineers International.

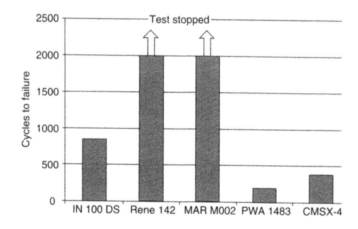

Figure 7.53 Dependence of EB-PVD TBC cyclic life on substrate alloy (U. Schulz, M. Menzebach, C. Leyens, and Y. Q. Yang, Influence of substrate material on oxidation behavior and cyclic lifetime of EB-PVD TBC systems, *Surf. Coat. Technol,* 2001, 146-147, 117–123). Reprinted with permission from Elsevier.

fail by spalling either within the TGO or at the TGO–bond coat interface. The strain-tolerant columnar microstructure extends the spall life significantly compared with that of the plasma-sprayed counterparts, as shown in Fig. 7.52 (Meier and Gupta, 1994).

The substrate alloy has a significant effect on the TBC life. Alloys containing oxygen REs such as yttrium or hafnium exhibit higher TBC life, as shown in Fig. 7.53 (Schulz et al., 2001). The bond coat for all samples was Ni20Co20Cr12Al with 0.1 to 0.2Y. The best performing alloys, René 142 and Mar-M 002, both contain 1.5%Y. Some of the alloying elements in the substrate migrate to the bond coat and the TGO and influence the TGO spallation through various mechanisms covered in the chapter on oxidation.

Failure Mechanisms and Life Modeling

Several life prediction methodologies have been proposed for the EB-PVD TBC, starting from the simplest empirical approach utilizing the Manson–Coffin fatigue model introduced in the plasma sprayed coating, Section 7.4, to the fracture mechanics model with the bond degradation model in between. Of these models, the Manson–Coffin model is presently the most utilized methodology for life prediction. The fracture mechanics model is the current focus of significant development. The rudiments of these models are discussed next.

- *Empirical approach using the Manson–Coffin type fatigue damage model:* This model relates cyclic life to total inelastic strain range and cumulative bond coat oxidation. However, it does not provide any insight into the failure process (Meier et al., 1992). In most applications, TBCs experience simultaneous thermal and externally applied mechanical strains. The thermal strains originate from the thermal expansion mismatch between the TGO and the metallic substrate. The strains are of the order of ~0.8% compression when cooled from 1135 to 77°C (2075 to 170°F) (Wright, 1998). The applied and thermal strains could be either in-phase (maximum compression at maximum temperature of the cycle) or out-of-phase (maximum compression at minimum temperature of the cycle). Thermomechanical fatigue tests (Fig. 7.54) indicate that the number of cycles to failure, N_f, can be related to total strain range through a relationship of the type

$$N_f = p(\Delta\varepsilon_{tot})^a,$$

where $\Delta\varepsilon_{tot}$ is the total strain (applied + thermal), and a and p are constants. The data reported by Wright (1998) were generated under the following types of test conditions: (1) samples cycled between T_{max}1135°C (2075°F) and T_{min}77°C (170°F) while under in-phase tensile loading; (2) samples for which there was no thermal strain; and (3) samples for which T_{min} was higher than 82°C (180°F). Although there is large scatter in the data typical of such analysis, the general trend gives the value of constant a between −3 and −4.

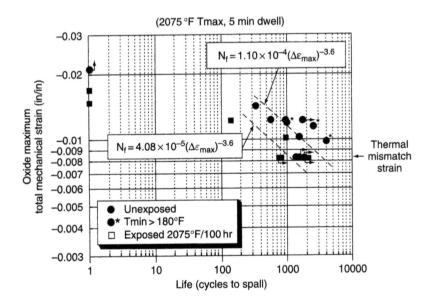

Figure 7.54 Maximum (hoop or axial) TGO scale total strain versus EB-PVD TBC life in TMF test (P. Kennard Wright, Influence of cyclic strain on life of a PVD TBC, *Mater. Sci. Eng.*, 1998, A245, 191–200). Reprinted with permission from Elsevier.

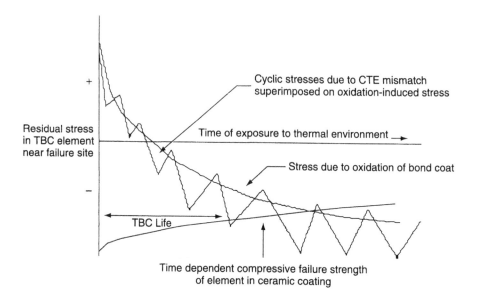

Figure 7.55 Failure occurs when stress equals bond strength (S. Bose and J. DeMasi-Marcin, Thermal barrier coating experience in gas turbine engines at Pratt & Whitney, *J. Therm. Spray Technol.*, 1997, 61(1), 99–104). Reprinted with permission from ASM International.

- *Residual stress and bond strength degradation model:* This model estimates life by equating out-of-plane stress at the TGO–bond coat interface with the interfacial bond strength (Fig. 7.55) (Bose and DeMasi-Marcin, 1997; Gell et al., 1999).

Both the stress and the bond strength are strong functions of cyclic thermal exposure. The out-of-plane stress increases while the bond strength decreases with continued cyclic thermal exposure. The TBC fails when they become equal. Although conceptually this model is simple to understand, it does not provide insight into the details of the failure process.

- *Fracture mechanics–based model:* This model utilizes such concepts as time-dependent interface toughness and strain energy release rate (Evans et al., 1997; He et al., 1998a and b). It provides some insight into individual events leading to the final loss of the TBC.

We have emphasized that failure of TBC deposited by the EB-PVD process generally occurs by spallation within the TGO or at the TGO–bond coat interface. The durability of the TBC is controlled by the energy density in the TGO and the formation or presence of imperfections in the vicinity of the TGO (Evans et al., 2001). A simple way to analyze the phenomenon is to calculate the total elastic strain energy per unit area. This is given by (Wright and Evans, 1999)

$$G_0 = (1 - \nu_{\mathrm{ox}}) h_{\mathrm{ox}} \, \sigma_{\mathrm{ox}}^{\,2} / E_{\mathrm{ox}},$$

where G_0 is the strain energy and h_{ox} is the TGO thickness, with other parameters having their usual meaning. At failure, the strain energy is consumed in inducing decohesion at the interface. Thus, at failure, the interface fracture toughness, Γ_0, should equal the strain energy G_0. The interface fracture toughness has been measured by Mumm and Evans (2000) with values between 56 and 80 J/m^2. The most likely scenario is that only part of the strain energy is consumed in delaminating the coating at the interface, the remainder being used for additional degradation processes.

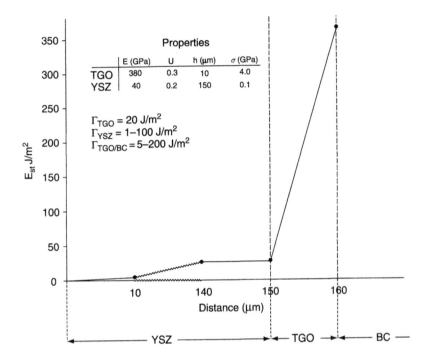

Figure 7.56 Cumulative strain energy of 150 μm TBC with 10 μm TGO as a function of distance from free surface of TBC (N. M. Yanar, G. Kim, S. Hamano, F. S. Pettit, and G. H. Meier, Microstructural characterization of the failures of thermal barrier coatings on Ni-base superalloys, *Mater. High Temp.*, 2003, 20(4), 495–506). Reprinted with permission from *Science and Technology Letters.*

In order to understand fracture at other locations within the TBC, cumulative elastic strain energy as a function of location needs to be considered. Figure 7.56 shows a plot of cumulative elastic strain energy versus distance from the surface of a 150-μm-thick TBC with 10 μm TGO (Yanar et al., 2003).

The strain energy within the TGO is significantly higher than in the ceramic coating, rising rapidly with TGO thickness. At a thickness of 5 μm of the TGO, the strain energy is $\sim 200\,J/m^2$. The magnitude of the strain energy is equivalent to the largest fracture energy reported for a metal–alumina interface. One, therefore, expects failure when the TGO grows to about 5 μm, not unlike the data observed in the field. However, failure could occur at other locations at lower strain energy due to defects and deviant microstructure.

Spall Mechanism and Fracture Mechanics Model

Thin oxide films including the TGO under residual biaxial compressive stress fail by spalling from the substrate. The fracture mechanics approach analyzes the failure as consisting of a series of events, each of which has to meet stress–strain and size criteria. The series of events leading to failure consists of the following sequence:

(1) A discrete separation forms at the interface between the TGO and the substrate (Fig. 7.57). This separation may form by one or more of the following mechanisms:

(a) It preexists.

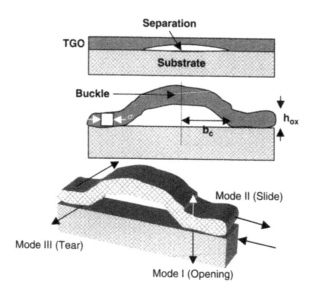

Figure 7.57 Buckle formation at the interface between the TGO and the bond coat.

(b) It forms by coalescence of voids at the interface. Voids are the result of unbalanced diffusional fluxes of metal ions involved in the oxidation process crossing the interface in the opposite direction. Unbalanced diffusion leads to creation of vacancies which, when high in concentration, condense to form voids.

(c) Creep and grain boundary sliding of the underlying substrate (Chutze, 1988) lead to decohesion of the oxide film to form the separation at the interface.

(d) Waviness associated with interfaces exhibits out-of-plane tensile stresses across the "hills" and compressive stresses across the "valleys." In the presence of defects such as voids and inclusions, the tensile stresses at the "hills" would induce separation.

(2) Once the separation reaches a critical radius with an associated stress (He et al., 1998a and b; Hutchinson et al., 1992), a blister called a "buckle" forms (Fig. 7.57).

   The required critical compressive stress and the radius of the separated area to develop a buckle are given by (Timoshenko and Gere, 1962; Hutchinson et al., 1992)

$$\sigma_c = [1.2235 \, E_{ox}/(1 - \nu_{ox}^2)](h_{ox}/b_c)^2,$$

which gives

$$b_c = 1.1062 h_{ox}[E_{ox}/(1 - \nu_{ox}^2)\sigma_c]^{1/2},$$

where h_{ox} is the oxide scale thickness and b_c is the smallest separation radius that buckles.

   Assuming that the stress in the scale is solely due to thermal expansion mismatch of the order of 3.5 GPa and the scale thickness is 5 μm, the calculated critical separation size to form a buckle comes out to be ∼50 μm, an order of magnitude larger than the TGO thickness. Such large flaw sizes are unlikely to preexist in the oxide scale or the TGO. However, existence of perturbations in the form of undulations of the oxide metal interface creates local tensile stresses at the convex tips, aiding in the buckling. Additionally, large growth stresses increase σ_c, thus reducing b_c.

(3) Compressive stress within the TGO leads to tensile stress at the periphery of the buckle (Fig. 7.57). The buckle so formed will remain stable and not propagate unless the stress intensity at the periphery of the buckle exceeds the critical stress intensity factor for fracture at the interface. Another way to look at the process is in terms of strain energy and its release rate (Wang and Evans, 1998). The extension of the buckle creates additional fracture surfaces that requires strain energy release rate G from the TGO,

$$G = G_0 \; g\{1 - (\sigma_c^2/\sigma^2)\},$$

where $G_0 = [(1 - \nu_{ox})h_{ox}\sigma^2]/E_{ox}$ is the elastic energy stored in the TGO, and

$$g = [1 + 0.9(1 - \nu_{ox})]^{-1}.$$

The strain energy release rate is related to the stress intensity factors through the relation (Hutchinson et al., 1992)

$$G = (K^2_{ox} + K^2_m)/E, \quad 1/E = (1/E_{ox} + 1/E_m),$$

where K is the stress intensity factor, E is plain strain ($h_{ox} << h_m$) Young's modulus, and subscripts ox and m refer to the TGO and the bond coat.

The buckle starts to propagate, leading to eventual spalling of the TGO and the failure of the TBC if $G \geq \Gamma_0$, where Γ_0 is the toughness of the interface. The buckle propagation may shift from one mode (Mode I, opening or tensile mode) to another (Mode II, sliding or shear mode, and Mode III, the tearing mode) through mixed modes in between. The value of Γ_0 depends on the particular mode of propagation. For propagation involving Mode I separation, the critical stress for the buckle to propagate is given by

$$\sigma_p = \phi[E_{ox}\Gamma_0/(1 - \nu_{ox})h_{ox}]^{1/2}.$$

For Mode I propagation along the interface, $\phi = 2.5$ and Γ_0 is the Mode I interface toughness. If the interface toughness is relatively high, that is, $\Gamma_0/\Gamma \geq 0.56$, where Γ is the oxide scale toughness, the buckle tends to propagate into the film (kink) and spallation results. The spall size is the same as the buckle size. The stress at which this occurs is given by the foregoing equation with $\phi \sim 1.7$.

Following He et al. (1998a and b), some rough order of magnitude calculations can be made with these assumptions: TGO modulus 400 GPa; $h_{ox} = 1 \mu m$; and interface fracture energy ranging between $1 J/m^2$ for a highly segregated Ni–alumina interface, and $10 J/m^2$ for a diffusion-bonded Ni/alumina interface (He and Hutchinson, 1989). Incorporation of the foregoing values into the buckle formation and spallation equation gives a critical stress in the range between 2 and 6 GPa for a TGO thickness of $1 \mu m$. These calculated stresses are comparable with the thermal expansion mismatch stresses measured within the oxide scale. In order to achieve the same critical buckling stresses at higher interfacial toughness values such as the 56 to $80 J/m^2$ reported by Mumm and Evans, and 80 to $100 J/m^2$ reported by others, h_{ox} needs to be between 7.5 and $10 \mu m$, which is in the range of TGO thickness observed in the field at TBC spallation.

The buckling equation clearly shows that the critical parameters controlling spallation are the stress σ, TGO thickness h_{ox}, and the interface and oxide scale toughness related to bond strength Γ.

Thermal Properties of TBC

The efficiency of a TBC is measured by the extent to which it reduces the temperature of a component while meeting durability demanded in its application. For a given heat flux and thickness, this efficiency is inversely proportional to the thermal conductivity of the ceramic layer, as discussed in Section 7.1. The indirect benefits of reduced temperature of the underlying metal are increased component life and reduced need for cooling air. Extensive research is, therefore, underway in industry and academia (Maloney, 2001; Clarke, 2003; Levi, 2004; Nicholls et al., 2002; Zhu et al., 2001; Zhu and Miller, 2004; Hass et al., 2001; Alperine et al., 1998) to identify and develop TBCs with thermal conductivities lower than that of the current industry standard, 7YSZ. For rotating components such as turbine blades, lower conductivity ceramic provides an added bonus in reducing coating thickness and thereby parasitic weight. The other important thermal properties include coefficient of thermal expansion and specific heat of the ceramic layer and the TGO. Thermal expansion of the ceramic layer and the TGO relative to the metallic bond coat determine the residual stresses in each layer and contribute to cracking and spallation.

Thermal Behavior of 7YSZ

Thermal Expansion

Coefficient of thermal expansion (CTE) of EB-PVD 7YSZ is shown in Fig. 7.58 (Meier et al., 1992).

For comparison, APS 7YSZ is also included in the plot. Considering the experimental error of measurement (1% for orientation perpendicular to the plane of deposit and 10% orientation parallel to the plane of deposit), there is no significant difference between APS and EB-PVD 7YSZ as well as between perpendicular and parallel orientation for the columnar structure.

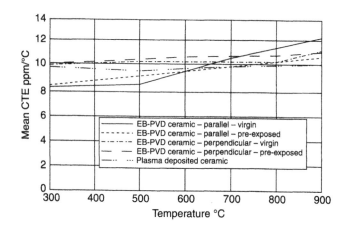

Figure 7.58 Coefficient of thermal expansion of 7YSZ as a function of temperature of measurement (S. M. Meier, D. M. Nissley, K. D. Sheffler, and T. A. Cruse, Thermal barrier coating life prediction model development, *J. Eng. Gas Turbine Power, Trans. ASME*, 1992, 114, 258–263). Reprinted with permission from American Society of Mechanical Engineers International.

Thermal Conductivity and Specific Heat

In order to develop low-conductivity ceramics for TBCs, the thermal properties of the baseline 7YSZ system need to be understood. A number of studies (references above; additional references include Klemens and Gell, 1998; Pawloski et al., 1984) have been conducted to measure these properties of both plasma sprayed and EB-PVD deposited 7YSZ.

Typical values of specific heat, both measured and calculated as functions of temperature, are shown in Fig 7.59 (Dinwiddie et al., 1996). The data show that heat capacity is the independent of microstructure, identical for plasma sprayed as well as EB-PVD 7YSZ. Thermal conductivity, however, depends strongly on the microstructural features, phase content, and the thermal exposure history of the coatings. The effect of microstructure will be discussed in a later section. The conductivity of the major phases of variously stabilized zirconia has been estimated from experimental data (Hasselman et al., 1987; Alperine et al., 1998). As shown in Table 7.7, the thermal conductivity of the three phases scales with the stabilizer content, which controls the defect concentration.

The available data on the thermal conductivity (Alperine et al., 1998; Meier et al., 1992; DeMasi et al., 1989; Miller et al., 1993; Eaton et al., 1994; Dinwiddie et al., 1996; Morrell

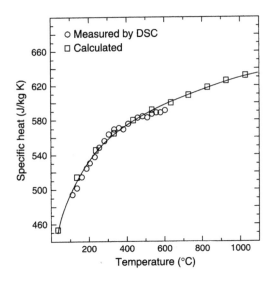

Figure 7.59 Specific heat of both plasma sprayed and EB-PVD 7YSZ (Ralph B. Dinwiddie, Stephen C. Beecher, Wallace D. Porter, and Ben A. Nagaraj, The effect of thermal aging on the thermal conductivity of plasma sprayed and EB-PVD thermal barrier coatings, *ASME 96-GT-982*, 1996). Reprinted with permission from American Society of Mechanical Engineers International.

Table 7.7 Thermal Conductivity of Various Phases of Zirconia

Phase	Thermal Conductivity, W/mK	
	Yttria Stabilized	Magnesia Stabilized
Monoclinic	4.2	5.2
Tetragonal	3.5	4.8
Tetragonal	2.3	1.8

and Taylor, 1985; Staniek and Marci, 1996; Portal, 1997) of yttria-stabilized zirconia are summarized in Table 7.8. Apart from the effect of the method of measurement, variation in conductivity for the same composition probably comes from differences in microstructure (porosity, microcrack density, and column orientation in EB-PVD deposited coating).

The primary effect of extended thermal exposure of a TBC is sintering and densification, which result in increased thermal conductivity. This is evident from Table 7.8 and more clearly demonstrated by NASA data (Fig. 7.60). The data were obtained by continually exposing the TBC ceramic surface to high heat flux generated by a 3.0 kW CO_2 laser of 10.6-μm wavelength. The metallic substrate side of the sample was air-cooled. Thermal conductivity was determined by measuring the ceramic surface temperature and monitoring the temperature of the back face as a function of time.

Table 7.8 Summary of Thermal Conductivity of YSZ

Deposition Process	Composition	Thermal Conductivity, W/mK	
		300 K	1300 K
Bulk	7 YSZ	2.9	2.7
APS	6 YSZ	1.1	
	8 YSZ	1.0	
	12 YSZ	0.6	
	7 YSZ (various porosity)	0.9–1.1	
	7 YSZ	0.8	0.8
	20 YSZ as sprayed (7 v/0 porosity)	0.4	0.55
	Heat treated at 1480°C	1.5	1.5
EB-PVD	7 YSZ	1.7	2.0
	7 YSZ	1.3	1.3
	8 YSZ as coated	1.5	1.3
	Heat treated	1.9	1.5

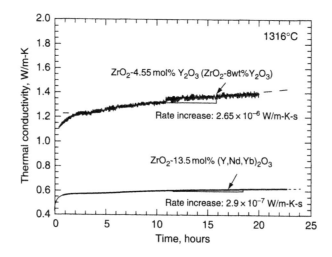

Figure 7.60 Effect of thermal exposure on thermal conductivity of plasma-sprayed TBCs (Dongming Zhu and Robert A. Miller, Development of advanced low conductivity thermal barrier coatings, *Int. J. Appl. Ceram. Technol.*, 2004, 1(1), 86–94). Reprinted with permission from Blackwell Publishing.

Increase in thermal conductivity on high-temperature cyclic furnace exposure has also been demonstrated on an 8YSZ TBC system using a method known as time-domain thermoreflectance (TDTR) (Zheng et al., 2005). The ceramic layer was deposited by the EB-PVD process on Pt aluminide bond coated René N5 superalloy. The measured thermal conductivity for the as-deposited TBC was 1.84 ± 0.08 W/mK in-plane and 1.7 ± 0.25 W/mK through thickness; that is, along the columns. After 500 furnace cycles from room temperature to 1135°C (2075°F), each cycle being 45 minutes at temperature and 15 minutes of cooling, the in-plane thermal conductivity increased to 2.2 ± 0.2 W/mK.

The increase in thermal conductivity of plasma-sprayed 8YSZ due to sintering has been shown to follow a "Larson–Miller" type relationship between temperature T in kelvins and time of exposure t in seconds (Zhu and Miller, 1999) (Fig. 7.61).

In most cases the increase in thermal conductivity is due primarily to sintering of the ceramic, which results in elimination of porosity and other defects, especially submicron-size microcracks.

The Effect of Microstructure

TBCs exhibit much lower thermal conductivity than their bulk counterparts (Fig. 7.62) (Nicholls et al., 2002), due to the presence of microcracks and porosity.

Additionally, plasma-sprayed TBCs exhibit thermal conductivity roughly half that of EB-PVD deposits for the same composition. A significant part of this difference is attributable to differences in microstructures between the coatings deposited by the two processes. The plasma-sprayed microstructure contains a network of microcracks. One set of these microcracks is parallel with the deposition surface and arises from the poor contact between the depositing splats. These microcracks provide discontinuities perpendicular to the path of heat conduction. The second set of microcracks is perpendicular to the deposition surface and arises from cracking of the individual splats due to thermal stresses generated on cooling. These perpendicular cracks do not contribute significantly to lower thermal conductivity. Additional features of the plasma-sprayed coatings that influence thermal conductivity are intersplat and

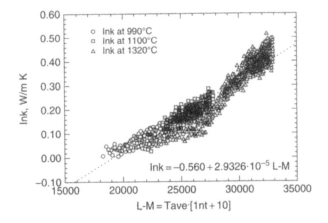

Figure 7.61 Change in thermal conductivity of plasma-sprayed 8YSZ as a function of thermal exposure plotted against "Larson–Miller (L-M)" parameter involving temperature T (K) for time t (s) (Dongming Zhu and Robert A. Miller, Thermal barrier coatings for advanced gas turbine and diesel engines, *NASA Report TM-1999-209453*, 1999). Courtesy of Robert A. Miller, NASA Glenn Research Center.

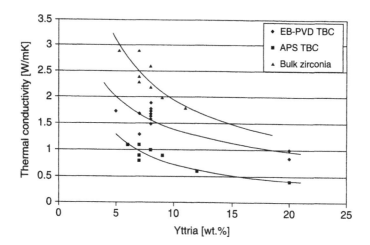

Figure 7.62 Thermal conductivity of plasma-sprayed, EB-PVD, and bulk zirconia as a function of yttria content (J. R. Nicholls, K. J. Lawson, A. Johnstone, and D. S. Rickerby, Methods to reduce thermal conductivity of EB-PVD TBCs, *Surf. Coat. Technol.*, 2002, 152–152, 383–391). Reprinted with permission from Elsevier.

intrasplat microporosity. The combination of the microstructural features has been successfully incorporated into thermal conductivity models (McPherson, 1984).

The microstructure of EB-PVD deposited TBCs exhibits fine porosity originating from the deposition process. However, it lacks the transverse microcracks typical of the plasma-deposited structure. The thermal conductivity reduction, therefore, comes only from the presence of fine porosity and thus is not as large as in the case of plasma spray. There are two types of porosity present in the structure, one residing within the columns, the other between them. The intracolumnar porosity, lying parallel to the deposition surface and perpendicular to the heat flow, reduces thermal conductivity. The intercolumnar porosity, being aligned perpendicular to the deposition surface, does not contribute significantly to the thermal conductivity reduction.

One of the characteristics of thermal conductivity of EB-PVD structure is its thickness dependence (Lawson et al., 1996). Reported values of thermal conductivity for coating thickness $<100\,\mu$m lie in the range 0.8–1.0 W/mK, while the mean value for average thickness of 250 μm is between 1.5 and 1.9 W/mK. In order to explain this behavior, a two-layer model has been proposed (Nicholls et al., 2002). The inner layer consists of fine grains with nanometer grain sizes. Calculations by Klemens (1997) show that for grain sizes of 5 nm, which is commensurate with the grains of the inner layer of EB-PVD 7YSZ, the thermal conductivity could be as low as 0.7 W/mK. In order to achieve the mean value of thermal conductivity for the total coating thickness, the outer layer needs to be closer to that of bulk, at around 2.2 W/mK.

Reduction of Thermal Conductivity

A number of approaches are available to create TBCs with thermal conductivity lower than that of the current standard 7YSZ. Some of these can be understood from an analysis of the fundamental equation of thermal conductivity of insulating solids, discussed in Section 7.4,

$$K = 1/3\ Cvl,$$

where C is the heat capacity per unit volume, v is the velocity of phonons, and l is the phonon mean free path. At temperatures $T > \theta$, where θ is the Debye temperature, the heat capacity C achieves a temperature-independent constant value $3k_B$ per atom, k_B being Boltzmann's constant. This is the well-known Dulong–Petit law. Thus, at high temperatures thermal conductivity per atom is given by

$$K = k_B v l.$$

Slack (1979) and later Clark (2003) simplified this equation in convenient forms, which allow one to calculate the minimum possible value of thermal conductivity, K_{min}. The mean phonon velocity v_{min} is approximated to the velocity of sound in the solid $\sim 0.87\,E^{1/2}/\rho^{1/2}$, and the minimum phonon mean free path l_{min} is equated to the dimension of the molecule (cube root of the molecular volume). The minimum thermal conductivity is now expressed as

$$K_{min} = 0.87\,k_B N_A^{2/3}(m^{2/3}\rho^{2/3}E^{1/2})/M^{2/3}.$$

Here N_A, E, M, m, and ρ are Avogadro's number, elastic modulus, molar mass, number of atoms per molecule, and density, respectively. This relationship clearly shows that thermal conductivity can be altered by alloying additions (M and ρ) and using novel oxide ceramics (M, m, E, and ρ). Because the foregoing relationship addresses intrinsic thermal conductivity only, it does not capture contributions by microstructural variations. If microstructural alteration is included, there are three avenues available to create low-conductivity TBCs, as discussed next.

Microstructural Modification

Because mean free paths for phonon scattering are smaller (~ 0.5 to $1\,nm$) than the typical grain size of TBC, 0.1 to $1\,\mu m$, grain size control has little effect except in the nanoscale ranges. More effort has, therefore, been focused on microcracking, porosity generation, and alignment of porosity. An example of the latter concept is the creation of zigzag-microstructure EB-PVD coatings (Haas et al., 2001) with reduction in conductivity by as much as 50%. A novel approach in microstructural modification to reduce thermal conductivity of EB-PVD deposited TBC is through the manipulation of deposition interfaces decorated with a second phase material. This concept has been successfully demonstrated and engine tested by Strangman (2006). The TBC structure on a platinum aluminide or MCrAlY bond coat consisted of the 7YSZ layers intermittently decorated by either Al_2O_3 or Ta_2O_5, the layer separation being of the order of 100–$200\,nm$. The structure was achieved by using the jumping beam technique from two ingots, one providing 7YSZ, while the other is the source of the decorating material. The structure so achieved exhibited a thermal conductivity of 0.7–$0.8\,W/mK$, compared with $1.8\,W/mK$ for 7YSZ. The reduction of thermal conductivity arises from the nanolaminate structure, with a large number of deposition interfaces and porosity stabilized by the second phase. The interfaces provide discontinuity to heat flow. The TBC deposited on turbine blades was successfully engine tested in an HTF 7000 engine.

Controlling thermal conductivity through microstructural manipulation should be treated with caution because of possible adverse effects on mechanical properties, particularly resistance to erosion.

One option to utilize the benefits of low thermal conductivity in the long run is to maintain the as-processed thermal conductivity while lowering susceptibility of the ceramic to sintering. This can be achieved through microstructural as well as compositional modification.

Alternate Alloying Additions

In addition to directly affecting the scattering mean free path and phonon velocity, appropriate alloying introduces vacancies due to valency difference with the host atoms. It also strains the lattice because of atomic size difference. Both of these phenomena reduce the phonon scattering mean free path. Ternary additions introduce additional complexity of structure and, therefore, further affect scattering. The success of these concepts is summarized in Fig. 7.63 (Nicholls et al., 2002).

It is evident that gadolinia is the most effective alloying constituent, followed by neodymia and ytterbia. Combinations of multiple alloying constituents such as gadolinia and ytterbia also exhibit low thermal conductivity over a wide temperature range due to the additional complexity of their structures.

Novel Oxide Ceramics

A number of oxide ceramics have been investigated to determine their suitability as TBC material. These include garnets of composition $Y_3Al_xFe_{(5-x)}O_{12}$, where x ranges between 0.0 and 5.0 (Padture and Klemens, 1997); perovskites of composition $SrZrO_3$ and $BaZrO_3$, and pyrochlore of composition $La_2Zr_2O_7$ (Vassen et al., 2000). All of these ceramics have complex structures, a requirement to increase phonon scattering. Additionally, they have high melting points. However, the perovskites either undergo destructive phase transitions or have poor thermal and chemical stability. The garnets in their bulk form in the temperature regime 296–1273 K (73–1832°F) exhibit thermal conductivity in the range 2.4 to 3.2 W/mK, comparable to that of bulk 7YSZ. With microstructural modification, this material shows potential to achieve lower thermal conductivity. The most promising candidate ceramic is $La_2Zr_2O_7$ of the pyrochlore family. The thermal conductivity is about 20% lower than that of 7YSZ. An additional benefit of lanthanum zirconate is its lower modulus of elasticity, which provides increased strain tolerance. Additional compounds with lower thermal conductivity (Wu et al., 2002) are

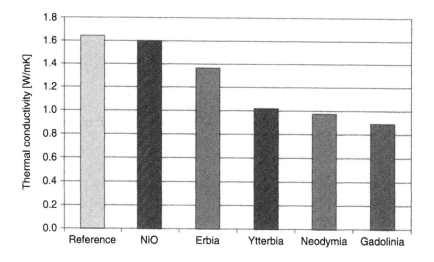

Figure 7.63 Thermal conductivity of zirconia as a function of alloying species (J. R. Nicholls, K. J. Lawson, A. Johnstone, and D. S. Rickerby, Methods to reduce thermal conductivity of EB-PVD TBCs, *Surf. Coat. Technol.*, 2002, 151–152, 383–391). Reprinted with permission from Elsevier.

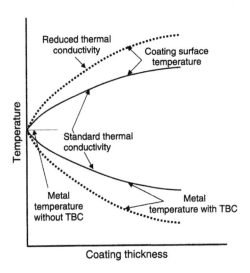

Figure 7.64 Schematic of metal and ceramic coating surface temperature as a function of ceramic coating thickness and thermal conductivity.

$Ga_2Zr_2O_4$ (both pyrochlore and fluorite structure) and $Sm_2Zr_2O_4$ (pyrochlore structure). They exhibit 30% lower thermal conductivity compared with their 7YSZ counterpart. An important issue with some of the low-conductivity novel oxide ceramics is that their susceptibility to hot corrosion is unknown and needs to be investigated. It is likely that ceramic compositions containing such oxides as BaO and SrO would be susceptible to sulfate salts in a way similar to MgO-containing zirconia.

Thermal Consequence of Increased Insulation

Based on what we have discussed so far, the total insulation of a TBC can be enhanced by increased coating thickness as well as reduced thermal conductivity of the ceramic coating. Increased thickness, however, has its associated weight and durability penalty. Although both the approaches reduce the metal temperature, the ceramic surface temperature tends to increase, as shown in the schematic of Fig. 7.64. The absolute magnitude of changes in temperature both for the metal and the ceramic coating can be calculated if the heat flux is known.

7.6   ENVIRONMENTAL BARRIER COATINGS (EBCs)

Processes by which the surface of coatings or otherwise protective scales on metals and alloys are defeated by environmental contaminants have been discussed in Chapter 5 on high-temperature corrosion. A similar phenomenon has also been observed in nonmetallic systems. A premier example is a ceramic matrix composite (CMC) consisting of a SiC matrix reinforced with SiC fibers. The system is usually known as SiC/SiC CMC. When exposed to an oxidizing environment at high temperature, SiC forms a slow-growing protective scale of SiO_2. However, in the presence of water vapor at high temperature, the SiO_2 scale slowly vaporizes. A number of reactions (More et al., 2000; Opila et al., 2006) have been proposed, which include

$$SiO_2 \text{ (solid)} + 2H_2O \text{ (vapor)} = Si(OH)_4 \text{ (gas)}.$$

Depending on the temperature and the partial pressure of the volatile species, the reaction results in slow recession of the oxide scale ($\sim 0.05\,\mu$m/hour at 1200°C [2057°F], 90% steam). The recession accelerates with increased gas velocity (Robinson and Smialek, 1999; Opila and Hann, 1997). Such surface recession has been observed in land-based gas turbine engines (Eaton et al., 2000; Miriyala and Price, 2000). For protection against surface recession in the presence of water vapor, an environmental barrier coating (EBC) is required. The general approach is to deposit a ceramic, which is stable in water vapor or has low silica activity, and is compatible with the CMC system. Two tested coating system designs consist of the following combinations (Spitsberg and Steibel, 2004):

1. SiC/SiC CMC–Silicon – Mullite/BMAS–BSAS

2. SiC/SiC CMC–Silicon – Mullite/BMAS–BSAS-graded zone–TBC (YSZ or new ceramic)

Mullite is $2Al_2O_3.SiO_2$; BMAS is barium magnesium aluminum silicate, and BSAS (barium strontium aluminum silicate) is 25 mol % BaO + SrO, 25 mol % Al_2O_3, 50 mol % SiO_2. The ceramic layers are usually deposited by plasma spraying followed by heat treatment. The intermediate mixed layers provide transition in thermal expansion. The TBC top coat is generally deposited by the EB-PVD process. Both the systems just described have been successfully evaluated in rig tests generating high surface temperature of 1650°C (~ 3000°F) and through thickness gradient 550°C (~ 1000°F).

Newer bond and ceramic coatings, innovative TBC designs, and associated processing techniques are constantly being developed to extend the useful life of the systems exposed to an aggressive high-temperature environment. However, the underlying principle and mechanisms seldom change. Hopefully, Chapters 6 and 7 help with understanding this.

REFERENCES

Alperine, S., M. Derrien, Y. Jaslier, and R. Mevrel, Thermal barrier coatings: The thermal conductivity challenge *AGARD-R-823*, paper 1, 1998.

Barkalow R. H., and F. S. Pettit, Mechanism of hot corrosion attack of ceramic coating material, in *Proc. 1st Conf. Advanced Materials for Alternate Fuel Capable Directly Fired Heat Engines*, pp. 704–714, Ed. J. W. Fairbanks and J. Stringer, U.S. Department of Energy, 1979.

Belov A. N., and G. A. Semenov, Thermodynamics of binary solid solutions of zirconium, hafnium, and yttrium oxides from high-temperature mass spectrometry data, *Russ. J. Phys. Chem.*, 1985, 59(3), 342–344.

Bengtsson, P., and T. Johannesson, Characterization of microstructural defects in plasma sprayed thermal barrier coatings, *J. Therm. Spray. Technol.*, 1995, 4(3), 245–251.

Bhatia, T., A. Ozturk, L. Xie, E. Jordan, B. M. Cetegen, M. Gell, X. Ma, and N. P. Padture, Mechanisms of ceramic coating deposition in solution-precursor plasma spray, *J. Mater. Res.*, 2002, 17(9), 2363–2372.

Borom, M. P., C. A. Johnson, and L. A. Peluso, Role of environmental deposits in spallation of thermal barrier coatings on aeroengine and land-based gas turbine hardware, *ASME 96-GT-285*, 1996.

Bose, S., and J. DeMasi-Marcin, Thermal barrier coatings experience in gas turbine engines at Pratt & Whitney, *J. Therm. Spray Technol.*, 1997, 6(1), 99–104.

Bruce, R. W., Development of 1232°C (2250°F) erosion and impact tests for thermal barrier coatings, *Tribol. Trans.*, 1998, 41, 399–410.

Cannon, R. M., and P. Y. Hou, Diffusion induced stress generation during oxidation, in *High Temperature Corrosion and Materials Chemistry*, Eds. M. McNallan et al., *Proc. Electrochem. Soc.*, 1998, 98/99, 594–607.

Chagnon, P., and P. Fauchais, Thermal spraying of ceramics, *Ceram. Int.*, 1984, 10(4), 119–131.

Chang, G. C., W. Phucharoen, and R. A. Miller, Behavior of thermal barrier coatings for advanced gas turbine blades, *Surf. Coat. Technol.*, 1987, 30, 13–28.

Cheng, J., E. H. Jordan, B. Barber, and M. Gell, Thermal/residual stress in thermal barrier coating system, *Acta. Mater.*, 1998, 46, 5839–5850.

Christensen, R. J., D. M. Lipkin, and D. R. Clarke, The stress and spalling behavior of the oxide scale formed on polycrystalline Ni_3Al, *Acta Mater.*, 1996, 44(9), 3813–3821.

Christensen, R. J., V. K. Tolpygo, and D. R. Clarke, The influence of the reactive element yttrium on the stress in alumina scales formed by oxidation, *Acta Mater.*, 1997, 45(4), 1761–1766.

Christensen, R. J., D. M. Lipkin, D. R. Clarke, K. Murphy, Nondestructive evaluation of the oxidation stresses through thermal barrier coatings using Cr^{3+} piezospectroscopy, *Appl. Phys. Lett.*, 1999, 69(24), 3754–3756.

Chutze, M., Stress and decohesion of oxide scales, *Mater. Sci. Technol.*, 1988, 4, 407–414.

Clarke, D. R., Materials selection guidelines for low thermal conductivity thermal barrier coatings, *Surf. Coat. Technol.*, 2003, 163–164, 67–74.

Demaray, E., Thermal barrier coatings by electron beam physical vapor deposition, *DOE contract DE-AC06-76 RL01830*, 1982.

DeMasi, J. T., K. D. Sheffler, and M. Ortiz, Thermal barrier coating life prediction model development, *NASA Contractor Report 182230*, p. 47, 1989.

DeMasi-Marcin, J. T., and D. K. Gupta, Protective coatings in gas turbine engine, *Surf. Coat. Technol.*, 1994, 68/69, 1–9.

DeMasi-Marcin, J. T., K. D. Sheffler, and S. Bose, Mechanisms of degradation and failure in plasma deposited thermal barrier coating, *ASME 89-GT-132*, 1989.

Dieter, G. E., *Mechanical Metallurgy*, p. 391, McGraw-Hill, New York, 1986.

Dinwiddie, R. B., S. C. Beecher, W. D. Porter, and B. A. Nagaraj, The effect of thermal aging on the thermal conductivity of plasma sprayed and EB-PVD thermal barrier coatings, *ASME 96-GT-982*, 1996.

Eaton H. E., and R. C. Novak, Particulate erosion of plasma-sprayed porous ceramic, *Surf. Coat. Technol.*, 1987, 30, 41–50.

Eaton. H. E., J. R. Linsey, and R. B. Dinwiddie, The effect of thermal aging on the thermal conductivity of plasma sprayed fully stabilized zirconia, *Thermal Conductivity*, 1994, 22, 289–300.

Eaton, H., G. Linsey, K. More, J. Price, J. Kimmel, and N. Miriyala, EBC protection of SiC/SiC composites in the gas turbine combustion environment, paper presented at the International Gas Turbine & Aeroengine Congress & Exhibition, Munich, Germany, May 8–11, 2000, *ASME2000-GT-631*.

Elsing, R., O. Knotek, and U. Balting, Calculation of residual thermal stress in plasma-sprayed coatings, *Surf. Coat. Technol.*, 1990, 43–44, 416–425.

Evans, A. G., M. Y. He, and J. W. Hutchinson, Effects of interface undulations on thermal fatigue of thin films and scales of metal substrates, *Acta Mater.*, 1997, 45(9), 3543–3554.

Evans, A. G., D. R. Mumm, J. W. Hutchinson, G. H. Meier, and F. S. Pettit, Mechanisms controlling the durability of thermal barrier coatings, *Prog. Mater. Sci.*, 2001, 46, 505–553.

Fauchais, P., A. Vardelle, and M. Vardelle, Recent developments of plasma sprayed thermal barrier coatings, *AGARD Report 823 Thermal Barrier Coatings*, pp. 3-1 to 3-12, April 1998.

Ferber, M. K., *Advanced Turbine Systems Annual Review*, Oak Ridge National Laboratory, Oak Ridge, TN, 1999.

Firestone, R. F., W. R. Logan, and J. W. Adams, Creep of plasma sprayed zirconia, *NASA CR 167868*, 1982.

Freborg, A. M., B. L. Ferguson, W. J. Brindley, and G. J. Petrus, Modeling oxidation induced stresses in thermal barrier coating, *Mater. Sci. Eng.*, 1998, A254(3), 182–190.

Gell, M., E. Jordan, K. Vaidyanathan, K. McCarron, B. Barber, Y. Sohn, and V. K. Tolpygo, Bond strength, bond stress and spallation mechanisms of thermal barrier coatings, *Surf. Coat. Technol.*, 1999, 120–121, 53–60.

Gell, M., L. Xie, X. Ma, E. H. Jordan, and N. P. Padture, Highly durable thermal barrier coatings made by solution precursor plasma spray process, *Surf. Coat. Technol.*, 2004a, 177–178, 97–102.

Gell, M., S. Sridharan, M. Wen, and E. H. Jordan, Photoluminescence piezospectroscopy: A multi-purpose quality control and NDI technique for thermal barrier coatings, *Int. J. Appl. Ceram. Technol.*, 2004b, 1(4), 316–329.

Groves, J. F., Directed vapor deposition, Ph.D. Dissertation, University of Virginia, 1998.

Groves, J. F., G. Mattausch, H. Morgner, D. D. Haas, and H. N. G. Wadley, Technology update-directed vapor deposition, *Surf. Eng.*, 2000, 16(6), 461–464.

Haas, D. D., A. J. Slifka, and H. N. G. Wadley, Low thermal conductivity vapor deposited zirconia microstructures, *Acta Mater.*, 2001, 49, 973–983.

Hasselman, D. P. H., L. F. Johnson, L. D. Bentsen, R. Syed, H. L. Lee, and M. V. Swain, Thermal diffusivity and conductivity of dense polycrystalline ZrO_2 ceramics: A survey, *Am. Ceram. Soc. Bull.*, 1987, 66(5), 799–806.

He, M. Y., and J. W. Hutchinson, Kinking of a crack out of an interface, *J. Appl. Mech.*, 1989, 56, 270–278.

He, M. Y., A. G. Evans, and J. W. Hutchinson, Effects of morphology on the decohesion of compressed thin films, *Mater. Sci. Eng.*, 1998a, A245, 168–181.

He, M. Y., A. G. Evans, and J. W. Hutchinson, Effects of morphology on the decohesion of compressed thin films, *Phys. Stat. Solid. A*, 1998b, 166(1), 5839.

Herman, H., Plasma-sprayed coatings, *Sci. Am.*, 1988, 259(3), 112–117.

Hou, P. Y., A. P. Paulikas, and B. W. Veal, Growth strains and stress relaxation in alumina scales during high temperature oxidation, paper presented at 6th Symposium on High Temperature Corrosion and Protection of Materials, Les Embiez, France, May 16–21, 2004.

Hutchinson, J. W., M. D. Thouless, and E. G. Liniger, Growth and configurational stability of circular, buckling-driven film delamination, *Acta Mater.*, 1992, 40(2), 295–308.

Johnson, C. A., J. A. Rudd, R. Bruce, and D. Wortman, Relationships between residual stress, microstructure and mechanical properties of electron beam-physical vapor deposition thermal barrier coatings, *Surf. Coat. Technol.*, 1998, 108–109, 80–85.

Jones, R. L., The development of hot-corrosion resistant zirconia thermal barrier coatings, *Mater. High Temp.*, 1991, 9, 228–236.

Jones, R. L., Thermogravimetric study of the 800C reaction of zirconia stabilizing oxides with SO_3-$NaVO_3$, *J. Electrochem. Soc.*, 1992, 139(10), 2794–2799.

Jordan, D. W., and K. T. Faber, X-ray residual stress analysis of a ceramic thermal barrier coating undergoing thermal cycling, *Thin Solid Films*, 1993, 235, 137–141.

Kaysser, W. A., M. Peters, K. Fritscher, and U. Schulz, Processing, characterization and testing of EB-PVD thermal barrier coatings, *AGARD Report 823*, 1998, pp. 9/1–9/11.

Klemens P. G., Thermal conductivity of nanophase ceramics, in *Chemistry and Physics of Nanostructures, TMS Proceedings*, Eds. E. Ma, B. Fultz, R. Shull, J. Morall, and P. Nash, TMS, 1997, 97–104.

Klemens, P. G., and M. Gell, Thermal conductivity of thermal barrier coatings, *Mater. Sci. Eng.*, 1998, A245, 143–149.

Kraemer, S., J. Y. Yang, C. A. Johnson, and C. G. Levi, Thermochemical interaction of thermal barrier coatings with molten CaO-MgO-Al_2O_3-SiO_2 (CMAS) deposits, Submitted to the Journal of the American Ceramic Society, 2006.

Lawson, K. J., J. R. Nicholls, and D. S. Rickerby, The effect of coating thickness on the thermal conductivity of CVD and PVD coatings, in *4th Int. Conf. on Advances in Surface Engineering*, Newcastle, UK, 1996.

Lelait, L., S. Alperin, and R. Mevrel, Alumina scale growth at zirconia–MCrAlY interface: a microstructural study, *J. Mater Sci.*, 1992, 27, 5–12.

Levi, Carlos G., Emerging materials and processes for thermal barrier systems, *Curr. Opin. Solid State Mater. Sci.*, 2004, 8, 79–91.

Leyens, C., I. G. Wright, and B. A. Pint, Hot corrosion of an EB-PVD thermal-barrier coating system at 950°C, *Oxid. Met.*, 2000, 54(5/6), 401–424.

Lipkin, D. M., D. R. Clarke, M. Hollatz, M. Robeth, and W. Pompe, Stress development in alumina scales formed upon oxidation of (111) NiAl single crystal, *Corr. Sci*, 1997, 39(2), 231–242.

Madejski, J., Solidification of droplets on cold substrates, *Int. J. Heat Mass Transfer*, 1976, 19, 1009–1013.

Maloney, M. J., Thermal barrier coating system and materials, U.S. Patent No. 6,284,323, September 2001.

Marple, B. R., J. Voyer, M. Thibodeau, D. R. Nagy, and R. Vassen, Hot corrosion of lanthanum zirconate and partially stabilized zirconia thermal barrier coatings, *J. Eng. Gas Turbine Power Trans ASME*, 2006, 128, 144–152.

McPherson, R., A model for the thermal conductivity of plasma sprayed ceramics, *Thin Solid Films*, 1984, 112, 89–94.

Meier, S. M., and D. K. Gupta, The evolution of thermal barrier coatings in gas turbine engine applications, *J. Eng. Gas Turbine Power Trans. ASME*, 1994, 116, 250–257.

Meier, S. M., D. M. Nissley, K. D. Sheffler, and T. A. Cruise, Thermal barrier coating life prediction model development, *J. Eng. Gas Turbine Power Trans ASME*, 1992, 114, 258–263.

Mercer, C., S. Faulhaber, A. G. Evans, R. Darolia, A delamination mechanism for thermal barrier coatings subject to calcium-magnesium-alumino-silicate (CMAS) infiltration, *Acta Mater.*, 2005, 53, 1029–1039.

Meyer, P., and S. Muehlberger, Historical review and update to the state of the art of automation for plasma coating processes, *Thin Solid Films*, 1984, 118, 445–456.

Miller, R. A., Current status of thermal barrier coatings—an overview, *Surf. Coat. Technol.*, 1987, 30, 1–11.

Miller, R. A., Life modeling of thermal barrier coatings for aircraft gas turbine engine, *J. Gas Turbines Power Trans, ASME*, 1989, 111, 301–305.

Miller, R. A., and C. E. Lowell, Failure mechanisms of thermal barrier coatings exposed to elevated temperatures, *Thin Solid Films*, 1982, 95, 265–273.

Miller, R. A., J. L. Smialek, and R. G. Garlick, Phase stability in plasma-sprayed partially stabilized zirconia–yttria, in *Advances in Ceramics*, Vol. 3, Eds. A. H. Heuer and L. W. Hobbs, American Ceramic Society, Cleveland, OH, 1981.

Miller, R. A., G. W. Leissler, and J. M. Jobe, Characterization and durability testing of plasma sprayed zirconia–yttria and hafnia–yttria thermal barrier coatings, Part I, *NASA TP 3295*, 1993.

Miriyala, N., and J. Price, The evaluation CFCC liners after field engine testing in gas turbine-II, paper presented at the International Gas Turbine & Aeroengine Congress & Exhibition, Munich, Germany, May 8–11, 2000, *ASME 2000-GT-648*.

More, K., P. Tortorelli, M. Ferber, and J. Kaiser, Observation of accelerated silicon carbide recession by oxidation at high water-vapor pressures, *J. Amer. Ceram. Soc.*, 2000, 83(1), 211–213.

Morrell, P., and D. S. Rickerby, Advantages/disadvantages of various TBC systems as perceived by the engine manufacturers, *AGARD Report 823*, 1998, 1088, 20/1–20/9.

Morrell, P., and R. Taylor, Thermal diffusivity of thermal barrier coatings of ZrO_2 stabilized with Y_2O_3, *High Temp High Press*, 1985, 17, 79–88.

Movchan, B. A., and A. V. Demchishin, Study of the structure and properties of thick vacuum condensates of nickel, titanium, tungsten, aluminum oxide, and zirconium dioxide, *Fiz. Metall. Metalloved*, 1969, 28, 83–90.

Mumm, D. R., and A. G. Evans, On the role of imperfections in the failure of a thermal barrier coating made by electron beam deposition, *Acta Mater.*, 2000, 48, 1815–1827.

Nagaraj, B. A., A. F. Maricocchi, D. J. Wortman, J. S. Patton, and R. L. Clarke, Hot corrosion resistance of thermal barrier coatings, paper presented at the International Gas Turbine and Aeroengine Congress and Exposition, Cologne, Germany, June 1–4, 1992.

Nicholls, J. R., Laboratory studies of erosion–corrosion processes under oxidizing and oxidizing/sulphidizing conditions, *Mater. High Temp.*, 1997, 14(3), 289–306.

Nicholls, J. R., Y. Jaslier, and D. S. Rickerby, Erosion and foreign object damage of thermal barrier coatings, *Mater. Sci. Forum*, 1997, 251–254, 935–948.

Nicholls, J. R., K. J. Deakin, and D. S. Rickerby, A comparison between erosion behavior of thermal spray and electron beam physical vapor deposition thermal barrier coatings, *Wear*, 1999, 233–235, 352–361.

Nicholls, J. R., K. J. Lawson, A. Johnstone, and D. S. Rickerby, Methods to reduce thermal conductivity of EB-PVD TBCs, *Surf. Coat. Technol.*, 2002, 151–152, 383–391.

Opila, E., and R. Hann, Jr., Paralinear oxidation of CVD SiC in water vapor, *J. Amer. Ceram. Soc.*, 1997, 80(1), 197–205.

Opila, E. J., N. S. Jacobson, D. L. Myers, and E. H. Copland, Predicting oxide stability in high-temperature water vapor, *JOM*, Jan. 2006, pp. 22–28.

Padture, N. P., and P. G. Klemens, Low thermal conductivity in garnets, *J. Am. Ceram. Soc.*, 1997, 80(4), 1018–1020.

Pawloski, L., D. Lombard, A. Mahlia, C. Martin, and P. Fauchais, Thermal diffusivity of arc plasma sprayed zirconia coatings, *High Temp. High Press.*, 1984, 16, 347–359.

Peng, X., and D. R. Clarke, Piezospectroscopic analysis of interface debonding in thermal barrier coatings, *J. Am. Ceram. Soc.*, 2000, 83(5), 1165–1170.

Portal, R., Etude de la conductivite thermique de couches minces de ZrO_2-Y_2O_3 deposees par EB-PVD, *Rapport de stage SNECMA*, 1997.

Reinhold, E., P. Boltzer, and C. Deus, EB-PVD process management for highly productive zirconia thermal barrier coating of turbine blades, *Surf. Coat. Technol.*, 1999, 120–121, 77–83.

Robertson, J., and M. I. Manning, Limits to adherence of oxide scales, *Mater. Sci. Technol.*, 1990, 6, 81.

Robinson, R. C., and J. L. Smialek, SiC recession caused by SiO_2 scale volatility under combustion conditions: I. Experimental results and empirical model, *J. Am. Ceram. Soc.*, 1999, 82(7), 1817–1825.

Ruckle, D. L., Plasma sprayed ceramic thermal barrier coatings for turbine vane platforms, *Thin Solid Films*, 1980, 73, 455–461.

Rudd, J. A., A. Bartz, M. P. Borom, and C. A. Johnson, Strength degradation and failure mechanisms of electron beam physical vapor deposited thermal barrier coatings, *J. Am. Ceram. Soc.*, 2001, 84(7), 1545–1552.

Safai, S., and H. Herman, Microstructural investigation of plasma-sprayed aluminum coatings, *Thin Solid Films*, 1977, 45, 295–307.

Sampath, S., and H. Herman, Rapid solidification and microstructure development during plasma spray deposition, *J. Therm. Spray Technol.*, 1996, 5(4), 445–456.

Sarioglu, C., M. J. Stiger, J. R. Blachere, R. Janakiraman, E. Schumann, A. Ashary, F. S. Pettit, and G. H. Meier, The adhesion of alumina films to metallic alloys and coatings, *Mater. Corr.*, 2000, 51, 358–372.

Scardi, P., E. Galvanetto, A. Tomasi, and L. Bertamini, Thermal stability of stabilized zirconia thermal barrier coatings prepared by atmosphere and temperature–controlled spraying, *Surf. Coat. Technol.*, 1994, 68–69, 106–112.

Scardi, P., M. Leoni, and L. Bertamini, Influence of phase stability in partially stabilized zirconia TBC produced by plasma spray, *Surf. Coat. Technol.*, 1995, 76–77, 106–112.

Schlichting, K. W., N. P. Padture, E. H. Jordan, and M. Gell, Failure modes in plasma-sprayed thermal barrier coatings, *Mater. Sci. Eng.*, 2005, A405, 313–320.

Schulz, U., K. Fritscher, and M. Peters, Thermocyclic behavior of variously stabilized EB-PVD thermal barrier coatings, *J. Eng. Gas Turbine Power Trans ASME*, 1997, 119, 917–921.

Schulz, U., K. Fritscher, C. Leyens, and M. Peters, Influence of processing and microstructure of EB-PVD thermal barrier coatings, *ASME 2000-GT-579*, 2000.

Schulz, U., M. Menzebach, C. Leyens, and Y. Q. Yang, Influence of substrate material on oxidation behavior and cyclic life time of EB-PVD TBC systems, *Surf. Coat. Technol*, 2001, 146–147, 117–123.

Schulz, U., B. Saruhan, K. Fritscher, and C. Leyens, Review on advanced EB-PVD ceramic topcoats for TBC applications, *Int. J. Appl. Ceram. Technol.*, 2004, 1(4), 302–315.

Schwingel, D., R. Taylor, T. Haubold, J. Wigren, and C. Gualco, Mechanical and thermophysical properties of thick PYSZ thermal barrier coatings: Correlation with microstructure and spraying parameters, *Surf. Coat. Technol.*, 1998, 108–109, 99–106.

Scott, H. C., Phase relationships in zirconia–yttria system, *J. Mater. Sci*, 1975, 10, 1527–1535.

Sevcik W. R., and B. L. Stoner, *NASA Contract. Rep. CR-135360*, 1978, Pratt & Whitney.

Shillington, E. A. G., and D. R. Clarke, Spalling failure of a thermal barrier coating associated with aluminum depletion in the bond coat, *Acta Mater.*, 1999, 47(4), 1297–1305.

Singhal, S. C., and R. J. Bratton, Stability of a $ZrO_2(Y_2O_3)$ thermal barrier coating in turbine fuel with contaminants, *J. Eng. Gas Turbine Power Trans ASME*, 1980, 102(10), 770–775.

Siry, Ch. W., H. Wanzek, and C.-P. Dau, Aspects of TBC service experience in aero engines, *Mat Wiss. U. Werkstofftech.*, 2001, 32, 650–653.

Slack, G., The thermal conductivity of nonmetallic solids, in *Solid State Physics: Advances in Research and Application*, Eds. H. Ehrenreich, F. Seitz, and D. Turnbull, Vol. 34, pp 1–73, Academic Press, New York, 1979.

Smialek, J. L., Moisture-induced delayed spallation and interfacial hydrogen embrittlement of alumina scales, *JOM*, Jan. 2006, pp. 29–35.

Smialek, J. L., and G. N. Morcher, Delayed alumina scale spallation on René N5 + Y: moisture effects and acoustic emission, *Mater. Sci. Eng.*, 2002, A332, 11–24.

Sohn, Y. H., R. R. Biederman, and R. D. Sisson, Jr., Microstructural development in physical vapor-deposited partially stabilized zirconia thermal barrier coatings, *Thin Solid Films*, 1994, 250, 1–7.

Sohn, Y. H., K. Schlichting, K. Vaidyanathan, E. Jordan, and M. Gell, Nondestructive evaluation of residual stress for thermal barrier coated turbine blades by Cr^{3+} photoluminescence piezospectroscopy, *Metall. Mater. Trans.*, 2000, 31A, 2388–2391.

Sohn, Y. H., E. Y. Lee, B. A. Nagaraj, R. R. Biederman, and R. D. Sisson Jr., Microstructural characterization of thermal barrier coatings on high pressure turbine blades, *Surf. Coat. Technol.*, 2001, 146–147, 132–149.

Spitsberg, I. and J. Steibel, Thermal and environmental barrier coatings for SiC/SiC CMCs in aircraft engine application, *Applied Ceram. Technol.*, 2004, 1(4), 291.

Standard Test Method for Adhesion or Cohesion of Flame Sprayed Coatings, *ASTM C 633-79*, American Society for Testing and Materials, West Conshohocken, PA.

Staniek, G., and G. Marci, in *Proc. Werkstoff-Kolloquium DLR*, pp. 50–53, Eds. M. Peters et al., 10 Dec. 1996.

Stecura, S., Optimization of NiCrAl–Y/ZrO_2–Y_2O_3 thermal barrier system, *NASA Tech. Memo. 86905*, NASA, Cleveland, OH, 1985.

Stiger, M. J., N. M. Yanar, F. S. Pettit, and G. H. Meier, Mechanism for the failure of electron beam physical vapor deposited thermal barrier coatings induced by high temperature oxidation, in *Elevated Temperature Coatings: Science and Technology III*, pp. 51–65, Eds. J. M. Hampikian and N. B. Dahotre, The Minerals and Materials Society, 1999.

Stott, F. H., D. J. de Wet, and R. Taylor, Degradation of thermal-barrier coatings at very high temperatures, *MRS Bull.*, Oct. 1994, pp. 46–49.

Strangman, T. E., Thermal barrier coatings for turbine airfoils, *Thin Solid Films*, 1985, 127, 93–105.

Strangman, T. E., Turbine coating life prediction model, in *Proc. 1990 Coatings for Advanced Heat Engines Workshop, Department of Energy Report CONF-9008151*, pp II35–II43, Castine, ME, 1990.

Strangman, T. E., Thermal strain-tolerant abradable thermal barrier coatings, *J. Eng. Gas Turbine Power Trans ASME*, 1992, 114, 264–267.

Strangman, T., Development of low conductivity nanolaminate thermal barrier coatings, Presentation GT2006-91448, at panel session WC-14-3, Turbo Expo—Power for Land, Sea and Air, Barcelona, Spain, May 10, 2006; U.S. Patent No. 6,482,537B1, Nov 19, 2002.

Szücs, F., Thermomechanische Analyse und Modellierung plasmagespritzer und EB-PVD aufgedampfter Wärmedammschicht-Systeme fur Gasturbinen, Serial 5, No 518, Fortschr.-Ber. VDI Verlag, Düsseldorf, Germany, 1998.

Tabakoff, W., Investigation of coatings at high temperatures for use in turbomachinery, *Surf. Coat. Technol.*, 1989, 39–40, 97–115.

Taylor, Raymond E., Heat-pulse thermal diffusivity measurements, *High Temp. High Press.*, 1979, 11, 38–58.

Thornton, J. A., High rate thick film growth, *Ann. Rev. Mater. Sci.*, 1977, 7, 239–260.

Thornton, J., D. Cookson, and E. Pescott, The measurement of strains within the bulk of aged and as-sprayed thermal barrier coatings using synchrotron radiation, *Surf. Coat. Technol.* 1999, 120–121, 96–102.

Tien, J. K., and J. M. Davidson, Oxide spallation mechanisms, in *Stress Effects and the Oxidation of Metals*, p. 200, Ed. J. V. Cathcart, AIME, New York, 1975.

Timoshenko, S., and J. M. Gere, *Theory of Elastic Stability*, 2nd ed., p. 390, McGraw-Hill, New York, 1961.

Tolpygo, V. K., and D. R. Clarke, Competition between stress generation and relaxation during oxidation of a FeCrAlY alloy, *Oxid. Met.*, 1998, 49, 187–212.

Tolpygo, V. K., J. R. Dryden, and D. R. Clarke, Determination of the growth stress and strain in α-Al_2O_3 scale during the oxidation of Fe-22Cr-4.8Al-0.3Y alloy, *Acta Mater.*, 1998, 46(3), 927–937.

Toriz, F. C., A. B. Thakker, and S. K. Gupta, Thermal barrier coatings for jet engines, *ASME 88-GT-279*, 1988.

Tortorelli, P. F., K. L. More, E. D. Specht, B. A. Pint, and P. Zschack, Growth stress–microstructure relationships for alumina scales, *Mater. High Temp.*, 2003, 20(3), 303–310.

Touloukian, Y. S. (Ed), *Thermophysical Properties of Matter*, Plenum Press, New York, 1977.

Trubelja, M. F., D. M. Nissley, N. S. Bornstein, and J. T. D. Marcin, *Proc. Adv. Turbine Syst. Ann. Prog. Rev. Meet.*, 1997, U.S. Dept. of Energy, Washington, DC.

Vassen, R., X. Cao, F. Tietz, D. Basu, and D. Stover, Zirconates as new materials for thermal barrier coatings, *J. Am. Ceram. Soc.*, 2000, 83(8), 2023–2028.

Walston, W. S., Coating and surface technologies for turbine airfoils, in *Superalloy 2004*, pp. 579–588, Eds. K. A. Green, T. M. Pollock, H. Harada, T. E. Howson, R. C. Reed, J. J. Schirra, and S. Walston, TMS, Warrendale, PA, 2004.

Wang, J. S., and A. G. Evans, Measurement and analysis of buckling and buckle propagation in compressed oxide layer on superalloy substrate, *Acta Mater.*, 1998, 46(4), 4993–5005.

Watchman, J. B., Jr., and D. G. Lam, Jr., Young's modulus of various refractory materials as a function of temperature, *J. Amer. Ceram. Soc.*, 1959, 42, 254.

Wellman, R. G., and J. R. Nicholls, On the effect of aging on erosion of EB-PVD TBCs, *Surf. Coat. Technol.*, 2004, 177–178, 80–88.

Wigren, J., D. Grevin, and J. F. DeVries, Effects of powder morphology, microstructure and residual stresses on the thermal shock life of thin thermal barrier coatings, *Proc. NTSC'96*, pp 855–861, Cincinnati, OH, 1996.

Wigren, J., and L. Pejryd, Thermal barrier coatings—why, how, where and where to, *Proc. 15th. International Thermal Spray Conf.*, pp. 1531–1542, Ed. C. Coddet, ASM International, 1998.

Wortman, D. J., B. A. Nagaraj, and E. C. Duderstadt, Thermal barrier coatings for gas turbine use, *Mater. Sci. Eng.*, 1989, A121, 433–440.

Wright, P. K., Influence of cyclic strain on life of a PVD TBC, *Mater. Sci. Eng.*, 1998, A245, 191200.

Wright P. K., and A. G. Evans, Mechanisms governing the performance of thermal barrier coatings, *Princeton Materials Institute Report PMI-99-11*, Princeton University, New Jersey, Feb. 1999.

Wu, J., X. Wei, N. Padture, P. G. Klemens, M. Gell, E. Garcia, P. Miranzo, and M. Osendi, Low-thermal conductivity rare-earth zirconates for potential thermal-barrier-coating application, *J. Am. Ceram. Soc.*, 2002, 85(12), 3031–3035.

Xie, W., K. P. Walker, E. H. Jordan, and M. Gell, Implementation of a viscoplastic model for a plasma sprayed ceramic thermal barrier coating, *J. Eng. Mater. Technol. Trans ASME*, 2003, 125, 200–207.

Yanar, N. M., G. Kim, S. Hamano, F. S. Pettit, and G. H. Meier, Microstructural characterization of the failures of thermal barrier coatings on Ni-base superalloys, *Materials at High Temp.*, 2003, 20(4), 495–506.

Youchison, D. L., M. A. Gallis, R. E. Nygren, J. M. McDonald, and T. J. Lutz, Effect of ion beam assisted deposition, beam sharing and pivoting in EB-PVD processing of graded thermal barrier coatings, *Surf. Coat. Technol.*, 2004, 177–178, 158–164.

Zheng, X., D. G. Cahill, and J. Zhao, Thermal conductivity imaging of thermal barrier coatings, *Adv. Eng. Mater.*, 2005, 7, 622–626.

Zhu, D., and R. A. Miller, Development of advanced low conductivity thermal barrier coatings, *Int. J. Appl. Ceram. Technol.*, 2004, 1(1), 86–94.

Zhu, D., N. P. Bansal, K. N. Lee, and R. A. Miller, Thermal conductivity of ceramic thermal barrier and environmental barrier coating materials, *NASA Report TM-2001-211122*, 2001.

Chapter 8

NONDESTRUCTIVE INSPECTION (NDI) OF COATINGS

The structural integrity of high-temperature metallic and ceramic coatings is of critical importance in meeting their intended performance and durability requirements. The integrity is adversely affected by the presence of defects, some of which may be in the as-deposited coatings as a result of processing aberrations, whereas others develop in use. Defect characteristics of metallic coatings include mechanical damage, contaminated interfaces, interfacial disbond, transverse and longitudinal cracks within the coatings, local lack of coatings, density and thickness variations, internal oxidation, oxidation and corrosion product accumulation at the interface, and excessive porosity and inclusions. In addition to some of these defects, ceramic coatings may exhibit discontinuity in deposition, misorientation of columnar structures, and infiltration by foreign materials. Whereas statistical process development and stringent quality control may eliminate the majority of these defects in as-processed coatings, the others need to be identified by routine destructive and nondestructive inspection (NDI). Destructive inspections tend to be expensive, particularly for gas turbine hardware, due to their high cost. Use of NDI techniques, therefore, is very attractive. However, in contrast to bulk structural alloys, the number of NDI techniques successfully applied to coatings is very limited.

8.1   NDI TECHNIQUES

The general principle of some of the conventional NDI techniques is well known (Bray and Stanley, 1989; *Metals Handbook,* 1999). These techniques belong to one of two categories. The first category is based on response of a material to stimulus from the environment in which it is operating or being tested or the response of existing defects in the material to externally imposed stimuli. An example is the ultrasonic NDI technique, in which an external stimulus in the form of an ultrasonic pulse is imposed on the material. The second category is not dependent on stimulus. This category is illustrated by radiography, which does not involve significant interaction with the material other than absorption or transmission. A few of the NDI techniques are briefly described next. A number of additional techniques at various stages of development for coatings are also reviewed.

Fluorescent Penetrant Inspection (FPI)

This technique is appropriate for detecting surface connected pores and cracks in a large class of materials, including coatings. The basic principle of the technique consists of wetting a clean and dry surface to be inspected by a liquid "penetrant" consisting of a dye in a carrier fluid. The dye has the property of fluorescing in the visible spectrum of light of wavelength between 4000 and 7000 Å when illuminated by ultraviolet light (wavelength in the range 3200 to 4000 Å), also known as black light. Because of surface tension forces, the dye penetrates into cracks. Excess dye is subsequently removed with the appropriate solvent remover. A developer is then spread over the surface to absorb, like a blotter, the remaining penetrant from around the cracks. The developer is usually a colored powder that draws penetrants in contact by capillary action. The color provides contrast for visual indication of cracks. Finally, the treated hardware is illuminated by ultraviolet light in a dark chamber and observed at a magnification of 3× to 5×. The cracks contain the dye fluoresce against a dark background, providing indications of locations, density, and dimensions of cracks and pores. One version of the process utilizes penetrants suitable for inspection under white light. The defect indications are observed by the variation of the concentration of color. An example of "black light" FPI inspection showing crack patterns on a turbine blade coated with platinum aluminide and exposed to 12,000 hours of service in a large industrial operation is shown in Fig 8.1 (Conner and Connor, 1994).

The FPI process has improved significantly by the introduction of laser scanning capability, in which the hardware is illuminated by rastering a laser beam over the sample.

Ultrasonic Inspection

In this technique an ultrasonic pulse, either of a fixed frequency or as a broad band, is introduced into the hardware by a piezoelectric transducer, which may also act as a detector probe. The transducer converts mechanical strains into electrical signal and vice versa. The ultrasonic frequency, generally in excess of 20 kHz and typically between 0.1 and 25 MHz, is such as to generate elastic waves in the solid material. The elastic waves propagate through the sample, strike defects including surfaces and interfaces, and undergo reflection known as echoes. The

Figure 8.1 FPI indication of extensive cracking on platinum aluminide coated blade exposed to 12,000 hours in large industrial gas turbine (Jeffrey A. Conner and William B. Connor, Ranking protective coatings: Laboratory vs. field experience, *JOM*, Dec. 1994, pp. 35–38). Reprinted with permission from The Minerals, Metals & Materials Society.

probe detects all waves and displays them on an oscilloscope as a spectrum of echoes arriving at different times. Analysis of the spectrum provides indication of shape, size, and location of the defects. Depending on the dimensionality of the scans, the evaluation method is designated as UT (ultrasonic testing) A, UT B, or UT C type scans. The UT A scan, usually used to measure thickness based on the incident signal and the signal reflected from a parallel surface, provides a one-dimensional display. The B scan consists of parallel sets of A scans, and the UTC scan provides a two-dimensional display of distribution of defects. If the detector registers the transmitted signal on the other side of the sample, the process is called "through-transmission" mode. If, on the other hand, the detector on the same side as the transducer senses the reflected signal, the process is in "pulse–echo" mode. Both the detection limit and the depth of penetration (thickness of samples) are controlled by the frequency of the ultrasonic signal. The ultrasonic NDI technique is applicable to metallic, nonmetallic, dielectric, magnetic, and nonmagnetic materials. For measurements at temperatures in excess of 100°C (212°F), high-temperature capable transducers with coupling materials should be used to avoid transducer contact with hot surfaces. Elastic properties such as modulus of elasticity and Poisson's ratio of the test materials can also be determined by ultrasonic techniques from a measurement of velocity of elastic waves. Thickness and elastic properties of ceramic and metallic coatings have been measured by a modified ultrasonic NDI technique that generates elastic surface waves, known as Rayleigh waves, with specially prepared focusing transducers (Kauppinen, 1997). Thickness, coating density, and microhardness of thermal spray coatings can be measured or estimated by ultrasonic NDE technique. The thickness is measured by introducing an ultrasonic signal into a coated component from a transducer. Time delay between signals reflected from the coating surface and the coating–substrate interface is recorded on a CRT display. Multiplying this delay by the velocity of sound in the coating gives twice the thickness of the coating. An ultrasonic technique using a control signal, the so-called CS technique, has been successfully used on thermally sprayed metallic coating (0.1% C steel) deposited on mild steel substrate to identify bond imperfections and measure porosity and thickness (Steffens and Crostack, 1981).

Eddy Current

This technique can only be used for electrically conducting materials. When an alternating electrical current flows through a coil known as the "probe," an alternating magnetic field is established along the coil axis. If the coil is brought into the proximity of a metallic surface such that the magnetic fields intersect the surface, eddy currents are generated in the metal in the form of circular loops (Fig. 8.2) (*Metals Handbook*, 1997). These currents create magnetic fields of their own, which tend to oppose the primary magnetic field of the coil. Defects in the metal, including thickness variation, reduce or interrupt the flow of eddy currents, which in turn increase the impedance of the coil. The probe, receiving an electrical signal that is electronically amplified with appropriate balancing and demodulation, monitors the impedance change. It is finally displayed on a computer monitor screen. The displayed signals are compared with those generated by reference standards of the same material with controlled flaws and cracks. The eddy current response from a metal is affected by several parameters that include the electrical conductivity of the metal as well as its magnetic permeability, the frequency of the inducing alternating current, the proximity of the probe to the surface, and lift-off. The last parameter is the signal generated by the movement of the probe over or off the surface. A variation of the technique just described is the frequency scanning eddy current technique.

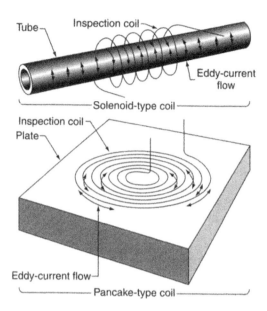

Figure 8.2 Eddy current generation by currents in two types of coils (Eddy current inspection, in *Metals Handbook*, Vol. 17, *Nondestructive Evaluation and Quality Control*, ASM International, 1997). Reprinted with permission from ASM International.

Frequency Scanning Eddy Current Technique (FSECT)

The frequency scanning eddy current technique (FSECT) is a variant of the process described earlier in which the eddy current probe operates within a range of frequency, typically between 100 kHz and 10 MHz. Impedance is measured as a function of the frequency. The applicability of this technique to coatings has been demonstrated in measuring thickness of antifretting coatings on Ti6Al4V alloy gas turbine compressor blades and MCrAlY coatings deposited on turbine blades made of Ni base superalloys IN 738 (Antonelli et al., 1997). In the latter case, the β NiAl content of the coating was also determined from the ratio between electrical conductivity and magnetic permeability as a function of engine exposure (Antonelli et al., 1998). The FSECT has successfully been used in situ to measure remaining coating thickness based on β-phase content on field-exposed turbine blades from frame-size industrial gas turbines (Rinaldi et al., 2006).

Another modification of the technique involves multifrequency eddy current meandering winding magnetometer (MWM) sensors (Zilberstein et al., 2003). The MWM sensor consists of a meandering primary winding that creates the magnetic field and, on the opposite side, a meandering secondary winding to monitor response. MWM sensors are highly reliable and repeatable. The capability of this technique has been demonstrated in measuring metallic and ceramic coating thickness to within $\pm 2\,\mu m$, determining electrical conductivity of metallic coating and surface roughness.

Infrared Imaging

Thermal imaging, also known as thermal wave imaging or infrared imaging, is a very attractive NDI technique for heat-insulating materials and has been evaluated for monitoring damage accumulation in TBCs (Ferber et al., 2000). The technique requires establishing a temperature gradient across the TBC-coated sample subjected to prior cyclic thermal exposure and monitoring the temperature of the ceramic coating surface with an IR camera. The temperature gradient may be established by placing the thermally cycled sample on a hot plate so that heat flows across the thickness. The damage created in the TBC due to cyclic thermal exposure generally consists of microcracks. These reduce heat transfer to the surface, which progressively appears cooler with continued thermal cycling.

Pulse–Echo Thermal Wave Infrared Imaging

A variant of the infrared imaging technique is the pulse–echo thermal wave infrared imaging (Newaz and Chen, 2005). The basic premise of the process is that defects generated in a TBC locally change the thermal conductivity. A flash lamp delivers a pulse of heat to the surface of a TBC-coated sample. The areas with underlying defects appear hotter because of slower dissipation of heat by conduction. The temperature also decays slowly. An IR camera connected to computers maps the temperature distribution. Analysis of the distribution of temperature and decay times provides information on the details of damage accumulation. This technique has been found to detect various types of flaws in the presence of delamination at the ceramic–bond coat interface (Zombo, 1997). An example of damage assessment of TBCs by the pulse–echo thermal wave infrared imaging technique is described by Chen et al. (2001). TBC-coated samples, consisting of 8YSZ ceramic, deposited by EB-PVD on bond-coated René N5 single-crystal superalloy, were thermally cycled in a furnace between 200°C (392°F) and 1177°C (2150°F). The cycle consisted of 9 minutes to heat up, 45 minutes hold at high temperature, and 10 minutes to cool down. Both thermal wave image and surface temperature profiles were measured on the TBC surface as a function of time. Progressively, the amplitude of the thermal signal increased until the TBC failed. The increase in thermal wave amplitude is attributed to the reduction in thermal conductivity due to continued damage accumulation. The reduction more than compensates for any increase in thermal conductivity due to ceramic sintering. The data show that the technique can be used for monitoring TBC health during thermal exposure. Thermal imaging is also frequently used to monitor corrosion and disbonding in structural components (Han et al., 1997).

Phase of Thermal Emission Spectroscopy

This method (Bennett and Yu, 2005) has been used to measure thermal properties of TBCs and detect delamination within the ceramic or at the interface without any special sample preparation. At the core of this method lies a periodically modulated laser beam, such as from a CO_2 laser (wavelength 10.6 μm), which heats up the TBC. The modulation frequency lies between 10 and 1000 Hz. The frequency determines the depth of thermal penetration. The heated ceramic coating emits thermal radiation, which is intercepted by a detector. The detector measures the difference in phase between the incident heating signal and the emitted signal. A mathematical model relates the phase difference to thermal properties of the TBC system and contact thermal resistance at delaminations. Bennett and Yu demonstrated the use of the technique on 7YSZ deposited by EB-PVD on FeCrAlY bond coat. The evaluation gave average thermal conductivity of EB-PVD deposited 7YSZ 1.44 ± 0.05 W/m K, average specific heat 2.10 ± 0.05 J/cm^3 K, and average depth of penetration of the laser 40 ± 3 μm.

Delamination, when present, is reflected as a thermal contact resistance. As mentioned earlier, one of the advantages of this method is the fact that no sample preparation is required.

Acoustic Emission

When defects form or grow in materials, internal stresses are redistributed, setting in stress waves that are known as acoustic emission. Piezoelectric sensors can detect these stress waves. The electrical signals from the sensors can be amplified, filtered, displayed on computer screens, and interpreted. Coated samples, when subjected to tensile or compressive loading, release acoustic emissions. Thermal barrier coated samples have been monitored by acoustic emission (Voyer et al., 1998) during thermal cycling under an infrared quartz lamp heat source. Initiation and growth of vertical cracks in the ceramic coating and delamination at the ceramic–bond coat interface are correlated with significant acoustic emission activity.

Photoacoustic Technique

In this process (Aithal et al., 1984) (Fig. 8.3), the surface of the coating sample is heated by a laser beam, which is chopped at a certain frequency ν. The sample is held in a gas-filled cell with a quartz window. The hot sample surface loses heat to the interior and to the surrounding gas, creating pressure or acoustic waves. Microphones held in the cell detect the waves. The signals generated in the microphones are amplified and displayed appropriately. The amplitude and phase of the acoustic waves depend on the surface temperature of the sample. The surface temperature in turn depends on the thickness, subsurface thermal properties, and presence of defects. The penetration depth of the acoustic wave is of the order of $2\pi\mu$, given by

$$2\pi\mu = 2\pi(\mathrm{K}/\pi\mathrm{C}\rho\nu)^{1/2},$$

where μ, K, C, ρ, and ν are respectively the thermal diffusion length, thermal conductivity, heat capacity, density, and laser chopping frequency. Changing the chopping frequency, generally

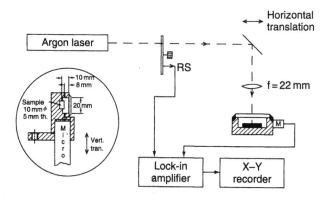

Figure 8.3 Schematic of photoacoustic technique (S. Aithal, G. Rousset, L. Bertrand, P. Cielo, and S. Dallaire, Photoacoustic characterization of subsurface defects in plasma-sprayed coatings, *Thin Solid Films*, 1984, 119, 153–158). Reprinted with permission from Elsevier.

between 4 and 300 Hz, to access the interior of the coating including defects and the interfaces varies the depth of penetration of the acoustic waves. The signals are affected by defects.

Aithal et al. have demonstrated that this technique can detect artificially introduced interfacial defects in plasma-sprayed coatings of thickness of the order of a few microns.

Mid-infrared Reflectance

This technique has been evaluated for TBCs based on yttria-stabilized zirconia (YSZ) (Eldridge et al., 2003). YSZ is translucent to mid-infrared (MIR) waves of wavelengths in the range 3 to 5 μm. When the surface of the TBC, particularly the air plasma sprayed version, is irradiated by MIR, the waves are reflected. The reflectance has three components: reflection from the external surface, scattering within the coating, and reflection from the interface between the ceramic and metallic bond coat. With continued thermal cycling of the TBC, cracks develop and link up within the ceramic coating near the interface (DeMasi-Marcin et al., 1990). The cracks introduce air gaps with large changes in index of refraction from 2 to 1. This results in an increase in reflectance. Eldridge et al. found that the TBC spalled when reflectivity reached 71%. The reflectance measurements can be conducted in a Fourie-r transform infrared (FTIR) spectrometer with an attachment to integrate reflected waves over a hemisphere. The reflectivity profile of as-coated samples can be used for NDI of defects, whereas the signals from thermally cycled coatings can be used to estimated remaining life.

Electrochemical Impedance Spectroscopy (EIS)

This technique is applicable to ionic conductors such as zirconia and oxides formed as corrosion products on metals. It is based on electrochemical principles (Sawyer and Roberts, 1974). The technique utilizes an electrolyte in contact with the test material such as the ceramic layer of TBCs (Ogawa et al., 1999; Jayaraj et al., 2004). The setup generally consists of a three-electrode electrochemical cell held in contact with an electrolyte at constant temperature. A convenient electrolyte is 0.01 molar potassium ferri- and ferrocyanide, $K_3Fe(CN)_6/K_4Fe(CN)_6.3H_2O$ (Zhang and Desai, 2005). The TBC acts as the grounded working electrode. Usually a calomel electrode is used as the reference, while platinum in the form of wire mesh provides a counterelectrode. A voltage is applied between the working and the counterelectrode. Electrical impedance of the system and phase angle are the parameters measured as a function of frequency of the applied voltage, within a range between a few millihertz and 100 kilohertz. The electrical impedance is a complex number with two parts, a real part containing resistance and an imaginary part containing capacitance, and requires vector plots for analysis. The data are analyzed by modeling the microstructural characteristics of the TBC system as an equivalent electrical circuit (Fig. 8.4). The circuit consists of a number of resistances R and capacitances C. In this figure, the subscripts S, YSZ, P, TGO, and T correspond respectively to contributions from the electrolyte, ceramic coating, porosity, TGO, and the ceramic–metallic coating interface. From the plots of resistance and capacitance as a function of frequency, the microstructural parameters such as TGO thickness, ceramic porosity, pore shape, and the presence of interfacial delamination are assessed without destroying test samples. Both the electrochemical resistance and capacitance of the ceramic coating and the TGO can be independently determined from the EIS spectra. Figure 8.5 (Sohn et al., 2004) is an example that shows the variation of these parameters as a function of

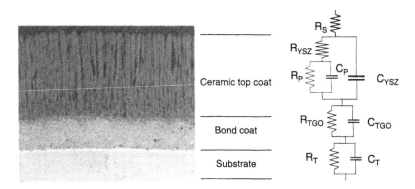

Figure 8.4 AC circuit equivalent of the microstructure of TBC consisting of EB-PVD ceramic, TGO at interface, and metallic bond coat (B. Jayaraj, S. Vishweswaraiah, V. H. Desai, and Y. H. Sohn, Electrochemical impedance spectroscopy of thermal barrier coatings as a function of isothermal and cyclic thermal exposure, *Surf. Coat. Technol.*, 2004, 177–178, 140–151). Reprinted with permission from Elsevier.

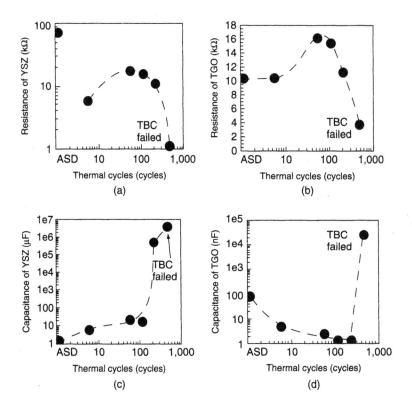

Figure 8.5 Variation of resistance and capacitance of 7YSZ and TGO layers as a function of thermal cycling: 10-minute heat up, 40-minute hold, and 10-minute forced air cool; hold temperature 1121°C, 2050°F (Y. H. Sohn, B. Jayaraj, S. Laxman, B. Franke, J. W. Byeon, and A. M. Karlsson, The nondestructive and nano-microstructural characterization of thermal-barrier coatings, *JOM*, Oct. 2004, pp. 53–56). Reprinted with permission from The Minerals, Metals & Materials Society.

thermal cycling of a TBC consisting of a EB-PVD 7YSZ ceramic coating and a NiCoCrAlY bond coat on IN738 superalloy substrate. It is clear that the evolution of both resistance and capacitance is a strong function of the number of cycles. This correlation indicates that the EIS can be used as an NDI process to assess extent of damage due to thermal cycling. More recent studies (Jayaraj et al., 2006) have shown that sintering due to high-temperature exposure as well as macrocrack formation due to thermal quenching in monolithic 7YSZ, parabolic growth of TGO, and imminent failure of EB-PVD TBC can be detected by the EIS technique.

Photoluminescence Piezospectroscopy (PLPS)

The basic principle of photoluminescence piezospectroscopy (PLPS) is illustrated in Fig. 8.6 (Gell et al., 2004). Typical alloys and metallic bond coats on which TBCs are applied always contain Cr either as part of the composition or as an impurity. The Cr atoms are incorporated within the aluminum oxide structure of the aluminum oxide TGO in the form of Cr^{3+} ions. On irradiation with a laser beam of appropriate frequency (for example, an argon laser of 5149 Å wavelength), Cr^{3+} ions fluoresce. At the atomic scale, what happens is the following: The outer electron in orbit around the Cr ion core absorbs the incident laser radiation, which raises the electron to a higher energy level. The electron subsequently falls to a lower energy level, releasing energy known as fluorescent radiation or luminescence. Since the process is stimulated by radiation in the visible spectrum, it is called photostimulated luminescence. The wavelength λ and frequency ν are related by $\nu = c/\lambda$, where c is the velocity of light. The photostimulated luminescence from the Cr^{3+} ions is spectrographically analyzed. The spectrum consists of two strong lines, R_1 and R_2 (Fig. 8.7), located at 14,402 and $14,432\,cm^{-1}$ frequencies for a stress-free aluminum oxide, such as sapphire. Additional lines corresponding to other aluminum oxide phases such as θ may also be present. The columnar microstructure of EB-PVD deposited yttria-stabilized zirconia (YSZ) acts as wave guides and does not significantly attenuate the intensity of the incident laser beam for YSZ thickness up to about $300\,\mu m$. However, air plasma sprayed YSZ with laminar microcracks and porosity greatly attenuates the incident beam, limiting YSZ thickness for analysis to below $75\mu m$. Stress in the TGO results in shifting of the lines relative to stress-free aluminum oxide. The frequency shift is related to the residual stress through the relation

$$\Delta\nu = 2/3\Pi_{ii}\sigma_{av},$$

where $\Delta\nu$ is the shift in frequency, Π_{ii} is the piezospectroscopic tensor, and σ_{av} is the residual stress within the TGO. Spectrographic analysis of the frequency shifts of R_1 and R_2 lines as well as the position of θ-phase aluminum oxide have been used to characterize TGO stress, to assess quality, and to detect the presence of interfacial delamination. For TBC systems that exhibit monotonic variation of stress with thermal exposure (which depends on a combination of processing parameters and bond coat characteristics) or where delamination at the interface allows luminescence from the stress-relieved TGO to be detected, PLPS could be used as an NDI technique. It has the potential to assess damage and estimate life remaining in actual use. Figure 8.8 is an example showing the evolution of luminescence spectra with thermal cycling of EB-PVD TBC. The ceramic is deposited on a platinum aluminide bond coat (Sohn et al., 2004). In the test, the TBC survived 420 cycles. At around 350 cycles signs of delamination are detectable as the TGO becomes practically stress-free.

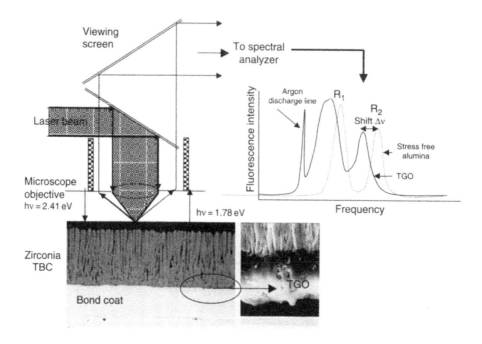

Figure 8.6 Principle of piezospectroscopic measurement of residual stress (Maurice Gell, Swetha Sridharan, Mei Wen, and Eric Jordan, Photoluminescence piezospectroscopy: A multi-purpose quality control and NDI technique for thermal barrier coatings, *Int. J. Appl. Ceram. Technol.*, 2004, 1(4), 316–329). Reprinted with permission from Blackwell Publishing.

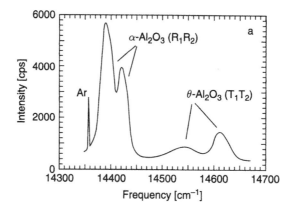

Figure 8.7 Fluorescence spectra from TGO containing both alpha and theta alumina (V. K. Tolpygo and D. R. Clarke, *Mater. High Temp.*, 2000, 17(1), 59–70). Reprinted with permission from *Science and Technology Letters.*

Interferometric Techniques

In this group of techniques, the principle of optical interference plays the key role. Two monochromatic beams of light originating from the same source are used to create an

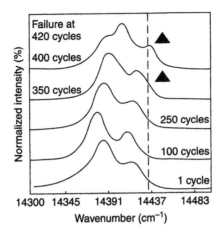

Figure 8.8 Evolution of PLPS spectra on thermal cycling from a TBC consisting of EB-PVD 7YSZ on a platinum aluminide bond coat. After 350 cycles, luminescence from stress-relieved TGO is visible. Failure occurred after 420 cycles (Y. H. Sohn, B. Jayaraj, S. Laxman, B. Franke, J. W. Byeon, and A. M. Karlsson, The non-destructive and nano-microstructural characterization of thermal-barrier coatings, *JOM*, Oct. 2004, pp. 53–56). Reprinted with permission from The Minerals, Metals & Materials Society.

interference pattern. One of the beams makes a journey to and from the object under inspection and meets the other beam coming directly from the source. An interference pattern forms that contains information about the object. Analysis of the interference patterns provides the details of various features of the object, including defects. Several types of interferometric techniques are available, that differ in their details.

Holography

Holography is a process in which a three-dimensional image of an object is created by using lasers (Fig. 8.9) (*Optical Holography, ASM Handbook*, 1997). It consists of two steps. The first step is to create a hologram by optically separating a laser beam into two beams using a beam splitter. One of the beams, known as the reference beam, is projected on a high-resolution holographic film after expanding and filtering the beam to cover the film uniformly. The second beam, similarly expanded and filtered, is reflected off the object. This beam, carrying all the information about the topography of the object, meets the reference beam at the holographic film, creating an interference pattern, that is captured on the film as the hologram. The second step is to illuminate the finished hologram by a beam akin to the original reference beam. This process faithfully reconstructs the object. This reconstructed image may be compared with the actual object at a later time to assess changes took place in such parameters as shapes, dimensions, and texture.

Thermal Wave Interferometry

This process (Almond et al., 1987) has been demonstrated on plasma-sprayed wear coatings as well as TBCs. It uses a laser beam modulated at a frequency between 5 and 800 Hz by a

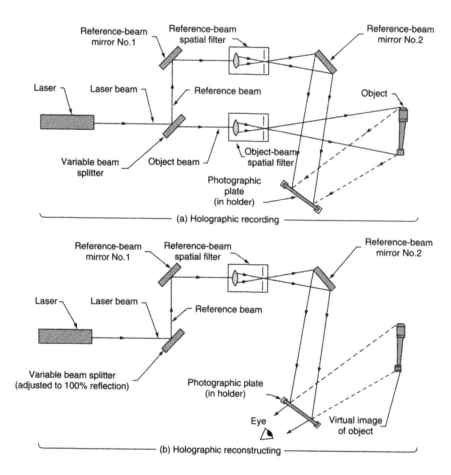

Figure 8.9 Schematic illustration of creation of a hologram and reconstruction of the object profile (Optical Holography, *Metals Handbook*, Vol. 17, Nondestructive Evaluation and Quality Control, ASM International, 1997). Reprinted with permission from ASM International.

mechanical chopper. The beam is scanned on the surface of the coating. The alternate heating of the coated surface generates a thermal wave, that propagates into the interior down to the interface. At the interface, part of the thermal wave is reflected back. The reflected component meets the incident wave at the surface, forming interference patterns. An infrared detector monitors the thermal profile, including the amplitude and the phase of the thermal wave. The profile varies with the thickness of the coating and the presence of subsurface and interface defects.

Shearography and Electronic Speckle Pattern Interferometry

These techniques are also based on holographic interferometric principles (Silva Gomes et al., 2000). However, they use two lasers to create interference patterns that appear as speckles. The processes have successfully been evaluated in inspecting delamination in TBCs.

REFERENCES

Aithal, S., G. Rousset, L. Bertrand, P. Cielo, and S. Dallaire, Photoacoustic characterization of subsurface defects in plasma-sprayed coatings, *Thin Solid Films*, 1984, 119, 153–158.

Almond, D. P., P. M. Patel, H. Reiter, The testing of plasma sprayed coatings by thermal wave interferometry, *Mater Eval.*, 1987, 45(4), 471-475.

Antonelli, G., M. Ruzzier, and F. Necci, Thickness measurement of MCrAlY high-temperature coatings by frequency scanning eddy current technique, *ASME 97-GT-1*, pp. 1–11, 1997.

Antonelli, G., M. Ruzzier, and F. Cernuschi, A calibration free electromagnetic technique for NDT of metallic coatings, *NDT.net*, 1998, 3(11), 145.

Bennett, T. D., and F. Yu, A nondestructive technique for determining thermal properties of thermal barrier coatings, *J. Appl. Phys.*, 2005, 97, 013520-1 to 013520-12.

Bray, D. E., and R. K. Stanley, *Nondestructive Evaluation—A Tool for Design, Manufacturing, and Service*, McGraw-Hill, New York, 1989.

Chen, X., G. Newaz, and X. Han, Damage assessment in thermal barrier coatings using thermal wave imaging technique, in *Proc. ASME International Mech. Eng. Congress and Exposition*, Nov. 11–16, 2001, pp. 239–2460.

Conner, J. A., and W. B. Connor, Ranking protective coatings: Laboratory vs. field experience, *JOM*, Dec. 1994, pp. 35–38.

DeMasi-Marcin, J. T., K. D. Sheffler, and S. Bose, Mechanism of degradation and failure in plasma-deposited thermal barrier coating, *J. Eng. Gas Turbines Power Trans ASME*, 1990, 112, 521–526.

Eldridge, J. I., C. M. Spuckler, J. A. Nesbitt, and K. W. Street, Health monitoring of thermal barrier coatings by mid-infrared reflectance, *Ceram. Eng. Sci. Proc.*, 2003, 24, 511–516.

Ferber, M. K., A. A. Wereszczak, M. Lance, J. A. Haynes, and M. A. Antelo, Application of infrared imaging to study controlled failure thermal barrier coatings, *J. Mater. Sci.*, 2000, 35, 2643–2651.

Gell, M., S. Sridharan, M. Wen, and E. Jordan, Photoluminescence piezospectroscopy: A multi-purpose quality control and NDI technique for thermal barrier coatings, *Int. J. Appl. Ceram. Technol.*, 2004, 1(4), 316–329.

Han, X., L. D. Favro, T. Ahmed, Z. Ouyang, L. Wang, X. Wang, F. Zhang, P. K. Kuo, and R. L. Thomas, in *Review of Progress in Quantitative NDE*, Vol. 16, pp. 353–356, Eds. D. O. Thompson and D. Chimenti, Plenum, New York, 1997.

Jayaraj, B., S. Vishweswaraiah, V. H. Desai, and Y. H. Sohn, Electrochemical impedance spectroscopy of thermal barrier coatings as a function of isothermal and cyclic thermal exposure, *Surf. Coat. Technol.*, 2004, 177–178, 140–151.

Jayaraj, B., S. Vishweswaraiah, V. H. Desai, and Y. H. Sohn, Changes in electrochemical impedance with microstructural development in TBC, *JOM*, Jan. 2006, pp. 60–63.

Kauppinen, P., The evaluation of integrity and elasticity of thermally sprayed ceramic coatings by ultrasonics, p. 130, *VTT Publication 325*, Technical Research Center of Finland, Espoo, Finland, 1997.

Metals Handbook, Vol. 17, *Nondestructive Evaluation and Quality Control*, Eds. J. R. Davis et al., ASM International, 1997, Eddy Current Inspection.

Newaz, G., and X. Chen, Progressive damage assessment in thermal barrier coatings using thermal wave imaging technique, *Surf. Coat. Technol.*, 2005, 190, 7–14.

Ogawa, K., D. Minkov, T. Shoji, M. Sato, and H. Hashimoto, NDE of degradation of thermal barrier coating by means of impedance spectroscopy, *NDT&E Int.*, 1999, 32, 177–185.

Optical Holography, ASM Handbook, Vol. 17, *Nondestructive Evaluation and Quality Control*, ASM International, 1997.

Rinaldi, C., V. Bicego, and P. P. Colombo, Validation of CESI blade life management system by case histories and in situ NDT, *J. Eng. Gas Turbines Power Trans. ASME*, 2006, 128, 73–80.

Sawyer, D. T., and J. L. Roberts, Jr., *Experimental Electrochemistry for Chemists*, Wiley, New York, 1974.

Silva Gomes, J. F., J. M. Monteiro, and M. A. P. Vaz, NDI of interfaces in coating systems using digital interferometry, *Mech. Mater.*, 2000, 32, 837–843.

Sohn, Y. H., B. Jayaraj, S. Laxman, B. Franke, J. W. Byeon, and A. M. Karlsson, The non-destructive and nano-microstructural characterization of thermal-barrier coatings, *JOM*, Oct. 2004, pp. 53–56.

Steffens H.-D., and H.-A. Crostack, Methods based on ultrasound and optics for the non-destructive inspection of thermally sprayed coatings, *Thin Solid Films*, 1981, 83, 325–342.

Voyer, J., F. Gitzhofer, and M. I. Boulos, Study of the performance of TBC under thermal cycling condition using an acoustic emission rig, *J. Therm. Spray Technol.*, 1998, 7(2), 181–190.

Zhang, J., and V. Desai, Evaluation of thickness, porosity and pore shape of plasma sprayed TBC by electrochemical impedance spectroscopy, *Surf. Coat. Technol.*, 2005, 190, 98–109.

Zilberstein, V., I. Shay, R. Lyons, N. Goldfine, T. Malow, and R. Reiche, Validation of multi frequency eddy current MWM sensors and MWM arrays for Coating production quality and refurbishment assessment, Paper No. GT2003-38170 in *Proc. ASME Turbo Expo 2003*, June 2003, pp. 581–590.

Zombo, P. J., Developing NDE methods for coated combustion turbine components, *TBC Workshop 1997*, pp 127–137, NASA Lewis Research Center, 1997.

Chapter 9

COATINGS REPAIR

Repair of high-value components in industrial processes characterized by elevated temperature and aggressive environment has become economical with the advent of robust repair processes. Turbine components in gas turbine engines are prime examples. If the cost of the repair process is within 70% of the price of a new part, the repair is considered economical (Underhill, 1985). Generally the economics of repair is assessed by the economic repair factor (ERF), given by

$$\mathrm{ERF} = (P_{\mathrm{rp}}/P_{\mathrm{nw}})(t_{\mathrm{nw}}/t_{\mathrm{rp}}),$$

where P and t respectively represent price and life, and the subscripts nw and rp refer to new part and repair process. For coated components both the life and price factors are significantly affected by the degradation and replacement cost of coatings.

Components requiring repair generally have significant exposure in the field with degradation due to such causes as substrate alloy degeneration, oxidation, corrosion, cracking due to thermal and low cycle fatigue, TBC spall, erosion, and foreign object damage. The repair cycle is generally adjusted to coincide with the overhaul of the complete equipment, such as gas turbine engines, to eliminate expenses for additional disassembly. New components may also require repair if they do not meet quality requirements, due primarily to manufacturing aberration and handling damage.

9.1 LIMITS TO COATINGS REPAIR

The total life of a component consists of a number of segments between prescribed overhauls of the whole equipment. Within each set of these overhauls, the repaired component needs to have acceptable integrity to meet the performance goals. In the case of coatings, this requirement imposes limits to the number of repair cycles. This arises because each time the old coating is stripped, there is a finite loss of the wall thickness due to the removal of the diffusion zone, which is roughly half the thickness of the coating. The interdiffusion zone of overlay coatings is usually thinner and, therefore, the wall thickness loss on coating stripping is expected to be less than that for diffusion coatings. Bell (1985) provides an example of an air-cooled turbine blade, made of Ni base superalloy Nimonic 108 and protected by a simple aluminide coating. The blade operated for 750 hours, predominantly in a marine military mission, exhibiting significant coating diffusion into the substrate material. The coating penetration was such that poststripping wall thickness loss would have limited the full life of a recoated blade to 1500 hours, which was half the required theoretical creep life of 3000 hours. An obvious solution to this problem to meet the remaining creep life was to upgrade the coating to platinum-containing aluminide, which typically lasted about three times longer than plain aluminide. The life extension is primarily due to the improved protective alumina scale adherence and consequent reduced rate of consumption of aluminum.

9.2 THE REPAIR PROCESS

For a component, the repair of coatings is generally one of the many steps (Mattheij, 1985; Bettridge and Ubank, 1986; Siry et al., 2001) in the overall repair process, as summarized in Table 9.1.

Component Cleaning

Prior to the removal of the old coating, the component needs to be cleaned to get rid of all the debris deposited during exposure in the operating environment. This step is typically done by washing and grit blasting. It should be certain that no residue of the cleaning medium is left behind.

Consequence of Remnants of Old Coating

Complete removal of the old coating is essential prior to the deposition of the new coating. The remnants of old coating with associated oxides, corrosion products, porosity, and active element depleted zone do not allow good adhesion between the new coating and the substrate, as illustrated by Fig. 9.1 (Schneider et al., 1985). In this example, remnants of old coating did not allow the new Ni–Cr–Si coating to adhere well to the substrate alloy. Complete metallic coating removal may be checked by etching or heat tinting. The underlying principle of heat tinting is that alumina or chromia scale-forming coatings tint differently than base alloys when oxidized at high temperature. From the tint of the surface scale formed, one can determine if the surface has any remnant coating. For TBC-coated components, complete ceramic coating removal is generally ascertained by visual inspection.

Table 9.1   Steps in the Overall Repair Process

Inspection of external surface and internal passages
Cleaning of component
Cleaning of internal passages (alkali autoclave/ultrasonic)
Removal of old coating
Cleaning and drying
Inspection and dimensional measurement
Grinding, blending, and polishing
Welding
Brazing
HIPing
Masking
Deposition of new coatings
Laser drilling of cooling hole, if required
Heat treatment

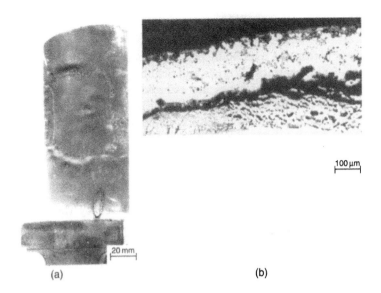

Figure 9.1 (a) An industrial gas turbine engine blade exhibiting coating flake-off after repair. (b) Inadequate removal of the old chromium coating prior to deposition of Ni–Cr–Si coating leads to 0flake-off (K. Schneider, B. Jahnke, R. Burgel, and J. Ellner, Experience of stationary gas turbine blades—view of a turbine manufacturer, *Mater. Sci. Technol.*, 1985, 1, 613–619). Reproduced with permission from Maney Publishing.

Removal of Ceramic Coatings

A number of processes have been evaluated for removal of ceramic coatings based on yttria-stabilized zirconia (YSZ) composition.

Grit Blasting

Abrasive grit blasting is an inexpensive process for removal of YSZ coatings (Anguelo and Anguelo, 1991). Generally, aluminum oxide is the blasting medium used with high-pressure air. Some of the issues with this process include the requirement of masking of areas where grit blasting would be detrimental and the lack of good control on material removal.

Alkali Solution

Aqueous alkali hydroxide solutions are effective in removing YSZ. The process is typically conducted in an autoclave at elevated temperature and pressure (Sangeeta, 1997). The added advantage of the process is the removal of some of the deposits within the internal passages and coating surface. Depending on the temperature, pressure, and geometry of the hardware, ceramic coating removal requires 1 to 6 hours. The alkali solution usually does not affect Ni base superalloy substrates. However, the extent of substrate damage, if any, needs to be ascertained before the process is used.

Molten Alkali

Digestion in molten potassium hydroxide (melting point 361°C, 682°F) has been found to accelerate removal of YSZ coatings (Warnes and Schilbe, 2001) compared with aqueous alkali solution in an autoclave. Removal times are reduced from hours to several minutes. The added advantage of the process is the accessibility of internal passages for coatings removal. The proposed mechanism for removal of YSZ is thought to be preferential reaction of alkali hydroxide with the aluminum oxide of the TGO,

$$2KOH + Al_2O_3 = 2KAlO_2 + H_2O.$$

That KOH does not react directly with YSZ is demonstrated by the release of large pieces of YSZ retaining the shape of the substrate such as a turbine blade airfoil and the lack of microscopic evidence of reaction.

Water Jet

A high-pressure water jet without abrasive grits has also been successfully used to remove the ceramic coating of TBCs (Sohr and Thorpe, 1993). The water jet used in metal cutting can be used without the abrasive garnet grits. The issues with this process are the lack of control and the requirement for expensive equipment.

Removal of Metallic Coatings

Acid Stripping

Metallic coatings are generally stripped chemically or electrochemically in acids such as hydrochloric or nitric acid at appropriate concentrations and temperatures. Degraded MCrAlYs with depleted aluminum are often difficult to acid strip (G. W. Goward, 2005, personal communication). Most of the Ni and Co base superalloys are reasonably resistant to acid attack, allowing complete removal of coating without loss of substrate material. For other substrate material, appropriate testing should be conducted and the stripping process should be closely controlled. Additional stripping processes reported in the literature include a high-pressure water jet.

9.3   RECOATING AND MATERIAL RESTORATION

Once the old coating is removed, and the component is cleaned and conditioned properly, a new coating can be deposited by the appropriate processes discussed in Chapters 6 and 7. Occasionally, coatings are used for dimensional restoration to compensate for lost substrate material. In such cases the coating material and the process should be selected so as to match density and reduce residual stress. At the completion of the coating process, the microstructure and the properties of the coating and the substrate are restored by appropriate heat treatment.

REFERENCES

Anguelo, M., and G. Anguelo, U.S. Patent No. 5,018,320, 1991.

Bell, S. R., Repair and rejuvenation procedures for aero gas-turbine hot-section components, *Mater. Sci. Technol.*, 1985, 1, 629–634.

Bettridge, D. F., and R. G. Ubank, Quality control of high-temperature protective coatings, *Mater. Sci. Technol*, 1986, 2, 232–242.

Mattheij, J. H. G., Role of brazing in repair of superalloy components—advantages and limitations, *Mater. Sci. Technol.*, 1985, 1, 608–612.

Sangeeta, D., U.S. Patent No. 5,643,474, 1997.

Schneider, K., B. Jahnke, R. Burgel, and J. Ellner, Experience of stationary gas turbine blades—view of a turbine manufacturer, *Mater. Sci. Technol.*, 1985, 1, 613–619.

Siry, C. W., H. Wanzek, and C. P. Dau, Aspects of TBC service experience in aero engines, *Mat.-Wiss. Werkstofftech.*, 2001, 32, 650–653.

Sohr, J. M., and M. L. Thorpe, *Aerospace Eng.*, 1993, 13, 19–23.

Underhill, J. W., Refurbishment of superalloy components for gas turbines—scope and reward, *Mater. Sci. Technol.*, 1985, 1, 604–607.

Warnes, B. M., and J. E. Schilbe, Molten hydroxide removal of thermal barrier coatings, *Surf. Coat. Technol.*, 2001, 146–147, 147–151.

Chapter 10

FIELD AND SIMULATED FIELD EXPERIENCE

High-temperature metallic and thermal barrier coatings (TBCs) are used for many industrial applications for aggressive environments, as described in Chapter 1. Although a wealth of information exists in the public domain devoted primarily to historical perspectives and the processing and properties of these coatings, literature covering assessment of their application and performance in the field is very scarce, rarely up to date, and in most instances, closely guarded by the industry for competitive business reasons. One of the primary users of high-temperature coatings is the gas turbine engine industry for commercial and military aircraft, industrial power generation, and marine applications. The experience illustrated by selected examples described in this chapter is, therefore, somewhat dominated by these applications. Experience in simulated tests is also included in this chapter because it has many of the attributes of actual field exposure.

10.1 GAS TURBINE ENGINE APPLICATION

Metallic Coatings

Coating Oxidation in Aircraft Engines

Overlay coatings, discussed in detail in sections starting at 6.4, provide the flexibility of tailoring the composition of the coating to the specifics of each application. Typical high-temperature overlay coatings have the general composition of NiCoCrAlY. One of the early methods of deposition of overlay coatings was by the EB-PVD process described in Section 6.6 and in greater detail, for ceramic coatings, in Section 7.5. An example of a coating deposited by this process is PWA 270, which has seen application in Pratt & Whitney–built gas turbine engines. The EB-PVD process relies on the vapor phase to form the coatings. Consequently, the vapor pressure of coating constituents is an important parameter. Retaining minor quantities of key reactive elements such as Hf and Y, which have very low vapor pressure compared with that of the other constituents, therefore, becomes an issue in EB-PVD NiCoCrAlYs. With the advent of the LPPS process, additional flexibility is afforded in controlling the composition and microstructure of overlay coatings. However, initial comparative studies (Pennisi et al., 1981) showed that NoCoCrAlYs deposited by EB-PVD on Hf-containing columnar-grain (DS) Mar-M200 performed better than the LPPS deposited counterpart. The adherence of the alumina scale on coating by the EB-PVD process had less to do with the coating itself and more to do with the alloy on which it was deposited. It benefited from the presence of Hf in the DS Mar-M200 alloy. It is well known that the reactive elements in the alloy substrate get incorporated at the interface between the alloy or the coating and the oxide scale, improving adherence of the oxide. For alloys without the reactive elements, this advantage is not available. For these alloys, the LPPS process provides the flexibility because the coating composition can be tailored and is independent of vapor pressure. In subsequent evaluation by Gupta and Duvall (1984), LPPS deposited Ni20Co18Cr12.5Al0.6Y0.4Si0.25Hf (composition in weight percent) was tested side by side against EB-PVD deposited Ni20Co18Cr12Al0.2Y. The coatings were deposited on nickel-base single-crystal superalloy PWA 1480 (does not have reactive elements) and evaluated in JT9D, as well as PW2037, experimental engines at Pratt & Whitney. These engines power Boeing 747 and 757 aircraft, respectively. After 375 hours/2500 cycles, the EB-PVD deposited NiCoCrAlY coating was defeated by oxidation (Fig. 10.1b), while LPPS deposited NiCoCrAlY + Si + Hf exhibited little distress (Fig. 10.1a). By 1984, the Hf- and Si-containing coating had accumulated more than 54,000 endurance cycles in several engines. The excellent performance is attributable to the presence of Hf and Si in the coating. Because of its success in the engine tests, this coating was introduced into production for application on turbine blades and vanes on various aircraft engine models at Pratt & Whitney.

Experience with diffusion aluminides goes back much further than with MCrAlYs. The diffusion aluminides were the first high-temperature coatings to be used to protect copper, iron, and later Ni and Co base superalloys against oxidation (Goward and Cannon, 1988) dating as far back as 1914. The first application on turbine blades was by the process of hot dipping for Allison and Curtiss Wright engines in 1952 (Nichols et al., 1965). Current applications include simple aluminides for protection against oxidation of external and internal surfaces of turbine blades and vanes, chromized aluminides to fight oxidation and type II hot corrosion, and platinum-containing aluminides for enhanced protection against oxidation and type I hot corrosion. One of the major limitations of diffusion coatings is the lack of flexibility

(a)   (b)

Figure 10.1 PWA 1480 single-crystal superalloy turbine blade coated by LPPS process with NiCoCrAlY + Si + Hf (a) and by EB-PVD with NiCoCrAlY (b). The coating in (b) had failed after 2500 cycles in a JT9D test engine, whereas the coating in (a) performed very well (D. K. Gupta and D. S. Duvall, A silicon and hafnium modified plasma sprayed MCrAlY coating for single crystal superalloys, in *Superalloys 1984*, Eds. M. Gell, C. S. Kortovich, R. H. Bricknell, W. B. Kent, and J. F. Radavich, Metallurgical Soc. of AIME, PA, USA, 1984, pp. 711–720). Reproduced with permission from The Minerals, Materials & Materials Society.

in controlling the composition. Connor (1992) reported testing of several Pt aluminides on directionally solidified, as well as single-crystal, blades in test engines at General Electric Aircraft Engine (GEAE). Some of the coating details are summarized in Table 10.1.

In general terms, the engine evaluation was characterized as a "very hot engine test." Posttest analysis of the coatings showed that the plain aluminides were severely defeated with significant base alloy attacks at leading and trailing edges of the blade. In areas where coating remnants were still intact, they were completely transformed from βNiAl to γ'Ni$_3$Al. All Pt aluminide–coated blades showed intact coatings after the engine test, indicating that considerable coating life remained. This observation is in line with the expected benefits of incorporation of Pt in the aluminides. Additional features observed after the engine test included significant growth of coating thickness (Fig. 10.2), which essentially is the result of movement of the alloy coating interface due to diffusion of Pt and Al into the alloy, and of Ni into the coating. Rumpling of coatings was also observed, which was more prominent for Pt aluminides than for plain aluminides, the latter due to the loss of coating at the locations where rumpling was observed. Some of the coatings exhibited thermal fatigue cracking limited to the outer "additive" layer. Another important factor to consider, which has led to rumpling observed by others, is the martensitic phase change (Chen et al., 2003) in the coating. However, without detailed microstructural analysis, the actual mechanism is difficult to identify. The

Table 10.1   Platinum Aluminide Coatings Tested in GEAE Test Engine

Coating	Source	Substrate
LDC 2E (Pack)	Howmet	SX Ni base
RT 22 (Pack)	Chromalloy	SX Ni base
RT 22G (Above Pack)	Chromalloy	SX & DS Ni base
MDC 150(CVD)	Howmet	SX & DS Ni base

SX = Single crystal; DS = Directionally solidified

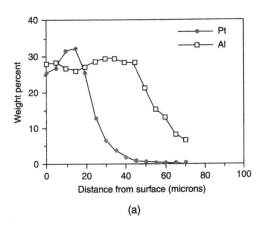
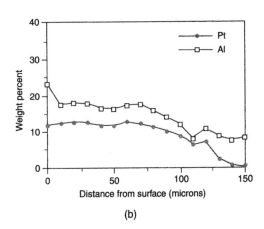

(a)   (b)

Figure 10.2   Concentration profile for RT-22 platinum aluminide coating on single-crystal Ni base alloy (a) before engine test, (b) after engine test (J. A. Connor, Evaluation of simple aluminide and platinum aluminide coatings on high pressure turbine blades after factory engine testing — round II, *ASME 92-GT-140*, 1992). Reprinted with permission from American Society of Mechanical Engineers.

rumpling phenomenon is not limited to the aluminides. Overlay NiCoCrAlY-coated high-pressure turbine blades in Pratt & Whitney's JT9D engine occasionally exhibited rumpling in the field, primarily at hot spots. The problem was alleviated by the use of patches of TBC, which reduced local temperatures and associated creep of the coating.

Coating Cracking in Aircraft Engines

The total reservoir of aluminum in coatings can be increased either by enhancing the aluminum content in the composition or by increased thickness of the coating. Both of these options increase oxidation resistance of the coated hardware. However, increased coating thickness raises the risk of cracking of the coating and crack propagation into the substrate alloy. Such cracking and propagation have been observed in the field. This has also been demonstrated by Meetham (1986) on aluminide-coated Nimonic 105 alloy turbine blades in a Spey engine test at Rolls-Royce, (Fig. 10.3). The cracking, observed at the leading edge of the blades, occurs at temperatures below the ductile-to-brittle transition temperature (DBTT), which is around 700°C (1292°F) for the coatings used. As Fig. 10.3 shows, the susceptibility of the coating to cracking and to propagation of the crack into the alloy, demonstrated by the number of cracked blades, is reduced significantly by postcoating diffusion heat treatment, which decreases the aluminum content from between 37% and 40% to between 25% and 30%. Reduced aluminum content results in hardness reduction of the coating from 900–1000 HV to 500–700 HV, making the coating "less brittle." One possible explanation of the thickness dependence in coating cracking and crack propagation is the relationship between stress intensity factor K_c and crack length, $K_c \sim \sigma \sqrt{\pi c}$, where σ is the stress in the coating and c is the crack length, which, for a cracked coating, equals the thickness of the coating. Thus, a crack in a thicker coating will experience higher stress intensity. When the stress intensity exceeds a critical value called the critical stress intensity factor, which is a materials property, the crack extends and may penetrate the substrate.

Oxidation and Hot Corrosion in Aircraft and Marine Engine Simulation Test

The performance of several diffusion and overlay coatings has been evaluated by the National Aerospace Laboratory NLR in The Netherlands in simulated service conditions. The test

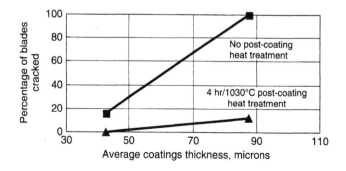

Figure 10.3 Effect of coating thickness on cracking propensity (based on data in G. W. Meetham, Use of protective coatings in aero gas turbine engines, *Mater. Sci. Technol.*, 1986, 2, 290–294). With permission from Maney Publishing.

intended to simulate the turbine conditions for Pratt & Whitney's F100 military engine, which powers the F-15 and F-16 fighter aircraft; General Electric's J85, which powers fighter training aircraft T-38; and General Electric's CF6-50 engine, which is used on both commercial and military transport aircraft (Mom and Hersbach, 1987; Wanhill et al., 1989; Mom and Boogers, 1986). The details of various coatings tested are summarized in Table 10.2. The simulated test facility consisted of two burner rigs. The first rig, operated at gas velocities of Mach number up to 0.75 with gas temperature within the range 600–1070°C (1112–1958°F), accepted nonrotating cascade assemblies of blades and vanes with a test area of 20 cm × 5 cm. The second rig was built for specimens in the form of cylindrical bars rotated in a carousel. The gas temperature range for this rig was 500–1650°C (932–3002°F). The mission profiles of military, commercial, and marine engines for both external surface and internal passages were simulated by controlling the time–temperature profile and injection of salt and SO_2 gas in the hot or cold gas streams. The posttest evaluation consisted of visual and microscopic observation. Relative ranking of coatings based on the result of the simulation test is shown in Table 10.3. The overall rankings, in general, validated the benefits of Pt, Si, Y, and Cr incorporation in the coating in fighting high-temperature corrosion. Also, tailored combinations of coatings fared better than single coating. The combination provided an avenue to incorporate multiple species, which may not be feasible in single coatings because of solubility and processing constraints. The absolute

Table 10.2 Details of coatings tested in rig simulating F100 and CF 6 Turbine Environment

Coating Type	Coating Process	Engine/Turbine Component on Which Coating Evaluated
PWA 73	High-activity pack aluminide	F100 1B, 2B, 1V, J85 1B
PWA 273	Low-activity pack aluminide	F100 1B
PWA 275	Low-activity over-the-pack aluminide	F100 1B, CF6 1B
HI 275	High-activity over-the-pack aluminide	CF6 1B
BB	Rh pack aluminide	F100 1B, 1V
RT 22A	Pt pack aluminide	F100 1B, 1V, J85 1B, CF6 1B
PWA 270	NiCoCrAlY overlay	F100 1B, 2B
Codep B	High-activity pack aluminide	J85 1B, CF6 1B
Cr–Al (l)	Low-activity Cr–Al pack	J85 1B
Cr–Al (h)	High-activity Cr–Al gas phase	CF6 1B
RB 505	Low-activity Cr–Al pack	J85 1B
RB 505 + Cr	Low-activity Cr–Al pack + Cr surface enrichment	J85 1B
SermaLoy J	Slurry Al–Si + diffusion treatment	J85 1B
RT 100	NiCoCrAlY LPPS	J85 1B
LCO-22	NiCoCrAlY LPPS	CF6 1B
C1A	Two-step Low-activity Cr–Al Pack	CF6 1B
C 30	Low-activity over-the-pack	CF6 1B
MDC 43	High-activity over-the-pack	CF6 1B
Pulse aluminide	High-activity over-the-pack with pulse argon	CF6 1B

B, blade; V, vane; 1, 2, 1st, 2nd stage. F100 Alloy tested: DS Mar-M200 + Hf; J85 and CF6 Alloy tested: René 80

Table 10.3 Relative Rankings of Coatings in Simulation Test

Engine	Coatings Ranked Low to High from Top to Bottom for Each Engine	Property Evaluated	Relative Ranking
F100	PWA73 BB PWA 275 PWA 273 RT22A PWA 270	Oxidation and corrosion	Increasing resistance to Oxidation and Corrosion ↓
J85	Cr–Al (l) SermaLoy J RB 505 RT 100 Codep B PWA 73 RB 505 + Cr RT 22A	Corrosion	Increasing resistance to Corrosion ↓
CF6-50 Internal Coating	SermaLoy J HI 275 Cr–Al (h) Pulse aluminide C 30 MDC 43	Corrosion	Increasing resistance to Corrosion ↓
External Coating	Codep B HI 275 Codep B + MDC 43 Codep B + Pulse aluminide Codep B + C 30 Codep B + Sermaloy J	Corrosion	Increasing resistance to Corrosion ↓

ranking of or between specific coatings should be considered with great caution because the simulation tests seldom reproduce the important nuances of actual engine and field tests.

Hot corrosion has also been observed in small engines such as those in helicopters and business jets, depending on the environment in which they operate and the alloys used. Eliaz et al. (2002) reported a case study of hot corrosion analysis of Mar-M 200 and IN 713C turbine blades, the former in a PT6 engine. In this and other tests, it is found that although degradation may be initiated by hot corrosion, final failure is likely to be due to other processes, such as low- or high-cycle fatigue.

Oxidation and Hot Corrosion in Industrial Gas Turbines

In the United States, documentation of field experience on coatings and alloys is more readily available for industrial gas turbine (IGT) engines than for aircraft engine applications, primarily because of the involvement of Electric Power Research Institute (EPRI), which caters to a consortium of power utilities and manufacturers in many areas of technology. EPRI-sponsored testing provides independent assessment of coatings for gas turbine power plants. An example of coatings performance assessment (McMinn et al., 1988) is EPRI's testing of coatings in Pratt & Whitney built FT4 turbine at Long Island Lighting Co (LILCO). The details of the materials tested are summarized in Table 10.4.

The testing period spanned August 1979 to August 1983, totaling field exposure of 812 hours with two intermediate visual examinations. The power plant was used as a "peaker," meaning it ran in cycles rather than at constant power. The objective of the test was to evaluate hot corrosion resistance of the coatings. Based on posttest microstructural evaluation, a relative ranking of the coatings was generated (Table 10.5). The benefits of Pt and a high concentration of Cr in fighting high-temperature corrosion are again confirmed by the test results.

Table 10.4 FT4 Coatings Evaluation at LILCO

Fuel Composition: No. 2 grade home heating oil S 0.25%, Pb 0.32 ppm, Na 0.45 ppm, Ca 0.62 ppm, K 0.50 ppm, Mn 233 ppm

Substrate Alloy: Mar-M 509, (Co10Ni23.3Cr7W3.5Ta0.2Ti0.5Zr)

Coatings Evaluated: Plain aluminide, Rh–aluminide, Pt–Rh aluminide, Pd–aluminide, Co21Cr12Al0.3Y, Ceramic overlay

Table 10.5 Coating Ranking Based on Performance in FT4 Industrial Gas Turbine Engine (1 High, 5 Low)

Coating	Location of Worst Coating Attack	Maximum Coating Thickness, μm (mil)	Coating Ranking Based on Test	Comments
CoCrAlY	Outer shroud	56 (2.2)	1	No substrate attack, isolated shallow penetration of coating
Pt–Rh–aluminide	Lower platform	102 (4.0)	1	No substrate attack, isolated penetration of coating, general thinning, voids in coating
Rh–aluminide	Lower platform	64 (2.5)	2	Shallow attack on substrate, coating penetrated at several locations, protective elsewhere
Rh–aluminide (HIPed)	Outer shroud	74 (2.9)	2	Shallow attack on substrate, coating penetrated at several locations, protective elsewhere
Aluminide	All locations	48 (1.9)	4	Complete local coating penetration, uniform coating attack, some substrate attack
Aluminide (HIPed)	Outer shroud, Mid-span	48 (1.9)	5	Complete coating defeat at many locations, some substrate attack
Pd–aluminide	Lower platform	173 (6.8)	5	Extensive coating penetration, some debonding

Several overlay and aluminide coatings have been field tested in Solar Turbines Mars T-14000 industrial gas turbines (Kubarych and Aurrecoechea, 1993). These coatings included Chromalloy produced simple aluminide RT-21 of average thickness 75 μm (3 mil), platinum aluminide RT-22 of thickness 65–115 μm (2.5–4.5 mil), and two overlay coatings designated VPS and 1386 of average thickness 100 μm (4 mil). The coatings were deposited on two different nickel base superalloys, the polycrystalline Mar-M 247 and single-crystal CMSX-4. The engine ran on natural gas for 4333 hours with 44 starts before inspection. Natural gas is generally a clean fuel, does not contain any significant amount of sulfur, and therefore, does not induce hot corrosion. After removal of some of the blades, the engine continued with replacements for more than 8000 hours with the goal to eventually accumulate a total of 20,000 hours of exposure. The estimated temperature over most of the airfoil was moderate, less than 980°C (1800°F). Without cooling holes, the temperature at the blade tip rose close to 1090–1120°C (2000–2050°F). Posttest visual observation as well as microstructural analysis showed:

- All four coatings successfully protected the substrate alloy.

- Both RT-21 and RT-22 degraded at the leading edge because of aluminum depletion leading to the transformation of β(NiAl) to γ'(Ni$_3$Al).

- For the MCrAlYs, the transformation of β(NiAl) to γ'(Ni$_3$Al) at the leading edge was practically complete.

- RT-22 was slightly better than RT-21 with less γ' formation on degradation.

The use of natural gas in the test preempted any hot corrosion. Thus, the degradation mode was predominantly oxidation. RT-22 faring better than RT-21 in degradation to γ' validates the benefit of Pt.

Large industrial gas turbine engines, used in power generation, require significantly higher component lives than do smaller power plants and aircraft engines. For example, turbine blades in IGTs are expected to last 70,000 to 100,000 hours, whereas lives of typical aircraft engine turbine blades are 15,000 to 20,000 hours. Oxidation, coating and substrate cracking, and hot corrosion, therefore, have additional significance for IGT. A number of alloys and coatings have been developed particularly for IGT applications. A few aircraft engine alloys and coatings have also been modified to accommodate IGT requirements. Schilke et al. (1992) have reviewed material issues in General Electric's large frame power plants. The review provides a glimpse of coatings used in IGTs in GE machines and their performance, summarized in Table 10.6.

Several points are evident from Table 10.6. As expected, platinum aluminide, both as single-phase and in two-phase coatings, provides protection against hot corrosion. This result, when combined with a 1.5× improvement of corrosion life by overlays over Pt aluminide, indicates that the hot corrosion is present in both type I and type II form. Also, the success of aluminizing over MCrAlYs indicates that the Al reservoir in the overlay coating is inadequate to meet the oxidation life requirements of the IGT parts.

High-temperature oxidation, as well as hot corrosion of LPPS-deposited CoCrAlY and CoNiCrAlY, have been reported by Kameda et al. (1997, 1999) on René 80 superalloy turbine blades of land-based gas turbines with various exposure times ranging between 8946 and 22,000 hours. Coating thickness varied between 120 and 200 μm (4.8–8 mil) for CoCrAlY coating tested in an engine burning liquefied natural gas (LNG). The CoNiCrAlY coating, tested with combined LNG and kerosene fuel, had thickness between 140 and 250 μm (5.6–10 mil). The CoCrAlY coating, exposed to a turbine environment burning LNG, underwent severe oxidation and loss of ductility after 21,000 hours of exposure. No hot corrosion was observed, obviously

Table 10.6   General Electric's Coatings for IGT Turbine Application

Coating	Performance characteristic	Comments
Platinum aluminide (single phase solid solution or solid solution + PtAl$_2$)	2X corrosion life relative to uncoated IN-738	Used between late 1970 and mid 1983
LPPS Overlays	1.5X corrosion life improvement relative to platinum aluminide	Introduced early 1980
GT-29	Baseline MCrAlY	
GT-29+	Aluminized over MCrAlY Used for firing temperature 1065°C (1950°F) in air cooled first blade and for firing temperature 955°C (1750°F) in uncooled first blade	Standard coating since 1990 Overaluminizing improves oxidation resistance
GT-29 In+	Used on internal surfaces in vanes	
GT-20	Used on vanes	Developed for aircraft engine

because of the inherent cleanliness of LNG. The use of combined fuel, however, induced not only oxidation, but also grain boundary sulfidation for CoNiCrAlY. The reduction in ductility was much more severe.

The data of Kameda et al. do not afford one-on-one comparison between CoCrAlY and CoNiCrAlY for fighting hot corrosion because they were tested using different fuels. However, the presence of sulfides in posttest analysis indicates that the corrosion is most likely to be type I. Life of the alloys and coatings in the IGT environment is influenced not only by the type of contaminants, but by their quantitative levels in the fuel, as shown by Schilke et al. (1992) (Fig. 10.4). For every 1 ppm increase in equivalent sodium, the life debit is approximately 50%. The source of sodium is likely to be in the form of a combination of chloride and sulfate. The latter is directly involved in hot corrosion. The chloride works in two ways. It produces

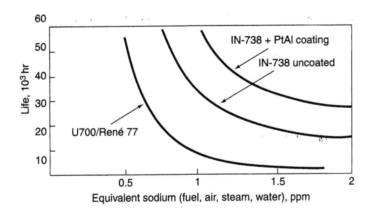

Figure 10.4 Alloy and coating hot corrosion life as a function of contaminant measured in sodium equivalent (P. W. Schilke, A. D. Foster, J. J. Pepe, and A. M. Beltran, Advanced materials propel progress in land-based gas turbines, *Adv. Mate. Proc.*, 1992, 141(4), 22–30). Reprinted with permission from ASM International.

sulfate salts in the combustor by reacting with SO_2 formed from oxidation of sulfur in the fuel. Additionally, sodium chloride accelerates the spallation of the alumina scale, which protects the alloys and the coatings from further oxidation.

Coating Cracking in Industrial Gas Turbine Engines

Local transient strains tend to crack coatings at temperatures below the DBTT. Such cracking has been found to be more prevalent in industrial gas turbines (Conner and Connor, 1994) than in aircraft engines as well as laboratory rig testing. Figure 10.5a shows cracking in a platinum aluminide coating after 12,000 hours of service in General Electric–built large industrial gas turbines. The cracks have penetrated the coating–alloy diffusion zone, which has started to exhibit signs of oxidation. An underlayer of LPPS-deposited MCrAlY coating with an aluminide overlayer (Fig. 10.5b) seems to provide a solution to cracking of the coating. The cracks still appear in the aluminide layer but do not penetrate the overlay. The overlay coating appears to blunt the cracks so that they do not propagate into the substrate even after 12,000 hours in the same engine as in Fig. 10.5a. One possible reason for the crack blunting behavior of the overlay coating is the presence of the ductile γ phase, which can undergo plastic deformation, unlike the brittle β phase.

The cracking resistance of the overlay coatings of $\gamma + \beta$ microstructure relative to the predominantly β phase of aluminides is also demonstrated in the field in other industrial gas turbine engines. For example, turbine blades (buckets) of nickel-base superalloys GTD 111 and IN-738 coated with platinum aluminide (22.1% aluminum in the outer layer), overaluminided CoCrAlY (16.1% aluminum in the outer layer), CoCrAlY (6.6% aluminum), and CoNiCrAlY (10.5% aluminum) were field exposed (Yoshioka et al., 2006) for up to 38,171 hours in frame 7E and 9E industrial gas turbine engines. Although the oxidation capability of the coatings correlated with aluminum content, the severity of coating cracking was in the

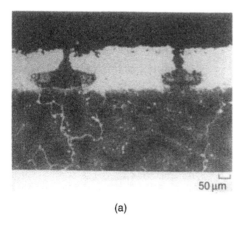
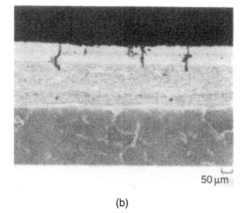

(a)   (b)

Figure 10.5 (a) Cracking in platinum aluminide coated turbine blade after 12,000 hours of service in large industrial gas turbine. Cracks have penetrated the diffusion zone of the substrate, which exhibits signs of onset of oxidation. (b) Crack arrested at the MCrAlY–aluminide interface after 12,000 hours in the same engine (J. A. Conner and W. B. Connor, Ranking protective coatings: Laboratory vs. field experience, *JOM*, Dec. 1994, pp. 35–38). Reprinted with permission from The Minerals, Metals & Materials Society.

following order: Pt aluminide > overaluminided CoCrAlY > CoCrAlY > CoNiCrAlY. Detailed posttest characterization showed that DBTT as well as minimum strain to cracking measured at elevated temperature correlated well with the cracking propensity, with cracks penetrating the alloys in the case of both platinum aluminide and overaluminided CoCrAlY. We know that for the same class of coating, the DBTT generally correlated with aluminum content, higher aluminum having higher DBTT.

Marine Application

Marine gas turbine engines are unique in that they are mobile, subjected to severe environments including terrenes, open oceans, and arctic surroundings. Some of these conditions may be present simultaneously. As a result, materials in marine applications experience severe corrosion. Alloys and coatings, therefore, need to be protected accordingly. Typically, high-chromium alloys and coatings fare better (Conde and McCreath, 1980). A number of coatings have been evaluated in the GE marine engine LM2500 (Grisik et al., 1980; Grossklaus, et al., 1986) which powers a large class of ships. This engine evolved from a military version, TF39, and the commercial counterpart, CF6. The testing was done across several engine builds aboard Gas Turbine ship, *Adm. Wm. Callaghan*. In addition to periodic removal of the engine, frequent boroscope inspection at roughly 400-hour intervals was also conducted. Further evaluation was done recently in a rainbow rotor in LM2500 powering Australian Navy ship, *HMAS Darwin*. Based on all the tests just described, the overall coating matrix and test results in terms of projected lives are summarized in Table 10.7.

Table 10.7 Coating Composition, Process, Supplier, and Results of Rainbow Test

Coating	Process (Supplier)	Test Result
BC21 Co22.5Cr10.5Al0.3Y (3.5–4.5 mil) (Standard production coating for US Navy's LM2500)	PVD (3.5–4.5 mil) (Chromalloy)	12000 hrs life

BC23
Co26Cr12Al1Hf5Pt (4.5–5.5 mil)
	PVD CoCrAl + Codeposition of
	    Al, Hf in pack + Pt electroplate
	    followed by diffusion heat treat
		13000–20000 hrs Life

BC23 **BC22** Co26Cr10.Al2.5Hf	LPPS LPPS	Baseline Between BC21 and BC23
BC23 (plasma) + Cr (5–6 mil) Co26Cr10.5Al2.5Hf5Pt & 0.5 mil Cr pack (25% near BC 23 to 40% near diffusion zone)	LPPS + Pack Chromizing	Better than BC23
Hi Cr CoCrAlY Co30Cr12Al0.5Y	LPPS	Eq to BC21
BC21 + PVD 20YSZ TBC	LPPS + EB-PVD	Erosion, some spall, no hot corr
Pt aluminide	Pack (Chromalloy)	~12000 hrs life

Some of the coatings listed in Table 10.7 have seen field exposure in the GE marine engine LM2500 (Wortman, 1985). Postexposure microstructural analysis revealed that type II hot corrosion was the primary mode of degradation due to the formation of low-melting eutectic $CoSO_4–Na_2SO_4$. In the field, both the three-layer and the plasma sprayed single-layer BC23 performed better than the baseline BC21 (Fig. 10.6). The test data, again, validate the benefits of Pt, Cr, and the reactive element Hf. They also demonstrate the potential barrier effects of TBC, which is chemically stable against sulfates.

TBC

Frit enamel coatings were used for thermal insulation in aircraft engine components as early as in the 1950s (Cannistraro, 1958). The concept slowly evolved into the modern design of the TBC. The first gas turbine engine application of a two-layer TBC system consisting of a bond coat on which a ceramic top coat is deposited was the "afterburner" of a Pratt & Whitney military engine (Soechting, 1995). The development and application of TBCs in gas turbine engines built by Pratt & Whitney has been reviewed by Bose and DeMasi-Marcin (1997) and others (Sheffler and Gupta, 1988; DeMasi-Marcin and Gupta, 1994; Meier and Gupta, 1994). Similar reviews on TBC application in GE aircraft engines (Wortman et al., 1989; Maricocchi et al., 1997), Rolls-Royce engines (Rhys-Jones and Toriz, 1989; Toriz et al., 1988, 1989; Bennett et al., 1987), and industrial gas turbines (Mustasim and Brentnall, 1997) show extensive application of and experience in this technology in the world of gas turbines.

The general progression of the use of TBCs in the turbine section of gas turbine engines started with the combustion chamber to reduce oxidation and thermal fatigue damage of float wall panels, followed by vane (also called nozzle) platforms to alleviate cracking due to high temperature and thermal strains, subsequently leading to the airfoil surfaces of the vanes to

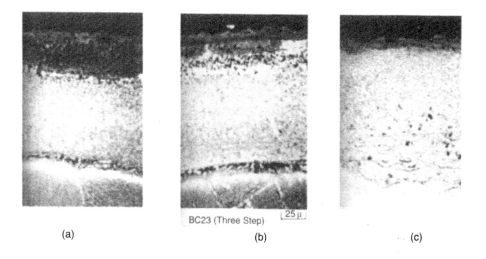

(a)   BC23 (Three Step) ⌞25 μ⌟   (c)
                                                        (b)

Figure 10.6 Microstructures of three metallic coatings at 10% span on the leading edge of second-stage turbine blade on LM2500 at the end of 5884 hours of service showing various degrees of type II hot corrosion: (a) BC21 with more attack than (b) three-layer BC23, and (c) plasma sprayed single-layer BC23 (D. J. Wortman, Performance comparison of plasma spray and physical vapor deposition BC23 coatings in the LM2500, *J. Vac. Sci. Technol. A,* 1985, 3(6), 2532–2536). Reproduced with permission from American Institute of Physics.

eliminate oxidation-induced burning. Included in between are turbine seals providing thermal insulation and abradability. A more recent application is on one of the most thermally and structurally demanding components in the turbine, the turbine blade. Figure 10.7 shows the coated components and the location in the turbine (DeMasi-Marcin and Gupta, 1994) of a gas turbine engine.

Plasma-Sprayed TBC

TBC in the Combustor

In commercial applications TBCs were introduced on Pratt & Whitney JT9D combustor panels to address increased temperature requirements. The earliest TBC for this application was air plasma sprayed magnesia stabilized zirconia (MSZ) on air plasma sprayed binary Ni–Al or Ni–Cr bond coat. MSZ-based TBC has also been used in combustors by several other gas turbine engine makers. For example, Rolls-Royce ran the coating with lives exceeding 8000 hours in civil aircraft engines and with twice that life in industrial RB211 engines (Meetham, 1986). The use of MSZ is limited to about 980°C (1796°F) because of phase destabilization on exposure to higher temperatures. It is also susceptible to hot corrosion, as discussed in Section 7.5. The MgO stabilizer is leached out by reaction with sulfate salts. The bond coat and the ceramic have subsequently been replaced by air plasma sprayed NiCoCrAlY and 7YSZ, respectively, to meet higher temperature requirements. This change resulted in a manifold improvement in TBC life (Meier and Gupta, 1992). Over the years this TBC design has accumulated hundreds of millions of on-wing flight hours/cycles in most modern gas turbine engines. Failure of this TBC, when it occurs, is limited within the oxide formed on the bond coat. With continued exposure in the engine, the oxide scale formed on the air plasma sprayed NiCoCrAlY at the

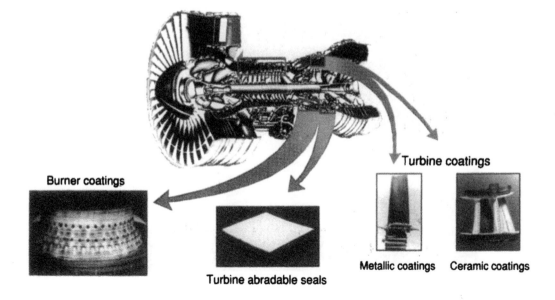

Figure 10.7 Coatings in the combustor and turbine section of gas turbine engine (J. T. DeMasi-Marcin and D. K. Gupta, Protective coatings in the gas turbine engine, *Surf. Coat. Technol.*, 1994, 68/69, 1–9). Reprinted with permission from Elsevier.

bond coat–ceramic interface fails to remain protective because of the presence of internal oxides in the as-coated bond coat. These oxides tie up enough aluminum so that over a period of time continued oxidation results in nonprotective oxides such as NiO, which is voluminous, weak, and prone to fracture and delamination.

A more challenging application of TBC is on high-pressure turbine vanes because of the higher temperature, higher heat flux, and larger thermal strain involved.

TBC on Vane Platform

An example of the early application of plasma-sprayed TBCs is on the vane platform of Pratt & Whitney's JT9D first-stage turbine vane. The vane TBC is distinguished from that on the combustor by the replacement of the bond coat. For the combustor application, the bond coat is plasma sprayed in air. Such bond coats do not meet the requirements of the more oxidizing vane environment. The bond coats are, therefore, replaced with more oxidation-resistant low-pressure plasma-sprayed NiCoCrAlY. A comparative engine test showed (Sheffler and Gupta, 1988) that a vane platform coated with 7YSZ was in good condition after 2778 endurance cycles while an adjacent vane platform with only oxidation-resistant metallic coating exhibited significant damage, including cracking and severe oxidation within only 1500 cycles in the same engine. Figure 10.8 shows a vane with a platform coated with pplasma-sprayed 7YSZ in excellent condition after 9300 hours of airline flight service.

GE Aircraft Engines tested a plasma-sprayed 8YSZ ceramic/BC52 (Ni10Co18Cr 6.5Al0.3Y) bond coat on the platform of René 80 cast second-stage turbine vane doublets (Wortman et al., 1989). The airfoil surfaces had plain aluminide coating. The 300 μm (12 mil) ceramic coating on the platform was hand polished to improve aerodynamic flow. The vanes were tested for 687 hours/2750 simulated engine cycles in a high-bypass engine CFC-80C, in proximity to the Adriatic Sea coast. Previous experience of exposure at the same site had

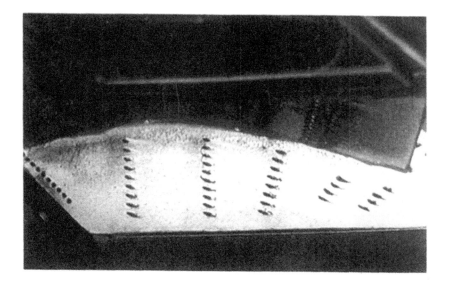

Figure 10.8 TBC-coated vane platform after 9300 hours of airline flight engine service in a JT9D-7R4G engine (K. D. Sheffler and D. K. Gupta, Current status and future trends in turbine application of thermal barrier coatings, *ASME 88-GT-286*). Reprinted with permission from American Society of Mechanical Engineers International.

shown a preponderance of hot corrosion. Posttest metallographic analysis indicated that neither the BC52 bond coat nor the ceramic coat on the platform had any indication of hot corrosion, whereas the aluminide-coated airfoil had significant hot corrosion attack. Rolls-Royce also tested a plasma-sprayed TBC on platforms of intermediate-pressure turbine vanes in the RB211-22B engine. A combination of bond coats and ceramic top coats were evaluated. The 7YSZ-coated TBC was found to perform well beyond an 11,000 hours test. The TBC has been incorporated on the high-pressure turbine vane platform in many Rolls-Royce engines (Meetham, 1986). Failure of the TBC used on vanes usually occurs within the ceramic but near the ceramic–bond coat interface. The failed surfaces still have remnant adherent ceramic. Because of the failure within the ceramic, the appearance is white, which is quite distinct from that of the combustor TBC. The latter typically appears black because of failure within NiO formed on the air plasma sprayed bond coat.

TBC on Vane Airfoil

The success of the plasma-sprayed TBC on vane platforms led to its application on the airfoil surfaces as well. Plasma-sprayed TBC, however, has several drawbacks relative to its EB-PVD deposited counterpart. These include susceptibility to erosion and significantly lower strain tolerance. Figure 10.9 shows a plasma-spray TBC-coated vane from an engine belonging to

Figure 10.9 Plasma-spray TBC-coated high-pressure turbine vane with severe erosion of coating after 17,260 hours (3006 cycles) on an engine from Lufthansa Airlines (Ch. W. Siry, H. Wanzek, and C. -P. Dau, Aspects of TBC service experience in aero engines, *Mat.-Wiss. Werkstofftech*, 32, 2001, pp. 650–653, © Wiley-VCH Verlag GmbH, D-69451, Weinheim, 2001). Reprinted with permission from John Wiley & Sons, Inc.

Lufthansa (Siry et al., 2001) with 17,260 hours (3006 cycles). The TBC is severely eroded. Where higher resistance to erosion is required or increased strains are involved, such as on relatively hot-running vanes and turbine blades for aircraft engines, TBC by the EB-PVD process is preferred to meet the required service life.

EB-PVD TBC

TBC on Blades and Vanes

An oxidation-resistant NiCoCrAlY metallic coating deposited by the LPPS process, a TBC with a patch of air plasma sprayed 7YSZ on LPPS NiCoCrAlY bond coat, and a TBC with a patch of EB-PVD 7YSZ on peened NiCoCrAlY bond coat were evaluated side by side at Pratt & Whitney on turbine blades in JT9D engines (Bose and DeMasi-Marcin, 1997). While the blades with only metallic coating and plasma-sprayed TBC patches exhibited severe distress, including oxidation and spall of the TBC, the blade with the EB-PVD TBC patch performed extremely well. These behaviors were confirmed in JT9D revenue engines (Fig. 10.10), in which the EB-PVD coated TBC ran in excess of 15,000 hours without any distress, while a metallic coated blade severely oxidized after only 3000 hours in the same engine. The strain tolerance capability of the EB-PVD structure relative to plasma spray is validated in these engine tests. Engine testing of a TBC consisting of EB-PVD 7YSZ and platinum aluminide bond coat on airfoil surfaces of turbine blades as well as vanes was pursued successfully by General Electric Aircraft Engines (Maricocchi et al., 1997). The test demonstrated improvement in component durability. Similar results were also reported by Rolls-Royce (Bennett et al., 1987) on vane airfoils in flight engines.

Although the major benefit of the use of TBCs is the reduction in component temperature and associated increase in durability, there are some spectacular indirect benefits. One such benefit is the elimination of pressure-side bulge in high-pressure turbine blades (Bose and

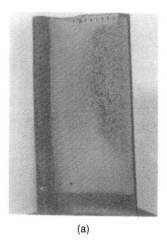

(a)   (b)

Figure 10.10 JT9D engine test result. (a) 15,000 hours with a patch of EB-PVD TBC, (b) 3000 hours with metallic coating showing severe oxidation resulting in scrap. (J. T. DeMasi-Marcin and D. K. Gupta, Protective coatings in the gas turbine engine, *Surf. Coat. Technol.*, 1994, 68/69, 1–9). Reprinted with permission from Elsevier.

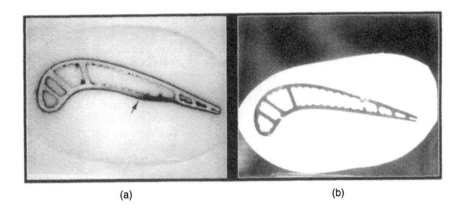

(a) (b)

Figure 10.11 (a) Pressure-side bulging of turbine blade cross-section, (b) bulge eliminated by the application of a TBC (Figs. 5 and 6, p. 103, in S. Bose and J. DeMasi-Marcin, Thermal barrier coating experience in gas turbine engines at Pratt & Whitney, *J. Therm. Spray Technol.*, 1997, 61(1), 99–104). Reprinted with permission from ASM International.

DeMasi-Marcin, 1997) (Fig. 10.11). The bulge resulted from creep. The driving force was thermal stress resulting from an 85°C (153°F) temperature difference between the concave and the convex side of the blade in combination with a high metal temperature. Applying a patch of TBC over the location, which not only reduced the overall temperature but also evened out temperature distribution to a great extent, eliminated the bulge.

Another benefit of a TBC is the reduction or elimination of creep of metallic coating, particularly on turbine blades. Creep stresses arise from the centrifugal load on the blade as well as the mismatch of the coefficient of thermal expansion between the coating and the alloy. The coating creep manifests itself as rumpling of the surface (Fig. 10.12) (Bose and DeMasi-Marcin, 1997). Rumpling affects both the aerodynamic performance of the coated hardware and the

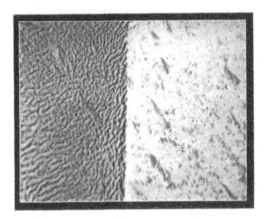

Figure 10.12 Rumpling of metallic coating on turbine blade (left) eliminated by the application of a patch of TBC (right). (S. Bose and J. DeMasi-Marcin, Thermal barrier coating experience in gas turbine engines at Pratt & Whitney, *J. Therm. Spray Technol.*, 1997, 61(1), 99–104). Reprinted with permission from ASM International.

Table 10.8 Industrial Gas Turbine Manufacturers and Models using TBCs

Manufacturer	Power Plant Model	TBC Coated Components
ABB	GT24/GT26	Vane
GE	7H	Blade & Vane
Siemens	V84/94.3A	Blade
MHI	501D/701D	Vane
	501F/701F	Blade & Vane
	501G/701G	Blade & Vane

durability of the coating. As Fig. 10.12 shows, the local patch of TBC with the associated lowering of temperature has eliminated the rumpling phenomenon.

TBC in Industrial Gas Turbine Engines

Successful performance of TBCs in aircraft engines has led to their application in power-generating gas turbine engines. Based on some promising rig test data, Mitsubishi Heavy Industries of Japan has introduced TBCs on turbine blades and vanes on most major models (Torigoe et al., 1994). The TBC, consisting of air plasma sprayed 8YSZ ceramic on an LPPS-deposited CoNiCrAlY bond coat, exhibited no distress such as cracking and TBC spalling for operations up to 24,000 hours with a maximum turbine inlet temperature of 1350°C (2462°F). The success of plasma-sprayed TBCs on turbine blades seems surprising, considering its poor performance in aircraft engines. However, from the magnitude of the turbine inlet temperature in the Mitsubishi test, it appears that the strain range experienced by the ceramic coating is significantly lower than that experienced in aircraft engines. Benefits of TBC on turbine blades have also been demonstrated in Russia (Osyka et al., 1995). The TBCs consisted of EB-PVD deposited 8YSZ on a CoCrAlY bond coat also deposited by EB-PVD. The TBC-coated blades ran in peak-load gas turbines GT-100 for more than 939 hours and 253 starts without any distress. Siemens-Westinghouse has also tested TBCs on industrial gas turbine engines. Blades having EB-PVD 7YSZ on MCrAlY bond coats with 27,000 hours exposure have been analyzed to map residual stress distribution (Sohn et al., 2000). Because of the significant improvement in durability exhibited by TBCs in a variety of engines, the major manufacturers of industrial gas turbines have incorporated TBCs on blades and vanes, as Table 10.8 indicates (Viswanathan and Scheirer, 1998)

10.2   OTHER APPLICATIONS

Coal Gasification Combined Cycle Power Plant

In the coal gasification process, pulverized coal is air separated and gasified. The resulting gas is cleaned of particulates. A wet cleanup process removes impurities, particularly sulfur, a universal constituent of coal. The coal gas is then burned as a fuel in a gas turbine engine. The exhaust heat from the engine is utilized to generate steam to power a steam turbine. The combined unit is used to generate electrical power. In many respects the power plant has the characteristics of gas turbine engines with the exception of the environment. While the wet cleanup process reduces sulfur and other impurities in the coal gas to very low levels, the gas temperature is lowered significantly with a consequent large reduction in efficiency of the power plant. A dry and hot sulfur cleanup process is, therefore, more attractive, with the downside that the contaminant level is likely to rise. Ni and Co base superalloys, overaluminided MCrAlY metallic coatings, and 8YSZ TBCs have been tested in such environments (Wada et al., 2000). In the test the turbine blade alloy was René 80H, a Ni base superalloy, whereas FSX414, a high-chromium Co base alloy, was used for turbine vanes. Coatings included Co29Cr5.8Al0.4Y and Co32Ni21Cr8Al0.5Y deposited by low-pressure plasma spray (LPPS) to a thickness of $200 \mu m$ (8 mil) followed by aluminiding. The tested TBC consisted an of air plasma sprayed $100 \mu m$ (4 mil) thick NiCoCrAlY bond coat and $200 \mu m$ (8 mil) thick 8YSZ ceramic coating, also deposited by air plasma spray. Cylindrical coated samples were evaluated in an actual coal gas environment at the Aioi Works of IHHI Co, Japan. Constituents in the coal ash within the gasifier created complex silicates of composition $Na_2O\ Al_2O_3\ xSiO_2$ ($x = 2$ and 6) at around 1000°C (1832°F). These silicates formed low-melting eutectics with melting points as low as 770°C (1418°F). Both CoCrAlY and CoNiCrAlY were attacked and defeated by the molted silicate. The TBC, however, successfully provided a physical barrier, inhibiting contact between the molten silicate and the metallic bond coat.

Fast Breeder Reactors

The environment of fast breeder reactors consists of moderate to high temperatures in the range between 200 and 625°C (392 and 1157°F), the presence of liquid metals such as sodium, and neutron radiation. Thermal fluctuations occur during sudden shutdowns of the reactors, with component temperature dropping from 625 to 400°C (1157 to 752°F) within the duration of a minute or so. The components have to survive about 240 such cycles in their service life. Most of the coatings in this environment are used for protection against wear, against corrosion due to contact with molten metal, and for trapping radionuclides (Johnson, 1984). One of the successful coatings used in this environment is diffusion aluminide deposited by a pack process on the Ni base superalloy Inco 718 and on stainless steels such as 304 and 316. The steels do not have an adequate reservoir of Ni to form βNiAl. Therefore, electrolytic or electroless Ni is deposited on steels first, followed by pack aluminiding (trademark TMT 2813, commercially available from Turbine Metal Technology, Burbank, CA, USA). The βNiAl coating has been found to be resistant to corrosion in molten sodium and to permeation by hydrogen and tritium. Coatings and alloys containing cobalt and boron are not acceptable for use in reactor environments because the cobalt isotope ^{60}Co releases high-energy gamma rays and has a long half-life, while boron exhibits a high neutron capture cross-section.

Coatings deposited by detonation gun and electro-spark deposition (ESD) have been found to perform better than equivalent coatings deposited by plasma spray, possibly due to lower oxide content and higher density.

Waste to Energy Plants

Environmentally friendly disposal of large quantities of waste by incineration with recovery of energy from the conversion has become an environmental and a social requirement of waste management all over the world. The incinerators require removal of harmful gases and solid ash. The heat generated in the process is used to produce steam at temperatures and pressures up to 500°C (932°F) and 100 atm (Kawahara, 1997), respectively. The environment within the incinerator is highly corrosive with the presence of Na, K, Ca, Zn, and Pb as oxides, chlorides, and sulfates in addition to gases such as CO, SO_2, H_2, O_2, and CO_2. The materials of construction of the waterwall and the superheater tubes in the incinerator require corrosion-resistant coatings. In studies done in Japan, flame-sprayed Al/80Ni20Cr has exhibited improved corrosion life (about 3 years). Recently, the HVOF process has been used to deposit NiCrSiB, which is expected to have better bond strength and, therefore, a longer life compared with the previous coating.

Diesel Engines

Diesel engines are used in transportation, marine, and utility industries. The demand for diesel power is projected to increase significantly in the future. However, to meet the demand, it has to satisfy efficiency, durability, and environmental requirements. TBCs are expected to help meet some of the challenges (Winkler and Parker, 1992; Yonushonis, 1997; Beardsley et al., 1999). TBCs reduce temperature so that the parts run cooler. They also improve thermal fatigue life by reducing large temperature distribution and the effect of transients. Additionally, they increase the heat conveyed by TBC surfaces into the combustion chamber, resulting in improved fuel burn. Pierz (1993) at Cummins Engine Company evaluated plasma-sprayed 20YSZ as well as mullite ($3Al_2O_3$ $2SiO_2$) as potential TBC systems for a heavy-duty four-stroke six-cylinder diesel engine. Mullite has a lower density as well as a lower coefficient of thermal expansion than zirconia, although the thermal conductivity is higher. Coatings of two thicknesses, 0.40 and 1.5 mm, were applied to aluminum pistons with crowns machined to restore geometry. Thin mullite TBC had the maximum durability, followed by thick mullite and 20YSZ.

Based on results of a 10-year study conducted by FCS Inc. of Connecticut, USA, under actual engine operating conditions, the benefits of TBCs for diesel engines have been found to include fuel savings (up to 11%), extended engine life by as much as 20%, increased power by 10% max, reduced emissions, reduced particulate formation, ability to use low-octane fuel, reduced engine noise, reduced part temperature by about 100°C (180°F), longer valve life, and reduced maintenance cost. Air plasma spray deposited TBCs have been tested on diesel engines used by the Connecticut Department of Transportation, Bridgeport Transit, and the U.S. Environmental Protection Agency. The various components on which the TBC is used, shown in Fig. 10.13, include pistons, piston crowns, exhaust valves, and cylinder heads. The fuel economy benefits over a 2-year period are shown in Fig. 10.14 (Winkler and Parker, 1992). One of the applications of TBCs in a demanding environment is on the diesel engines powering tugboats in marine use. The U.S. Department of Transportation has successfully tested plasma-sprayed TBCs on a twin-engine tugboat for more than 2 years (14,000 hours). The conditions of the TBCs were excellent when the project was completed.

The potential benefits of using TBCs in automobile engines have persuaded some segments of the industry such as racing cars to incorporate the technology into the combustion chambers, intake valve faces, and surfaces of the exhaust valves.

The selected examples presented in this chapter clearly show the tremendous benefits provided by high-temperature coatings in actual field applications in many industries. The manufacturers and users have been investing large resources focused on innovations and improvements in high-temperature coating technologies. The fundamental understanding of the behavior of the coatings has come a long way because of research and development carried out by industry and academic institutions, work that continues with increased fervor.

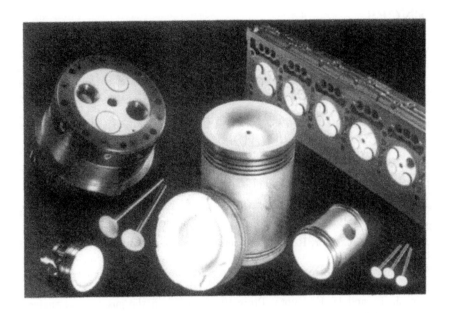

Figure 10.13 TBC applied to diesel engines part (M. F. Winkler and D. W. Parker, "Greener, meaner" diesels sport thermal barrier coatings, *Adv. Mater. Proc.*, 1992, 5, 17–22). Reprinted with permission from ASM International.

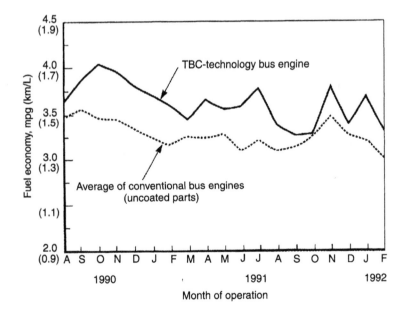

Figure 10.14 Fuel savings due to the use of TBC in a diesel engine based on 19-month transit company study (M. F. Winkler and D. W. Parker, "Greener, meaner" diesels sport thermal barrier coatings, *Adv. Mater. Proc.*, 1992, 5, 17–22). Reprinted with permission from ASM International.

REFERENCES

Beardsley, M. B., P. G. Happoldt, K. C. Kelley, E. F. Rejda, and D. F. Socie, Thermal barrier coatings for low emission, high efficiency diesel engine applications, *SAE Technical Paper* 1999-01-2255.

Bennett, A., F. C. Toriz, and A. B. Thakker, A philosophy for thermal barrier coating design and its corroboration by 10000 h service experience on RB211 nozzle guide vanes, *Surf. Coat. Technol.*, 1987, 32, 359–375.

Bose, S., and J. DeMasi-Marcin, Thermal barrier coating experience in gas turbine engines at Pratt & Whitney, *J. Therm. Spray. Technol.*, 1997, 6(1), 99–104.

Cannistraro, N. I., Where do we stand in ceramic coatings?, *Met. Prog.*, 1958, 74, 111–113.

Chen, M. W., R. T. Ott, T. C. Hufnagel, P. K. Wright, and K. J. Hemker, Microstructural evolution of platinum modified nickel aluminide bond coat during thermal cycling, *Surf. Coat. Technol.*, 2003, 163–164, 25–30.

Conde, J. F. G., and C. G. McCreath, The control of hot corrosion in marine gas turbines, *ASME 80-GT-126*.

Conner, J. A., and W. B. Connor, Ranking protective coatings: Laboratory vs. field experience, *JOM*, Dec. 1994, pp. 35–38.

Connor, J. A., Evaluation of simple aluminide and platinum modified aluminide coatings on HPT blades after factory engine testing—round II, *ASME 92-GT-140*.

DeMasi-Marcin, J. T., and D. K. Gupta, Protective coatings in the gas turbine engine, *Surf. Coat. Technol.*, 1994, 68/69, 1–9.

Eliaz, N., G. Shemesh, and R. M. Latanision, Hot corrosion in gas turbine components, *Eng. Fail. Anal.*, 2002, 9, 31–43.

Goward, G. W., and L. W. Cannon, Pack cementation coatings for superalloys: A review of history, theory, and practice, *J Eng. Gas Turbines Power Trans ASME*, 1988, 110, 150–154.

Grisik, J. J., R. G. Miner, and D. J. Wortman, Performance of second generation airfoil coatings in marine service, *Thin Solid Films*, 1980, 73, 397–406.

Grossklaus, W. D., Jr., G. B. Katz, and D. J. Wortman, Performance comparison of advanced airfoil coatings in marine service, in *High Temperature Coatings*, pp. 67–83, Eds. M. Khobaib and R. C. Krutenat, The Metallurgical Society, Warrendale, PA, 1986.

Gupta, D. K., and D. S. Duvall, A silicon and hafnium modified plasma sprayed MCrAlY coating for single crystal superalloys, in *Superalloys 1984*, pp. 711–720, Eds. M. Gell, C. S. Kortovich, R. H. Bricknell, W. B. Kent, and J. F. Radavich, Metallurgical Society of AIME, Warrendale PA, 1984.

Johnson, R. N., Coatings for breeder reactor components, *Thin Solid Films*, 1984, 118, 31–47.

Kameda, J., T. E. Bloomer, Y. Sugita, A. Ito, and S. Sakurai, High temperature environmental attack and mechanical degradation of coatings in gas turbine blades, *Mater. Sci. Eng.* 1997, A229, 42–54.

Kameda, J., T. E. Bloomer, and S. Sakurai, Oxidation/carbonization/nitridation and in-service mechanical property degradation of CoCrAlY coatings in land-based gas turbine blades, *J. Therm. Spray. Technol.*, 1999, 8(3), 440–446.

Kawahara, Y., Development and application of high-temperature corrosion resistant materials and coatings for advanced waste-to-energy plants, *Mater. High Temp.*, 1997, 14(3), 261–268.

Kubarych, K. G., J. M. Aurrecoechea, Post field test evaluation of an advanced gas turbine blade, *Proc. of ASM 1993 Mater. Congress Materials Week*, Pittsburgh, PA, 1993, 59–68.

Maricocchi, A., A. Bartz, and D. Wortman, PVD TBC experience on GE aircraft engines, *J. Therm. Spray. Technol.*, 1997, 6(2), 193–198.

McMinn, A., R. Viswanathan, and C. L. Knauf, Field evaluation of gas turbine protective coatings, *J. Eng. Gas Turbines Power Trans ASME*, 1988, 110, 142–149.

Meetham, G. W., Use of protective coatings in aero gas turbine engines, *Mater. Sci. Technol.*, 1986, 2(3), 290–294.

Meier, S. Manning, and D. K. Gupta, The evolution of thermal barrier coatings in gas turbine engine applications, *ASME 92-GT-203*, 1992.

Meier, S. Manning, and D. K. Gupta, The evolution of thermal barrier coatings in gas turbine engine applications, *J. Eng. Gas Turbines Power Trans ASME*, 1994, 116, 250–257.

Mom, A. J. A., and J. A. M. Boogers, Simulated service test behavior of various internal and external coatings applied on CF6-50 first stage turbine blades, in *High Temperature Alloys for Gas Turbines and Other Applications 1986*, Part II, pp. 1245–1264, Eds. W. Betz, R. Brunetaud, D. Coutsouradis, H. Fischmeister, T. B. Gibbons, I. Kvernes, Y. Lindblom, J. B. Marriott, D. B. Meadowcroft, and D. Reidel, Dordrecht, The Netherlands, 1986.

Mom, A. J. A., and H. J. C. Hersbach, Performance of high temperature coatings on F100 turbine blades under simulated service conditions, *Mater. Sci. Eng.*, 1987, 87, 361–367.

Mustasim, Z., and W. Brentnall, Thermal barrier coatings for gas turbine applications: An industrial note, *J. Therm. Spray. Technol.*, 1997, 6(1), 105–108.

Nichols, E. S., J. A. Burger, and D. K. Hanink, *Mech. Eng.*, March 1965, pp. 52–56.

Osyka, A. S., A. I. Rybnikov, S. A. Leontiev, N. V. Nikitin, and I. S. Malashenko, Experience with metal/ceramic coating in stationary gas turbines, *Surf. Coat. Technol.*, 1995, 76–77, 86–94.

Pennisi, F. J., and D. K. Gupta, Tailored plasma sprayed MCrAlY coatings for aircraft gas turbine application, *NASA Contract Rep CR 165234*, 1981.

Pierz, P. M., Thermal barrier coating development for diesel engine aluminum pistons, *Surf. Coat. Technol.*, 1993, 61, 60–66.

Rhys-Jones, T. N., and F. C. Toriz, Thermal barrier coatings for turbine applications in aero engines, *High Temp. Technol.*, 1989, 7(2), 73–81.

Schilke, P. W., A. D. Foster, J. J. Pepe, and A. M. Beltran, Advanced materials propel progress in land-based gas turbines, *Adv. Mater. Proc.*, 1992, 141(4), 22–30.

Sheffler, K. D., and D. K. Gupta, Current status and future trends in turbine application of thermal barrier coatings, *ASME 88-GT-286*, 1988.

Siry, Ch. W., H. Wanzek, and C.-P. Dau, Aspects of TBC service experience in aero engines, *Mat.-Wiss. Werkstofftech.*, 2001, 32, 650–653.

Soechting, F. O., A design perspective on thermal barrier coatings, *Thermal Barrier Coating Workshop*, *NASA Conference Publication 3312*, pp. 1–15, 1995.

Sohn, Y. H., K. Schlichting, K. Vaidyanathan, E. Jordan, and M. Gell, Nondestructive evaluation of residual stress for thermal barrier coated turbine blades by Cr^{3+} photoluminescence piezospectroscopy, *Metall. Mater. Trans. A*, 2000, 37, 2388–2391.

Torigoe, T., T. Kitai, I. Tsuji, H. Kawai, and Y. Kasai, Zirconia TBC application in power generating gas turbine, in *Advanced Materials and Coatings for Combustion Turbines*, pp. 131–134, Eds. V. P. Swaminathan and N. S. Cheruvu, ASM International, 1994.

Toriz, F. C., A. B. Thakker, and S. K. Gupta, Thermal barrier coatings for jet engines, *ASME 88-GT-279*, 1988.

Toriz, F. C., A. B. Thakker, and S. K. Gupta, Flight service evaluation of thermal barrier coatings by physical vapor deposition at 5200H, *Surf. Coat. Technol.*, 1989, 39/40, 161–172.

Viswanathan R., and S. T. Scheirer, Materials advances in land-based gas turbines, paper presented at *POWER GEN*, Orlando, FL, Dec. 9–11, 1998.

Wada, K., L. Yan, M. Takahashi, K. Takaishi, and T. Furukawa, Degradation of gas turbine blades in integrated coal-gasification combined cycle plant, in *Proc. International Symposium on High Temperature Corrosion and Protection 2000*, Science Reviews, pp. 329–336, Eds. T. Narita, T. Maruyama, and S. Taniguchi, H., Japan, September 17–22, 2000.

Wanhill, R. J. H., A. J. A. Mom, H. J. C. Hersbach, G. A. Kool, and J. A. M. Boogers, NLR experience with high velocity burner rig testing 1979–1989, *High Temp. Technol.*, Nov. 1989, pp. 202–211.

Winkler, M. F., and D. W. Parker, "Greener, meaner" diesels sport thermal barrier coatings, *Adv. Mater. Proc.*, 1992, 5, 17–22.

Wortman, D. J., Performance comparison of plasma spray and physical vapor deposition BC23 coatings in the LM2500, *J. Vac. Sci. Technol. A*, 1985, 3(6), 2532–2536.

Wortman, D. J., B. A. Nagaraj, and E. C. Duderstadt, Thermal barrier coatings for gas turbine use, *J. Mater. Sci. Eng.*, 1989, A121, 433–440.

Yonushonis, T. M., Overview of thermal barrier coatings in diesel engines, *J. Therm. Spray Technol.*, 1997, 6(1), 50–56.

Yoshioka, Y., D. Saito, H. Okamoto, S. Ito, and K. Ishibashi, Damage and degradation assessment of stage 1 bucket coatings in a 1100°C-class gas turbine, Paper GT2006-90748, presented at *ASME Turbo Expo 2006* Barcelona, 2006.

Appendix

Compositions of Selected Superalloys

Alloy	Approximate Nominal Composition in Weight Percent												
	Al	B	C	Co	Cr	Hf	Mo	Ni	Ta	Ti	W	Zr	Others
Nickel-Base Polycrystalline													
Waspaloy	1.2	0.005	0.07	13.5	19.5		4.2	Bal		3		0.09	Fe 15.8
Hastelloy X	2		0.15	1.5	22		9	Bal			0.6		Fe 15.8
IN 718	0.05		0.04		19		3	Bal		0.9			Fe 18, Nb 5, Cu 0.25
IN 792	3.2	0.02	0.2	9	13		2	Bal	4	4.2	4	0.1	Nb 2
IN 100	5.5	0.015	0.17	15	9.5		3	Bal		4.8		0.06	V 1
B-1900 + Hf	6	0.015	0.11	10	8	1.2	6	Bal	4.3	1		0.08	
René 80	3	0.015	0.17	9.5	14		4	Bal		5	4	0.03	
René 80 + Hf	3	0.015	0.08	9.5	14		4	Bal		4.8	4	0.02	Hf 0.75
IN 738	3.4	0.01	0.17	8.5	16		1.7	Bal	1.7	3.4	2.6	0.1	Nb 0.9
Mar-M247	5.5	0.016	0.15	10	8.4	1.4	0.6	Bal	3	1	10	0.06	
DS* Mar-M247	5.5	0.015	0.15	10	8.4	1.4	0.6	Bal	3	1	10	0.06	
Mar-M200	5	0.015	0.15	10	9			Bal		2	12	0.05	Nb 1
DS Mar-M200 + Hf	5	0.015	0.13	10	9	2		Bal		2	12	0.05	Nb 1

Nickel-Base
Single-Crystal

PWA 1480	5			5	10			Bal	12	1.5	4		
PWA 1483	3.7			9	12.8		1.9	Bal	4	4.2	3.8		
PWA 1484	5.7			10	5	0.1	1.9	Bal	8.7		5.9		Re 3
René N	3.7			7.5	9.3		1.5	Bal	4	4.2	6		
René N4	4.2	0.004		7.5	9.75	0.15	1.5	Bal	4.8	3.5	6		Nb 0.5
René N5	6.2			7.5	7	0.15	1.5	Bal	6.5		5		Re 3, Y 0.01
René N6	5.75			12.5	4.2	0.15	1.4	Bal	7.2		6		Re 5.4, Y 0.01
CMSX-2	5.6			4.6	8		0.6	Bal	6	1	8	0.06	
CMSX-4	5.6			9	6.5		0.6	Bal	6.5	1	6		
Cobalt Base													
Mar-M509			0.6	Bal	23.5			10	3.5	0.2	7	0.5	
FSX-414		0.01	0.25	Bal	29			10			7.5		Fe 1

DS, Directionally solidified

Melting Points °F (°C)

YSZ	NiCoCrAlY	NiAl	Ni-Base Superalloy	Alumina
~2690 (4874)	~1260 (2300)	~1638 (2980)*	1260–1343 (2300–2450)	2030 (3686)

* Interface melting at ~1296° C (2365° F).

TBC Properties

	Dense YSZ	APS 7YSZ	EB-PVD 7YSZ	NiCoCrAlY	NiAl	(Ni,Pt)Al
Density (g/cm³)	6.1	5	5.1	~7–7.5 (APS is ~6.5)	~5.9	~7 (PtAl₂ is 6.25)
Thermal Conductivity (W/m K at 1000°C)	2.5	0.6–1.1	1.5–2	20–35	~75	
Specific Heat (J/g °C)		0.582–0.603		0.628–0.712	0.64	
Emittance at 0.8–0.9 µm		0.25	0.325–0.35			
Debye Temperature °C	106					
Knoop Hardness		~4		~2–3		

Coefficient of Thermal Expansion/°C × 10⁻⁶

Temperature °C	NiCoCrAlY	NiAl	Pt Aluminide	Ni-Base Superalloy	Alumina	APS 7YSZ	EB-PVD 7YSZ (Perp. To Column)
25	13.3	14	13.6	12.2	8		
300	14.1	14.2	14.4	12.9	8.3	9.82	8.2
500	15	14.6	14.9	13.5	8.6	9.64	8.6
700	16.1	14.6	15.6	14.1	8.8	9.88	10.7
900	18	14.9	16.6	15.1	9.2	10.19	12.2

Young's Modulus

	APS 7YSZ	EB-PVD 7YSZ	NiCoCr AlY	Pt Aluminides	Super-alloys	Alumina	Plain Aluminide
E (GPa)						380–400	
Temperature °C(°F)							
24 (75)	2.5	35–40	170		200–220	400	100 (<100), 190 (<110)
538 (1000)	2.5						290 (<111)
871 (1600)	2.3						
982 (1800)	2					350	
1093 (2000)	1.8					340	
1204 (2200)	1.75					330	
Poisson's Ratio ν				0.3	0.3	0.25	

TGO Bond Coat Interface Fracture Toughness, J/m^2

Segregated interface	1
Diffusion bonded Ni–alumina	10
Cleaner interface	60–100

Approximate values of DBTT of Aluminides and MCrAlYs

Coating	Estimated DBTT, °C (°F)
NiAl	868–1060 (1594–1940)
(Ni,Pt)Al	> plain aluminide
CoAl	878–1070 (1612–1958)
Co18Cr9Al1Y	150–200 (302–392)
Co18Cr11Al1Y	250–300 (482–572)
Co20Cr12.5Al1Y	600–650 (1112–1202)
Co29Cr6Al1Y	700–800 (1292–1272)
Co27Cr12AlY	800–900 (1272–1652)
Ni20Cr9-11AlY	25–200 (77–392)
Ni38Cr11AlY	600–650 (1112–1202)
$PtAl_2$	870–1070 (1598–1958)
Commercial platinum aluminide (Temperature at which fractures at 3% strain)	~930 (1706)

DBTT, ductile-to-brittle transition temperature

Author Index

Subject Index

Printed and bound by CPI Group (UK) Ltd, Croydon, CR0 4YY

03/10/2024

01040333-0012